# Evolutionary Economics and Social Complexity Science

Volume 37

The Japan Association for Evolutionary Economics (JAFEE) always has adhered to its original aim of taking an explicit "integrated" approach. This path has been followed steadfastly since the Association's establishment in 1997 and, as well, since the inauguration of our international journal in 2004. We have deployed an agenda encompassing a contemporary array of subjects including but not limited to: foundations of institutional and evolutionary economics, criticism of mainstream views in the social sciences, knowledge and learning in socio-economic life, development and innovation of technologies, transformation of industrial organizations and economic systems, experimental studies in economics, agent-based modeling of socio-economic systems, evolution of the governance structure of firms and other organizations, comparison of dynamically changing institutions of the world, and policy proposals in the transformational process of economic life. In short, our starting point is an "integrative science" of evolutionary and institutional views. Furthermore, we always endeavor to stay abreast of newly established methods such as agent-based modeling, socio/econo-physics, and network analysis as part of our integrative links. More fundamentally, "evolution" in social science is interpreted as an essential key word, i.e., an integrative and /or communicative link to understand and re-domain various preceding dichotomies in the sciences: ontological or epistemological, subjective or objective, homogeneous or heterogeneous, natural or artificial, selfish or altruistic, individualistic or collective, rational or irrational, axiomatic or psychological-based, causal nexus or cyclic networked, optimal or adaptive, micro- or macroscopic, deterministic or stochastic, historical or theoretical, mathematical or computational, experimental or empirical, agent-based or socio/econo-physical, institutional or evolutionary, regional or global, and so on. The conventional meanings adhering to various traditional dichotomies may be more or less obsolete, to be replaced with more current ones vis-à-vis contemporary academic trends. Thus we are strongly encouraged to integrate some of the conventional dichotomies. These attempts are not limited to the field of economic sciences, including management sciences, but also include social science in general. In that way, understanding the social profiles of complex science may then be within our reach. In the meantime, contemporary society appears to be evolving into a newly emerging phase, chiefly characterized by an information and communication technology (ICT) mode of production and a service network system replacing the earlier established factory system with a new one that is suited to actual observations. In the face of these changes we are urgently compelled to explore a set of new properties for a new socio/economic system by implementing new ideas. We thus are keen to look for "integrated principles" common to the above-mentioned dichotomies throughout our serial compilation of publications. We are also encouraged to create a new, broader spectrum for establishing a specific method positively integrated in our own original way.

Ichiro Takahashi

# Microfoundations of Keynesian Economics

## An Evolutionary Approach to Stabilizing an Unstable Economy

Springer

Ichiro Takahashi
Visiting Research Fellow at the Institute of Economic Research
Chuo University
Tokyo, Japan

ISSN 2198-4204 ISSN 2198-4212 (electronic)
Evolutionary Economics and Social Complexity Science
ISBN 978-981-95-9389-7 ISBN 978-981-95-9390-3 (eBook)
https://doi.org/10.1007/978-981-95-9390-3



This Springer imprint is published by the registered company Springer Nature Singapore Pte Ltd.
The registered company address is: 152 Beach Road, #21-01/04 Gateway East, Singapore 189721, Singapore

*To my wife*

# Preface

My father ran a small business and lived under the constant threat of dishonored checks. I grew up watching my parents navigate daily uncertainty, sometimes hopeful and sometimes anxious about the future of the firm. When I was in high school, my father's company went bankrupt. That experience left a lasting impression. It made me wonder whether it is possible to build an economic system in which unemployment, bankruptcy, and persistent insecurity are not endemic features of everyday life.

The Japanese term "keizai" traces its origin to the classical Chinese expression jingguo jimin (經國濟民)—literally, "to govern the state and relieve the people"—reflecting an early conception of economic activity as inseparable from social welfare and governance. I came to believe that economics should be an academic discipline concerned not only with efficiency, but with the conditions under which people can live with dignity and security. This conviction led me to study economics and, ultimately, to devote my research to the problem of macroeconomic stability.

My intellectual path to this book has been indirect. I began by studying Keynesian economics and its microeconomic foundations, moved through game theory and equilibrium selection, and later returned to macroeconomics equipped with new analytical tools. Agent-based simulations and evolutionary models convinced me that macroeconomic outcomes are better understood as stochastic processes rather than as movements toward unique equilibria. In earlier work, I showed that macroeconomic dynamics can give rise to multiple quasi-steady states associated with different employment levels, with the economy wandering among them over very long horizons. This insight became the starting point for the present book.

The central argument of this book is that macroeconomic stability should not be understood as convergence to a pointwise equilibrium, but as the shape, support, and boundary behavior of the economy's long-run invariant distribution. Even when a neutral "anchor" exists, the economy may spend most of its time away from full employment, occasionally approaching it, but rarely remaining there. In this sense, unemployment is not an anomaly to be eliminated once and for all, but a structural feature whose magnitude and persistence depend on institutional and behavioral mechanisms.

The book revolves around three closely related questions. First, does a capitalist monetary economy possess inherent stabilizing mechanisms, or is instability its natural state? Second, does nominal wage rigidity stabilize or destabilize the economy in the long run? Third, should macroeconomic stability be pursued by stabilizing nominal anchors—such as the money supply—or by allowing active monetary adjustment over the business cycle?

These questions place the book at the intersection of several traditions. Classical and mainstream macroeconomic theories have long emphasized the stabilizing role of flexibility, arguing that adjustments in wages and prices restore full employment. In contrast, Keynes and the Post-Keynesian tradition stressed fundamental uncertainty, endogenous instability, and the possibility that flexibility itself may amplify fluctuations. This book revisits these debates from a different angle, using evolutionary game theory and stochastic processes to study how adjustment mechanisms shape invariant distributions rather than equilibrium points.

While the analysis developed in this book bears directly on policy debates—concerning wage rigidity, monetary regimes, and financial stability—its primary objective is theoretical. The aim is to clarify the mechanisms through which real anchors, nominal institutions, and boundary effects interact to shape long-run macroeconomic outcomes. Issues of institutional design and redistribution are addressed only to the extent that they help elucidate the economic content of the results.

At its core, the book advances a microeconomic reinterpretation of Keynesian economics. *The General Theory of Employment, Interest and Money* is famously difficult, but its central ideas can be sharpened and made more transparent through modern analytical tools. In this respect, my thinking has been influenced by a broad Keynesian tradition, including the work of Hirofumi Uzawa, Katsuhito Iwai, James Tobin, and Hyman Minsky, each of whom contributed in distinct ways to the understanding of investment, demand, and macroeconomic instability.

This book represents the culmination of my work in economics. The question that first arose in my youth—whether a stable economic system is possible—has accompanied me for more than 50 years. The answer developed here may not be definitive, but it is the most coherent one I have been able to formulate. If the path to it appears roundabout, it is because the problem itself demands patience. It is my hope that this book will serve as a useful foundation for readers exploring the microfoundations of macroeconomic stability and encourage further research in this direction.

The author utilized ChatGPT 5.5 during the preparation of this manuscript to refine the text's clarity and logical flow. All resulting revisions were reviewed and approved by the author, who assumes full responsibility for the content.

This research was supported by JSPS KAKENHI (Grants-in-Aid for Scientific Research, Grant Number: 22K01403), Ministry of Education, Culture, Sports, Science and Technology (MEXT), Japan.

The publication of this book took longer than I had originally expected. I would like to express my heartfelt gratitude to Yutaka Hirachi at Springer Japan for his patient and steadfast support throughout this process.

Tokyo, Japan Ichiro Takahashi

# Declarations

**Competing Interests** The authors have no competing interests to declare that are relevant to the content of this manuscript.

# Contents

# List of Figures

# List of Tables

# Chapter 1
# Introduction: Framework and Methodological Approach

> *I have come to believe that the plan, properly worked out and applied, is incomparably the best proposal ever offered for speedily and permanently solving the problem of depressions; for it would remove the chief cause of both booms and depressions, namely, the instability of demand deposits, tied, as they now are, to bank loans.*
>
> —Irving Fisher, *100% Money* (1935)

The modern monetary economy is prone to recurrent instability not because productive capacity fluctuates violently, but because the means of payment expand and contract with credit conditions. When demand deposits are created as the counterpart of bank lending, the payments system becomes inherently procyclical: expansions of credit fuel booms, while contractions amplify downturns. This book takes seriously the long-standing diagnosis, articulated most clearly by Irving Fisher, that the root of macroeconomic instability lies in the institutional coupling of money and bank credit. By separating payments balances from credit intermediation and embedding this structure in a Keynesian model of investment, expectations, and effective demand, the analysis that follows reexamines the dynamics of stability, fluctuation, and neutrality in a monetary economy.

This book develops a framework for analyzing the endogenous dynamics of a monetary market economy driven by decentralized private agents. Rather than reviving Keynesian fine-tuning of aggregate demand through discretionary policy, the aim is to understand how both instability and stability emerge from the interaction of firms, households, and banks operating under realistic institutional constraints. By combining Keynes's behavioral insights with Fisher's emphasis on balance-sheet consistency and monetary architecture, the framework explains how macroeconomic fluctuations can arise naturally—without reliance on exogenous shocks, representative-agent optimization, or discretionary government control. Accordingly, the analysis focuses not on policy-driven equilibrium selection, but on the mechanisms through which decentralized decisions generate persistent macroeconomic dynamics over time.



I. Takahashi, *Microfoundations of Keynesian Economics*, Evolutionary Economics and Social Complexity Science 37, https://doi.org/10.1007/978-981-95-9390-3_1

A central institutional theme is that macroeconomic stability depends not only on behavioral rules and real frictions but also on the monetary architecture through which bank lending is financed and payments are settled. This raises a foundational question that motivates the analysis throughout the book: can the deposit channel of financial instability be removed without blunting real investment?

This book revisits Irving Fisher's claim[1] that depressions are propagated by the instability of demand deposits tied to bank lending, but does so within a broader analytical framework of decentralized decision-making and balance-sheet consistency.

We begin by building a Keynesian short-run block with posted prices and posted wages, a split-bank payments architecture, and evolutionary investment dynamics. By construction, a 100%-reserve payments book eliminates deposit fragility; bank lending is funded by nonmonetary claims. Aggregate demand is stabilized not by discretionary fine-tuning, but by a simple *distribution rule* for capital income: firms are rebated immediately at the accounting instant, so households ultimately own the capital stock and consume out of both current income and wealth.

The resulting structure delivers three layers of stability. First, it yields *bounded short-run dynamics*: with prices and wages posted for the period, disturbances move quantities but do not generate explosive paths. Second, conditional on the capital stock, the short-run block admits a locally stable fixed point, delivering *intermediate-run* stability in the form of a *symmetric quasi-steady state* (SQSS). Third, investment and entry dynamics induce an ergodic Markov process for capital, implying a unique invariant distribution for the capital stock whose support is confined to a stability corridor (formalized in Chap. 8).[2]

The next two figures preview the book's long-run object: the transition probabilities of the capital stock and the induced invariant distribution concentrated within the stability corridor. Figure 1.1 plots the probabilities of expansion $p_+(K)$ and contraction $p_-(K)$; Fig. 1.2 plots the corresponding invariant distribution.

These two objects—$p_+(K)$, $p_-(K)$ and the induced invariant distribution—provide the book's basic summary of long-run macroeconomic dynamics.

## 1.1 Keynesian Economics as a Monetary Production Economy

The model's fundamental structure is rooted in Keynes' *The General Theory of Employment, Interest and Money*, which we distill into three interlocking features.

---

[1] While the Chicago Plan proposed by Benes and Kumhof (2012) offers several merits, including eliminating bank runs, enhancing monetary stability, reducing credit cycles, and ensuring more effective monetary policy, the focus of this book is on developing a framework that integrates these ideas with the principles of 100%-reserve banking.

[2] See Wicksell (1936, orig. 1898), Leijonhufvud (1973), and Dimand (1997).

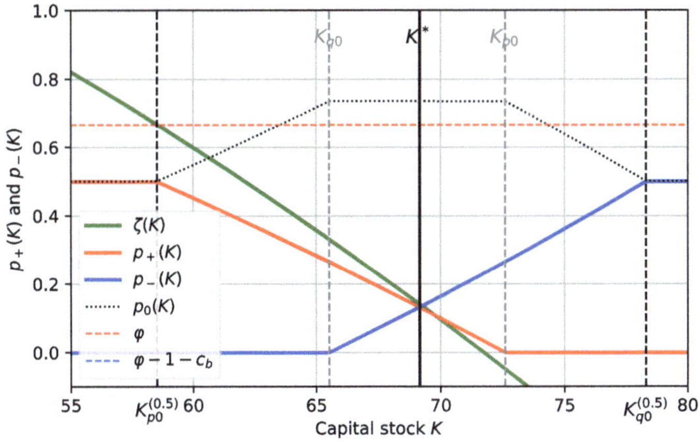


**Fig. 1.1** Transition probabilities: expansion $p_+(K)$ and contraction $p_-(K)$.
These probabilities summarize the entry/exit (mutation) dynamics and thereby define the Markov transition kernel for the capital stock $K_t$

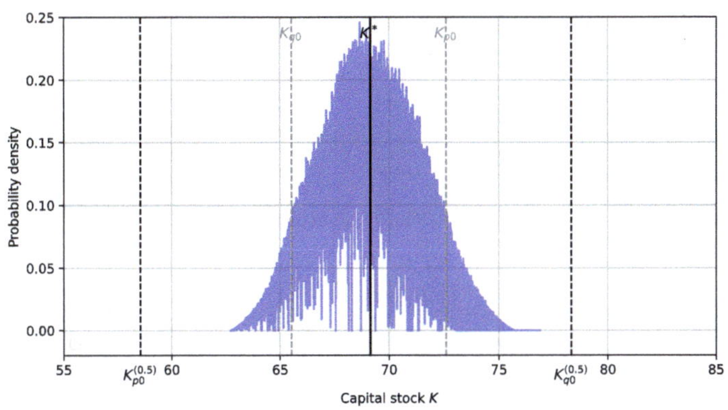


**Fig. 1.2** Invariant distribution of $K$.
The stationary density of the capital stock is concentrated within a central stability corridor, with most probability mass lying near intermediate values of $K$. Although the support extends beyond $[K_{q0}, K_{p0}]$, excursions outside this region are rare

Before turning to these three pillars, it is worth emphasizing an even more basic departure from much of contemporary macroeconomics: Keynesian analysis begins from the institutional separation between households and managerial firms. Firms are organized production units—embodying equipment, productive capital, and coordinated labor—whose central decision is how much capacity to operate and how much new capital to install. Households, by contrast, make consumption, saving, and portfolio choices. This division implies that investment and consumption are determined by distinct agents under distinct constraints, so aggregate demand need not adjust automatically to aggregate productive capacity. In this sense, Keynesian economics is not a Robinson–Crusoe economy in disguise; it is a theory of a monetary capitalist system in which coordination between firms' investment plans and households' spending plans becomes a central macroeconomic problem.

1. **Liquidity preference and effective demand failures.**
   Agents are not compelled to spend their money immediately. They may hoard liquidity for precautionary reasons or because money provides services that other assets do not. As a result, desired expenditure can fall short of productive capacity even in the absence of mechanical price rigidity, and unemployment and idle capacity can persist rather than being eliminated by automatic market clearing.
2. **Fixity of capital and the central role of investment.**
   Unlike labor or raw materials, capital goods are fixed in form and location once installed. This irreversibility makes investment decisions central for macroeconomic dynamics. Keynes's *marginal efficiency of capital* (MEC) captures how expectations—and in particular shifts in business confidence—govern whether firms expand capacity or retrench. Abrupt changes in confidence therefore translate into discontinuous changes in investment, generating downturns and upturns.
3. **The marginal propensity to consume and the multiplier.**
   Households typically consume only a fraction of marginal income, saving the remainder. This behavior implies a multiplier: investment disturbances are amplified into larger movements in income and employment.

## 1.2 Balance-Sheet Consistency: Linking Financial Wealth and Real Capital

A major weakness of many macroeconomic models—including some Keynesian ones—is the lack of an explicit linkage between financial wealth and the real capital stock. Traditional models often treat aggregate financial assets (money, bonds, equity) as detached from the physical stock of capital goods, creating inconsistencies that can mislead both theory and policy.

This book adopts the stock–flow consistent (SFC) discipline associated with Godley and Cripps (1983) and Godley and Lavoie (2007, 2012): transactions are recorded in sectoral matrices, flow identities and budget constraints hold exactly, and stocks evolve from flows. Where we depart from canonical SFC models is the monetary closure. Instead of loans-create-deposits with central bank accommodation, we impose a 100%-reserve payments system and let banks fund lending with nonmonetary liabilities. The accounting remains fully SFC; only the financial architecture and portfolio equations differ. In particular, household wealth equals the sum of money balances, bonds, and equity claims, which in turn correspond to the real capital stock held by firms.

### 1.2.1 Benefits of Balance-Sheet Consistency

1. **A richer consumption function.** Balance-sheet consistency yields a natural two-component consumption function

$$C = c_y \frac{Y}{P} + c_w \frac{\mathcal{W}}{P},$$

where $\mathcal{W}$ is the market value of household financial wealth. In the benchmark environment developed in this book, $\mathcal{W}/P$ is proportional to the real capital stock, so the consumption function can be written in reduced form as $C = c_y(Y/P) + c_w K$. This formulation links spending to both current income and accumulated wealth and helps reduce the short-run equilibrium to a one-dimensional mapping in $K$.
2. **Internal coherence.** A balance-sheet-consistent model prevents impossible states, such as wealth creation without real backing or unexplained losses. Every financial position is mirrored by a counterpart elsewhere in the system.
3. **Transparent feedback loops.** Because financial wealth is linked explicitly to real capital, the framework makes clear how capital accumulation feeds back into consumption, saving, and subsequent investment incentives.
4. **Clarifying the role of money and credit.** Under Fisher's 100%-reserve architecture, the model separates real sources of instability from money–credit amplification mechanisms by making the payments system and lending finance explicit.

## 1.3 From Short-Run Equilibria to Fluctuations

In the short run, given posted prices and wages, agents can reach a temporary equilibrium conditional on the predetermined capital stock and on expectations held within the period. This balance is fragile. When confidence or other determinants of investment incentives shift between periods, investment behavior changes, the capital stock evolves, and the economy transitions to a new short-run equilibrium. In this way, macroeconomic fluctuations arise endogenously from the dynamics of investment and expectations in a decentralized economy.

## 1.4 The Case for a Fisherian Institutional Setting

Irving Fisher, a contemporary of Keynes, proposed a 100%-reserve banking system in which banks hold full reserves against transactions deposits and do not create money via lending. Adopting this institutional setting allows us to strip out deposit fragility and isolate a cleaner baseline for studying real sources of instability—confidence shifts, investment frictions, and demand externalities—rather than banking-driven credit cycles.

A complementary line of argument emphasizes the advantages of fractional-reserve banking. By allowing banks to expand balance sheets beyond contemporaneous saving, such systems can relax financing constraints, support large and lumpy investment projects, and thereby foster rapid technological progress (see, e.g., Schumpeter 1934, Ch. 3). At the same time, this elasticity of credit may amplify fluctuations or enable large-scale expenditures without immediate increases in funding costs. The present analysis abstracts from these intermediation benefits in order to isolate the stabilizing role of a fixed-reserve monetary architecture; the comparison should therefore be understood as analytical rather than normative.

Against this backdrop, Fisher's 100%-reserve architecture and Keynes's demand-management program seem to pull in different directions. In fact, they share a common aim: dampening the amplitude of cycles to avoid the two pathologies that threaten real activity—cumulative inflation on the upside and deflationary slumps with persistent unemployment on the downside. Both acknowledged that modern economies oscillate around a moving equilibrium; modest, bounded fluctuations are not themselves the central problem. The threat of systemic instability is realized when an impulse drives the system outside its basin of attraction, a region metaphorically referred to as the "stability corridor."[3] Fisher (1933) extended this logic by identifying debt as a key amplifier of deflationary spirals.[4]

As Fig. 1.3 illustrates, both approaches can be read as targeting the same objective—keeping the economy inside a stability corridor where equilibrating forces prevail—so as to prevent cumulative inflationary or deflationary dynamics that arise once bounds are breached.

Keynes's "Harvey Road" view placed responsibility on informed fiscal–monetary policy to steer aggregate demand back toward balance,[5] whereas Fisher sought to embed an *automatic stabilizer* in the monetary architecture: a 100%-reserve payments system to weaken the credit–money feedbacks that amplify booms and busts. Different instruments—discretionary macroeconomic management in Keynes, institutional monetary design in Fisher—but the same target—macroeconomic stability (see Dimand 2005).

---

[3] Our use of "stability corridor" follows Wicksell (1898/1936) in spirit: when the market (money) rate is aligned with the natural rate, price movements are self-damping; when rates stray beyond those bounds, movements become cumulative (see Wicksell, 1936, Ch. 9). The term "corridor" follows Leijonhufvud (1973), who emphasizes that economies behave well within bounds but can exhibit coordination failures once displaced outside them.

[4] For Fisher's stabilization objective, see Fisher (1935), *100% Money: Designed to Keep Checking Banks 100% Liquid; to Prevent Inflation and Deflation*. For Keynes's demand-management aim, see Keynes (1933), Keynes (1936), Leijonhufvud (1968), and Blanchard and Fischer (1989). For discussion of their exchange and overlapping aims, see Allen (1993) on the 1944 Keynes–Fisher correspondence.

[5] Shorthand for Keynes's belief that capable public servants could steer demand through fiscal and monetary policy; "Harvey Road" alludes to the Cambridge address associated with the Keynes family.

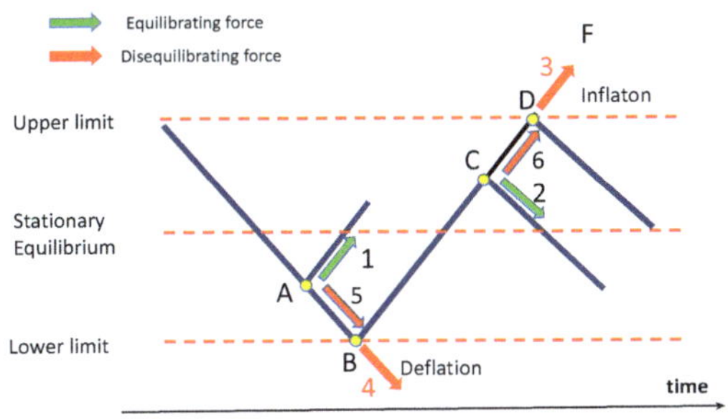


**Fig. 1.3** Stability corridor and cumulative dynamics outside the bounds. When activity stays inside the corridor, equilibrating forces dominate; breaching the *upper* (inflation) or *lower* (deflation) limit triggers disequilibrating cumulative processes. Arrows: 1 Recovery; 2 Slowdown; 3 Inflation; 4 Deflation; 5 Recession; 6 Overheating

## 1.5 Contributions of This Book

This book makes six contributions to Keynesian macroeconomic theory.

1. *An evolutionary model of investment dynamics.*
   Keynes's investment theory is reinterpreted through evolutionary game dynamics. Firms are modeled as *incumbents*, who replace depreciation by investing $I = \delta K$, and *mutants*, who temporarily deviate by expanding or halting investment. When the share of mutants exceeds a critical threshold, their behavior spreads through imitation, tipping the economy to a new stationary distribution of capital. This framework makes Keynes's notion of shifting confidence mathematically tractable and explains how small local deviations can cascade into systemic regime shifts.
2. *A balance-sheet-consistent consumption function.*
   By linking household wealth explicitly to the real capital stock, we derive a consumption function that depends on both income and capital, $C = c_y(Y/P) + c_w K$. This not only captures Keynes's insight about the marginal propensity to consume but also helps make the short-run equilibrium fully expressible as a function of $K$.
3. *Three interacting layers of uncertainty.*
   The framework rests on a three-layer structure—linking individual firms, industries, and the macroeconomy—that connects microlevel sentiment shocks to aggregate capital adjustment through sectoral heterogeneity. The intermediate industry layer is crucial: it allows the stochastic transition of aggregate capital to be interpreted not as a single macroshock, but as the aggregate outcome of heterogeneous sectoral decisions. In this way, firm-level sentiment shocks are aggregated through sectoral disturbances to generate smooth probabilistic

transitions, linking the law of large numbers at the industry level with the existence of a smooth invariant distribution at the macrolevel.[6]

4. *Reduction to a one-dimensional Markov process.*
   With consumption and investment both tied to $K$, macroeconomic dynamics can be summarized as a one-dimensional Markov process for the capital stock. This simplification enables rigorous probabilistic analysis while retaining economic content.
5. *Existence and uniqueness of a stochastic equilibrium.*
   We prove existence (and, under additional conditions, uniqueness) of an invariant probability distribution for the economy. Despite ongoing fluctuations, the system converges to a stable stochastic steady state in which the distribution of outcomes is time-invariant. This result provides a firm foundation for interpreting persistent macroeconomic volatility.
6. *New insights into long-run policy effects.*
   Unlike conventional macroeconomics, which often predicts long-run neutrality of money and wages, our framework shows how nominal institutions can have persistent effects by shifting the invariant distribution of $K$ and altering its proximity to boundary regions. Exogenous changes in money or wage-setting parameters therefore have long-run consequences for output and employment through their distributional effects.

More broadly, the present framework contributes to a growing strand of macroeconomic analysis that departs from representative-agent foundations.[7]

## 1.6 Main Findings (Preview)

The analysis establishes four headline results:

1. **Existence and confinement.** The economy admits a unique stochastic steady state: the capital stock follows an ergodic Markov process with a well-defined invariant distribution, whose support is confined to a stability corridor determined by real anchors and boundary regimes.
2. **Neutrality at the anchor, not in distribution.** Long-run monetary neutrality holds at the level of the stochastic anchor (the neutral point and associated real variables). However, nominal institutions (money supply and wage-setting rules)

[6] Excluding the industry layer would force the unrealistic assumption that the entire macroeconomy moves synchronously as a single unit. In that case, stochastic stability would become a purely statistical artifact rather than a reflection of propagation mechanisms.

[7] An alternative approach along these lines is developed by Aoki and Yoshikawa (2011), who derive aggregate demand from the distribution of sectoral productivities using methods inspired by statistical physics. While their framework emphasizes probabilistic aggregation across heterogeneous units, the present analysis focuses instead on evolutionary dynamics and coordination among agents.

matter by shifting the *distribution* of outcomes around that anchor and by altering how frequently the economy visits lower-boundary (slack) and upper-boundary (full-utilization) regions.

3. **Wage adjustment as automatic stabilization under a money cap.** With a fixed money supply, endogenous nominal wage adjustment stabilizes the economy against deflationary pressure: wage cuts reduce the price level and transactions demand, reduce the market interest rate, and move the invariant distribution away from the lower boundary.
4. **Keynesian boundary asymmetry.** Interior dynamics are locally self-stabilizing, whereas proximity to the boundaries can activate mechanisms that turn disturbances into cumulative inflationary or deflationary episodes. The practical objective of stabilization policy is therefore to keep the economy away from its real boundaries, rather than to fine-tune a point equilibrium outcome.

### A Conceptual Implication

Taken together, the first and fourth results imply a conceptual conclusion that runs counter to conventional intuition. The existence of a unique stochastic steady state does not, by itself, guarantee robustness: stability in this framework is an interior property, sustained by mechanisms that operate only when the economy is sufficiently far from its real boundaries. In particular, the Keynesian asymmetry identified above implies that self-stabilizing forces are active in the interior of the state space, while proximity to boundaries—such as those associated with capacity constraints, inflationary pressure, or deflationary slack—activates cumulative dynamics. This observation casts new light on the role of market structure. Under perfect competition, where markups vanish and long-run profits are driven to zero, the economy operates at the corridor boundary (profit sustainability or utilization boundary) rather than in the interior. As a result, the very conditions conventionally associated with long-run efficiency undermine local robustness, rendering the steady state fragile to small disturbances. In this sense, persistent zero-profit equilibria are not merely knife-edge in theory but structurally exposed to disturbances in stochastic environments.

### Methodological Stance

Unlike representative-agent models that derive consumption and labor supply from a single utility-maximization problem, this book treats consumption, money demand, and employment as distinct behavioral blocks. Labor demand is determined on the firm side, while household consumption depends on realized income rather than on a notional labor–leisure optimization problem. This modular structure reflects a Keynesian view in which employment is an outcome of effective demand and wage-setting institutions, not a decision variable chosen by households.

Accordingly, the absence of an explicit household labor-supply function is not a simplifying omission but a substantive modeling choice. The framework emphasizes that households respond to income and monetary conditions as they are realized, while employment is governed by firms' hiring decisions, financial constraints, and aggregate demand. This separation allows the analysis to focus on the endogenous propagation of demand fluctuations and wage dynamics without imposing the strong informational and optimization requirements embedded in joint consumption–labor choice.

**Roadmap**

For expositional clarity, the book can be read as consisting of four broad parts, although chapters are numbered sequentially. It begins by establishing the accounting structure and behavioral blocks of the monetary economy, then develops an evolutionary theory of investment and entry. Building on this foundation, the analysis turns to aggregate dynamics and the characterization of the stability corridor. The final part interprets the results in institutional and policy terms, and situates the framework relative to Keynesian and modern stabilization approaches.

## 1.7 Structure of the Book

The book proceeds in four parts:

1. **Foundations**
   - Model Structure: Actors, Timing, and Accounting
   - Fisher's 100%-Reserve Banking System and Balance-Sheet Consistency
   - Microfoundations of Consumption and Money Demand
   - Short-Run Dynamics: Demand, Finance, and Markups (PM–LM)
2. **Investment Theory**
   - Keynesian Uncertainty, Liquidity Preference, and the Fixity of Capital
   - The Mutant–Incumbent Framework for Evolutionary Investment Dynamics
   - Behavioral Frictions and Adjustment Thresholds
   - Conditions for Tipping and Regime Shifts
3. **Macroeconomic Dynamics**
   - Theory of Wage Setting
   - Reduction to a One-Dimensional Markov Process in Capital Stock
   - Analytical Results on Stability, Fluctuations, and Ergodicity
4. **Policy and Interpretation**
   - Long-Run Effects of Money and Wage Policies
   - Comparison with Keynesian Fine-Tuning and Modern Stabilization Approaches
   - Links to Sunspot Equilibria and Agent-Based Simulations

## 1.8 Scope, Minimality, and Closure

The aim of this monograph is to isolate the internal stabilizing mechanism of a market economy operating under stationary conditions. To that end, we adopt a deliberately minimal and closed benchmark environment: (i) income from firms and financial intermediaries is distributed equally across households, so that aggregate demand is not driven by heterogeneity in marginal propensities to consume;[8] (ii) there are no external shocks to technology or preferences; (iii) there is no foreign sector, no government sector, and no discretionary monetary intervention; and (iv) population and technology are stationary; (v) prices adjust to clear goods markets (the pricing–markup curve), while money demand comes from a pooled transactions sector (the LM curve).

We work with a stationary short-run benchmark we call the symmetric quasi-steady state (SQSS). In an SQSS, capital is fixed, fundamentals and population are constant, prices clear goods markets, and the interest rate adjusts to clear the payments market; hence, net investment is zero, and gross investment merely replaces depreciation, $I = \delta K$. This stripped-down SQSS environment is not a description of reality; it is a benchmark designed to make the stabilizing channel—the endogenous movement of the short safe rate and its implications for investment—as transparent as possible. This benchmark provides the foundation for the stochastic dynamics analyzed in later chapters. Extensions (e.g., heterogeneous MPCs by income, policy rules, foreign trade, trend growth) are left for future work.

### *1.8.1 Scope Disclaimer: Financial Markets and Fragility*

Following Uzawa (1984), two institutional features are central to Keynesian dynamics: (i) firms and households choose investment and saving independently, and (ii) corporate bonds and shares are highly marketable, so their prices reflect market expectations rather than firms' "real values." The tension between the fixity of productive capacity (with its organic structure) and the liquidity of financial claims can therefore generate financial fragility.

This monograph abstracts from that fragility channel. In our baseline:

- Asset markets do not feed back into investment via leverage, collateral, or valuation loops (no Minsky-style debt dynamics, no Tobin–$q$ amplification).
- Banks operate under a 100%-reserve payments architecture; inside-money creation and credit cycles are absent.

[8] In general, distribution matters: there are large differences in the marginal propensity to consume between the top 1% and the bottom 80%, so shifts in distribution can have macroeconomic consequences. See Stiglitz (2018, esp. discussion around inequality, aggregation, and MPCs).

- The confidence index $\mu$ shifts the marginal efficiency of investment directly; security prices are not modeled.
- Expectation formation—the Minskyan engine of instability—is central to the phenomena studied here, but an explicit treatment lies beyond the scope of this volume.

The payoff is a minimal, transparent environment that isolates the real-side mechanism we study: how confidence-driven fluctuations interact with the interest-rate stabilizer in the PM–LM block. Financial fragility is important, but it is left for future work.

Consistent with Keynes's own framing, we treat financial-market forces as amplifiers rather than first causes. In Keynes (1936, Ch. 22), "the essential character of the Trade Cycle ... is mainly due to the way in which the marginal efficiency of capital fluctuates," while it is "aggravated by" other short-period factors. In Keynes (1936, Ch. 12), he likewise warns that "the position is serious when enterprise becomes the bubble on a whirlpool of speculation," underscoring that market liquidity can magnify instability without constituting its fundamental source. This abstraction from explicit financial fragility is deliberate and complements the focus on boundary-driven real dynamics developed in the remainder of the book.

### *1.8.2 Conventions and Standing Assumptions (Money and Wages)*

For ease of reference, the remainder of this chapter records the global environment and standing assumptions used throughout the book; readers may skip this material on a first pass.

**Assumption 1.1 (Global Standing Assumptions (Used Unless Noted)) (M) Money stock.** The nominal money stock $M$ is treated as exogenous and constant in the baseline chapters. In the long-run equilibrium chapter (Chap. 8), we study the comparative-static effects of exogenous changes in $M$ on the invariant distribution—both graphically (simulation) and analytically. We do not endogenize $M$ anywhere in the book.

**(W) Money wages.** The nominal wage $W$ is predetermined within the period and held constant across time in the baseline chapters (PM–LM, consumption & money demand, SQSS). We endogenize wage setting only in the dedicated wage-negotiation chapter (Chap. 7), where a flex–fix mechanism is microfounded. In the long-run equilibrium chapter, we may also consider exogenous counterfactuals for $W$ as comparative statics.

**(C) Propensity parameter.** The composite demand parameter $c := c_w/(1 - c_y)$ is treated as structural in baseline analysis; Chap. 8 examines exogenous "$c$-changes" as robustness exercises.

#### 1.8.2.1 Global Environment and Standing Assumptions (GE)

**Assumption 1.2 (GE: Model skeleton)**

**GE1. Agents and market structure.** A unit mass of symmetric monopolistically competitive firms facing Dixit–Stiglitz demand with elasticity $\eta > 1$; a representative household sector; a payments/banking block that implements the LM locus. Firms set prices with a constant markup.

**GE2. Technology and depreciation.** $Q = F(K, N) = K^{\alpha} N^{1-\alpha}$ with $0 < \alpha < 1$. Capital depreciates at rate $\delta \in (0, 1)$.

**GE3. Capital fixity and gestation.** Within period $t$, installed capital is predetermined; investment has one-period gestation: $K_{t+1} = (1 - \delta)K_t + I_t$.

**GE4. Timing and pricing.** Given $(M, W, K_t, \bar{N}_t)$ at the start of $t$, solve PM–LM to pin $(r_t, \omega_t)$ and set $P_t = W/\omega_t$. Prices and the rate $r_t$ are committed for the rest of period $t$.

**GE5. Employment (quasi-fixity).** Employment used in production is preset: $N_t = \bar{N}_t$; utilization/intensity may vary intra-period. A reduced form is $Q_t = \bar{Q}_t + \kappa_q(X_t - \bar{Q}_t)$ with $\kappa_q \in [0, 1]$ and $\bar{Q}_t = F(K_t, \bar{N}_t)$.

**GE6. Money and transactions scale.** CES–MIU money demand yields LM:

$$\frac{M}{P_t} = \lambda^{1-\beta} C_t \Big(\frac{r_t}{1 + r_t}\Big)^{-\beta}, \qquad \omega_t = \frac{W}{P_t}.$$

The scale $C_t$ used for LM at the start of $t$ is the replacement-only baseline $\bar{C}_t = c_y \bar{Y}_t^r + c_w K_t$, where $\bar{Y}_t^r = \bar{Q}_t - \delta K_t$.

**GE7. Consumption standing result.**

$$C_t = c_y\, Y_t^r + c_w\, K_t, \qquad 0 < c_y < 1,\ c_w > 0,$$

with $Y_t^r = Y_t / P_t$ the real net income.

**GE8. Income and profits.**

$$Y_t^r = Q_t - \delta K_t = \omega_t N_t + r_t K_t + \Pi_t^F, \qquad \Pi_t^F \equiv Q_t - [\omega_t N_t + (r_t + \delta) K_t].$$

Unless otherwise noted, firm (and bank) profits are rebated lump-sum to households (they enter $Y_t^r$).

**GE9. Expectations.** Short-run beliefs (those used to set PM–LM and $C_t$) are held fixed within $t$; long-run beliefs may update across periods (confidence shocks are introduced later).

**GE10. Baseline (replacement only).** We anchor each period at $I_t = \delta K_t$; deviations (e.g., investment shocks) propagate within $t$ via $\kappa_q$; PM–LM are resolved only at the start of $t+1$.

These standing assumptions define the common environment maintained throughout the book. Chapter 2 now operationalizes this skeleton by specifying the actors, timing, and accounting structure in detail.

## References


Allen, W. R. (1993). Irving fisher and the 100 percent reserve proposal. *Journal of Law and Economics, 36*(2), 703–717. https://doi.org/10.1086/467295, https://ideas.repec.org/a/ucp/jlawec/v36y1993i2p703-17.html.

Aoki, M., & Yoshikawa, H. (2011). *Reconstructing macroeconomics: a perspective from statistical physics and combinatorial stochastic processes*. Cambridge University Press. https://doi.org/10.1017/CBO9780511510670.

Benes, J., & Kumhof, M. (2012). The Chicago Plan revisited. IMF Working Paper WP/12/202, International Monetary Fund. https://doi.org/10.5089/9781475505528.001.

Blanchard, O., & Fischer, S. (1989). *Lectures on macroeconomics*. MIT Press.

Dimand, R. W. (1997). Irving fisher and modern macroeconomics. *American Economic Review, 87*(2), 442–444.

Dimand, R. W. (2005). Fisher, keynes, and the corridor of stability. *American Journal of Economics and Sociology, 64*(1), 185–199.

Fisher, I. (1935). *100% money: Designed to keep checking banks 100% liquid; to prevent inflation and deflation; largely to cure or prevent depressions; and to wipe out much of the national debt*. Adelphi Company. https://books.google.com/books?id=O0tZLbBNlBUC, revised edition.

Godley, W., & Cripps, F. (1983). *Macroeconomics*. Oxford University Press.

Godley, W., & Lavoie, M. (2007). *Monetary economics: An integrated approach to credit, money, income, production and wealth*. Palgrave Macmillan.

Godley, W., & Lavoie, M. (2012). *Monetary economics: An integrated approach to credit, money, income, production and wealth* (2nd edn.). Palgrave Macmillan.

Keynes, J. M. (1933). *The means to prosperity*. Macmillan and Co. https://gutenberg.ca/ebooks/keynes-means/keynes-means-00-h.html. Pamphlet enlarged from four articles in *The Times*, March 1933.

Keynes, J. M. (1936). *The general theory of employment, interest and money*. Macmillan.

Leijonhufvud, A. (1968). *On Keynesian economics and the economics of Keynes* (Vol. 158). Oxford University Press.

Leijonhufvud, A. (1973). Effective demand failures. *The Swedish Journal of Economics, 75*(1), 27–48.

Schumpeter, J.A. (1934). *The Theory of Economic Development: An Inquiry into Profits, Capital, Credit, Interest, and the Business Cycle*. Harvard University Press.

Stiglitz, J. E. (2018). Where modern macroeconomics went wrong. *Oxford Review of Economic Policy, 34*(1–2), 70–106. https://doi.org/10.1093/oxrep/grx059.

Uzawa, H. (1984). *Keinzu "Ippan Riron" wo Yomu [Reading Keynes's "General Theory"]*. 7, Iwanami Shoten.

Wicksell, K. (1936). *Interest and prices: A study of the causes regulating the value of money*. Macmillan for the Royal Economic Society. Translated by R. F. Kahn; introduction by Bertil Ohlin. First published in German (1898) as *Geldzins und Güterpreise*

# Chapter 2
# Model Structure: Actors, Timing, and Accounting Framework

> *The difficulty lies, not in the new ideas, but in escaping from the old ones, which ramify, for those brought up as most of us have been, into every corner of our minds.*
>
> — John Maynard Keynes, *The General Theory*, Ch. 24

## 2.1 Introduction

The aim of this chapter is to fix the language, timing, and accounting conventions that underlie the short-run pricing–markup (PM) and liquidity–money (LM) blocks, as well as the investment dynamics studied in later chapters. The purpose is not to add complexity, but to ensure that the model's "moving parts" are fully understood before they are set in motion.

We therefore separate three layers that are often conflated: (i) *who holds what* (stocks and balance sheets, expressed in real units); (ii) *when things happen* within the period, culminating in an end-of-period book close—the *accounting instant*;[1] and (iii) *how the monetary plumbing works* (payments versus funding). The payoff of this discipline is that later, when we analyze short-run PM–LM behavior and the capital dynamics, each mechanism can be activated without ambiguity about what it touches.

The remainder of the chapter proceeds as follows. Section 2.2 outlines the basic structure of the model and introduces the principal actors and markets. Section 2.3 describes the stochastic structure of the economy and explains how micro-level industry dynamics aggregate into macroeconomic outcomes. Section 2.4 specifies the timing of events within each period and presents the transaction-flow matrix that summarizes the accounting relationships among sectors. Section 2.5 then introduces the monetary architecture of the model under a 100%-reserve banking system,

---

[1] The accounting instant is a zero-duration end-of-period close: loans and funding are redeemed and reissued, profits are distributed, and the payments ledger is cleared. It is a bookkeeping device that pins down stock–flow consistency.

 
I. Takahashi, *Microfoundations of Keynesian Economics*, Evolutionary Economics and Social Complexity Science 37, https://doi.org/10.1007/978-981-95-9390-3_2

distinguishing between the payments and funding books. Finally, Sect. 2.6 discusses the role of risk capital and the determination of the funding rate under a fixed money supply.

## 2.2 Model Outline

We begin with a bird's-eye view of the symmetric quasi-steady-state (SQSS) environment and the minimal *payments–funding* monetary architecture.

**Sectors** There are three sectors: (i) households, (ii) firms, and (iii) a split bank with a *payments book* and a *funding book*. The payments book provides par settlement, while the funding book provides credit intermediation. The economy uses five basic instruments or resources: goods, labor, money (payments balances), bank paper (funding liabilities), and productive capital stock. Unless stated otherwise, all balance-sheet entries are reported in *real* (goods) units at date-$t$ prices. Households ultimately own the real economy; at the accounting instant, installed capital is carried at replacement cost (accounting $q = 1$).

**Remark 2.1 (Two $q$'s Used in the Book)** Investment and entry are governed by

$$q_t^{\text{inv}} \equiv \frac{p_t^{K,\text{mkt}}}{p_t^I},$$

the market value of installed capital relative to replacement cost. In Chap. 5, we show that $q^{\text{inv}} = 1 + \frac{1}{\alpha(\eta-1)}$ under the baseline markup/elasticity assumptions. At the accounting instant, by contrast, installed capital is carried at replacement cost, so

$$q_t^{\text{acc}} \equiv \frac{p_t^{K,\text{book}}}{p_t^I} = 1.$$

This is a valuation convention used to close the accounts. Unless stated otherwise, "$q = 1$" in this chapter refers to $q^{\text{acc}}$.

**Households** Households are treated as the consolidated private owner of the economy. Before the accounting instant, they may hold payments deposits $D_t^H$, funding claims $F_t$ (the bank's nonmonetary paper), and—because firms have not yet paid out—a residual equity claim

$$E_t^F = K_t - L_t,$$

where $K_t$ denotes the capital stock installed at the start of period $t$.[2] At the accounting instant, firms distribute profits and retained earnings, closing the equity

[2] $K_t$ denotes the capital stock operational at the outset of period $t$.

position, so that private wealth is carried in the funding instrument. In this sense, aggregate household net worth equals the capital stock: $NW_{H,t} = K_t$ (see Lemma 2.1).

Households are homogeneous and passive in the short run: they supply labor, own firms, and receive an equal share of national income $Y_t$. Consumption $C_t$ is determined by a stable reduced-form rule[3] introduced later; for now, we use only that $C_t$ rises with contemporaneous income and with wealth, and that the marginal propensity to consume out of income is finite (strictly less than one). The nominal wage $W$ is fixed within the period (and endogenized only in a later chapter).

**Firms** A large number of price-setting firms produce differentiated varieties using a common technology. Each firm combines installed capital $K_i$ with variable labor $N_i$ to produce a single good. New investment ordered in period $t$ becomes productive only in $t+1$ (one-period gestation). Firms finance all investment with *external* funds through the bank's *funding book*; transactions deposits are not used to finance capital expenditure.[4]

**Remark** Firms finance all investment with external funds through the bank's funding book. This simplifying assumption is consistent with the full-payout convention imposed throughout the model, under which firms do not accumulate retained earnings internally. The assumption isolates the macroeconomic role of external finance and investment coordination. Allowing partial internal financing would likely smooth investment fluctuations by reducing firms' dependence on external funding conditions, but the underlying coordination mechanism would remain operative as long as investment decisions continue to depend on expectations and profitability.

**Bank: Two Books by Design** The bank is split into a *payments book*, which provides par settlement, and a *funding book*, which provides credit intermediation. The payments book is a custody ledger: within the period, it carries reserves $R_t$ matched one-for-one against deposits $D_t$; at the accounting instant, both are cleared and reissued as needed. The funding book finances loans with nonmonetary liabilities $F_t$; at the accounting instant, we impose zero bank equity, $E_t^B = 0$, so that funding equals loans, $F_t = L_t$. This *institutional split between payments and funding* is the backbone of the model.[5]

---

[3] A microfoundation—using CES (constant elasticity of substitution) preferences with money in utility and an overlapping-generations (OLG) structure—and the explicit reduced form for $C_t$ are developed in Chap. 3.

[4] For concise treatments of monopolistic competition in macro, see Solow (1998). Our symmetry construction follows Iwai (1981, 1987) and differs mainly in making installed capital $K$ explicit.

[5] The separation between a reserve-backed payments ledger and a credit-funding ledger reflects the long-standing distinction between settlement balances used for payments and funding instruments used for credit intermediation in monetary economics; see, e.g., (Fisher, 1935; Mehrling, 2011).

Together, these elements provide the mechanism through which decentralized heterogeneous behavior is translated into a smooth transition kernel and an invariant distribution studied later.

## 2.3 Industry-Level Dynamics and Micro–Macro Link

Before turning to the within-period timing and balance-sheet mechanics, we briefly describe the stochastic aggregation structure that links firm behavior to the evolution of aggregate capital.

The stochastic foundation of the capital dynamics rests on three layers—firms, industries, and the macroeconomy. The hierarchy is summarized schematically as

$$(\epsilon_i) \longrightarrow U \longrightarrow (\nu_t, K_t),$$

where idiosyncratic firm-level shocks $\epsilon_i$ are aggregated within industries, industry-level investment adjustments are driven by a random draw $U$, and the macrostate evolves as a Markov process in $(K_t, \nu_t)$.

In modeling the stochastic evolution of aggregate capital $K_t$, we treat the *industry* (rather than the individual firm) as the fundamental decision and aggregation unit. Each industry contains sufficiently many heterogeneous firms that idiosyncratic shocks $\epsilon_i$ average out within the industry. At the same time, industry-level investment behavior remains responsive to macrosentiment $\nu_t$ and to industry-specific disturbances captured by the draw $U$. This layered structure yields a disciplined aggregation path: firm-level variation maps into industry-wide investment adjustments, which in turn generate the state-dependent transition probabilities $p_\pm(K_t, \nu_t)$ and multiplicative capital increments $(1 \pm \epsilon^\pm)$ that govern the law of motion of aggregate capital. The setup preserves behavioral richness while keeping the aggregate dynamics analytically tractable within a Markovian framework.

Before laying out the within-period clock, we display balance sheets in real units *immediately before* and *immediately after* the end-of-period book close (the accounting instant) as shown in Table 2.1.[6] The two panels in Table 2.1 anchor the stocks that the timeline connects. Panel A reports the within-period snapshot, when deposits and reserves are active in the payments book and firms still carry equity. Panel B reports the post-close balance sheet, in which deposits and reserves have been cleared, bank equity is set to zero, and firms have paid out, leaving private wealth carried in funding claims. With the before/after stocks in view, the next section details the within-period sequence that takes the economy from panel A to panel B.

[6] For the plumbing assumptions and identities, see Sect. 2.5.1.

**Table 2.1** Balance sheets in real units before and at the accounting instant

Panel A: Before accounting instant (snapshot)

| Households (real) | |
|---|---|
| Assets | Liabilities + equity |
| Deposits (payments) $D_t^H$ | (No household liabilities) |
| Funding claims $F_t$ | |
| Equity claims on firms $E_t^F$ | |
| | Net worth $NW_{H,t}$ |

| Firms (real) | |
|---|---|
| Assets | Liabilities + equity |
| Fixed capital $K_t$ | Loans $L_t$ |
| | Equity $E_t^F = K_t - L_t$ |

| Bank—payments (real) | |
|---|---|
| Assets | Liabilities + equity |
| Reserves $R_t$ | Deposits $D_t$ |

| Bank—funding (real) | |
|---|---|
| Assets | Liabilities + equity |
| Loans to firms $L_t$ | Funding liabilities $F_t$ |
| | Bank equity $E_t^B = L_t - F_t$ |

Panel B: At the accounting instant (post-close)

| Households (real) | |
|---|---|
| Assets | Liabilities + equity |
| Funding claims $F_t$ | |
| | Net worth $NW_{H,t} = F_t$ |

| Firms (real) | |
|---|---|
| Assets | Liabilities + equity |
| Fixed capital $K_t$ | Loans $L_t$ |

| Bank—payments (real) | |
|---|---|
| Assets | Liabilities + equity |
| Reserves $R_t = 0$ | Deposits $D_t = 0$ |

| Bank—funding (real) | |
|---|---|
| Assets | Liabilities + equity |
| Loans to firms $L_t$ | Funding liabilities $F_t = L_t$ |
| | Bank equity $E_t^B = 0$ |

*Note.* At the accounting instant, the payments book is cleared, so $D_t = R_t = 0$ as a within-period snapshot. Bank equity in the funding book is set to zero, $E_t^B = 0$, so $F_t = L_t$. Firms have already paid out the residual equity claim, so household net worth is carried entirely in funding claims, $NW_{H,t} = F_t = L_t$. In equilibrium, $D_t \equiv D_t^H$ since households are the only deposit-holding sector

The accounting identities underlying the split-bank structure are developed in Sect. 2.5.1 (Lemma 2.1) and are taken as institutional primitives in what follows. The following assumption specifies how financial contracts are carried from one period to the next.

**Assumption 2.1 (One-Period Rollover)** At the accounting instant, last period's loans and funding $(L_t, F_t)$ are redeemed (principal and interest are settled), and new instruments $(L_{t+1}, F_{t+1})$ are issued for the next period.

## 2.4 Timing of Events Within the Period

**Within-period clock** We use the following timeline, which will later underpin the short-run pricing–markup (PM) and liquidity–money (LM) block in Chap. 4.

*T.1* *Contracts posted:* Firms post *prices* for their varieties and *wages* for the period, and banks post *interest rates*; all remain fixed intra-period. This reflects real-world price posting under uncertainty rather than Walrasian tâtonnement.[7]

*T.2* *Production and sales:* Firms employ $N_t$, produce, and sell. The payments book processes transactions at par; deposits and reserves move in lockstep ($D_t = R_t$) during the period.

*T.3* *Interest accrual:* Interest on funding claims $r_t^F F_t$ is paid to households; interest on loans $r_t^L L_t$ is paid by firms.[8]

*T.4* *Accounting instant (close): End-of-period book close.* Old $L_t$ and $F_t$ are redeemed and new $L_{t+1}$ and $F_{t+1}$ are issued; firms pay out so that $E_t^F = 0$; the bank transfers any residual spread so that $E_t^B = 0$; and the payments ledger is cleared so that $D_t = R_t = 0$ as a snapshot.

This timing cleanly separates *transactions* (handled by the payments book) from *intermediation* (handled by the funding book). It also explains why short-run PM–LM shocks primarily move *quantities* within the period: prices, wages, and rates reset only when the books are closed.

We allow loan and funding rates $(r_t^L, r_t^F)$ to differ within the period; their spread is institutionally sluggish and is closed at the accounting instant.

### 2.4.1 *Transaction-Flow Matrix*

As explained in Chap. 1, we adopt stock–flow consistent (SFC) accounting: transactions are recorded in sectoral flow matrices so that budget constraints and adding-up identities hold exactly, while stocks evolve as the accumulated flows.

Table 2.2 reports a minimal transaction-flow matrix (TFM) at the end-of-period *accounting instant* (real units). It records only interest flows and principal redemption/issuance among households, firms, and the bank's funding book. The table makes the split-bank closure transparent: payments clear ($D = R = 0$), bank equity closes at zero ($E^B = 0$), and firms pay out ($E^F = 0$); hence $F = L$. It also verifies that all column sums are zero. A full-period TFM (including wages, profits, and consumption) appears in Appendix section "Stock-Flow Consistency and Accounting Closure".

---

[7] See Iwai (1974) on price, wage, and employment adjustments by firms facing uncertainty; for the broader disequilibrium critique of tâtonnement, see Iwai (1981).

[8] We track these flows in the transaction-flow matrix (TFM); intermediation spreads adjust only sluggishly, matching observed institutional frictions.

**Table 2.2** Transaction-flow matrix at the accounting instant (real units)

| | Households | Firms | Bank (funding) |
|---|---|---|---|
| Interest on loans | | $-r_t^L L_t$ | $+r_t^L L_t$ |
| Interest on funding | $+r_t^F F_t$ | | $-r_t^F F_t$ |
| Redeem old loans ($t$) | | $-L_t$ | $+L_t$ |
| Redeem old funding ($t$) | $+F_t$ | | $-F_t$ |
| Issue new loans ($t$+1) | | $+L_{t+1}$ | $-L_{t+1}$ |
| Issue new funding ($t$+1) | $-F_{t+1}$ | | $+F_{t+1}$ |
| *Column sum* | 0 | 0 | 0 |

*Note.* The payments book is omitted at the accounting instant ($D_t = R_t = 0$). Bank equity in the funding book closes at zero ($E_t^B = 0$); hence $F_t = L_t$; firms pay out, so the residual equity claim closes ($E_t^F = 0$). The one-period rollover convention implies $L_{t+1} = (1 + r^L)L_t$ and $F_{t+1} = (1 + r^F)F_t$, so each column sum is zero

## 2.5 Monetary Architecture Under 100%-Reserve Banking

Under a 100%-reserve system, the bank is decomposed into a *payments book* (deposits $D$ fully backed by reserves $R$) and a *funding book* (loans $L$ financed by nonmonetary paper $F$). The payments book is a custody ledger for settlement, while the funding book intermediates credit. All variables are reported in real units (e.g., $D_t := \mathcal{D}_t/P_t$, $R_t := \mathcal{R}_t/P_t$, $C_t := \mathcal{C}_t/P_t$), where $\mathcal{D}_t$, $\mathcal{R}_t$, and $\mathcal{C}_t$ denote nominal deposits, reserves, and idle cash balances, respectively.

### 2.5.1 *Payments Book (Deposits and Settlement)*

In the split-bank 100%-reserve setup, the payments book implements par settlement: payments deposits are fully backed by reserves one-for-one. Let $i_*$ denote the competitive yield on perfectly liquid payments balances. In the baseline environment, reserves are fiat money issued by the government and the nominal reserve stock is fixed at $M^S$. Throughout the book, we normalize $i_* = 0$ unless noted.

Allowing $i_* > 0$ would not change any mechanism emphasized here: it simply shifts opportunity-cost terms. Formally, the formulas carry through by replacing $r_t^F$ with $r_t^F - i_*$ wherever an opportunity-cost term appears. The normalization $i_* = 0$ therefore isolates the role of portfolio choice and funding rates; none of the results relies on it.

To make the accounting transparent, we impose two conventions that pin down what holds *at the accounting instant* and thereby anchor stock–flow consistency.

**Assumption 2.2 (Settlement Convention)** Payments balances $D_t$ and reserves $R_t$ are *intra-period* settlement instruments: they expand as needed to clear transactions, and the payments ledger is cleared at the accounting instant, so that $D_t = R_t = 0$ at the book close. Any residual payments balances are rolled over at par into nonmonetary funding paper $F_t$; at the start of the next period, transactions balances are reconstituted from $F_t$ at par as needed.

**Assumption 2.3 (Payout at the Instant (Firm Equity Closes))** Firms distribute period profits at the accounting instant so that the residual equity claim closes at the book close: $E_t^F = 0$.

Under the settlement convention and payout at the instant, the split-bank closure implies a transparent consolidation: at the accounting instant, private net worth is carried in funding claims and corresponds (under $q^{\mathrm{acc}} = 1$) to the replacement-cost value of installed capital.

**Lemma 2.1 (Split-Bank Closing Identities)** *At the accounting instant, the following identities hold:*

Payments: $D_t = R_t$ *within the period, and the payments ledger is cleared at the instant, so* $D_t = R_t = 0$ *as a snapshot at close.*
Funding: *Bank equity closes at zero,* $E_t^B = 0$*; hence funding liabilities equal loans,*

$$F_t = L_t.$$

Payout: *Firms distribute earnings so that the residual equity claim closes at the instant,* $E_t^F = 0$.

*Hence, aggregate household net worth at the accounting instant is carried in funding claims:*

$$NW_{H,t} = F_t = L_t.$$

*Under the replacement-cost valuation convention for installed capital (*$q^{acc} = 1$*), aggregate household net worth corresponds to a claim on the installed capital stock, so* $NW_{H,t} = K_t$.

For later use, the remark below clarifies how the notion of real money balances in Chap. 3 differs from the deposit balance $D_t$ used here.

**Remark 2.2 (Within-Period Money Services vs. End-of-Period Stock)** Money demand in the consumption chapter is defined in terms of the *within-period* real services of money, denoted by $H_t$ (money-in-utility). This corresponds to the time average of real transactions balances during the period, not the post-close stock, which is zero by Assumption 2.2. Formally,

$$H_t \equiv \frac{1}{\Delta t} \int_t^{t+\Delta t} \frac{D(\tau)}{P_t} \, d\tau.$$

Having clarified how payments clear within each period under 100% reserves, we now turn to the complementary side of the balance sheet: the funding flows that enable investment and link bank liabilities to firms' capital formation.

### 2.5.2 Funding Book (Loans and Funding)

The funding book intermediates credit: It issues nonmonetary claims $F_t$ to households and extends loans $L_t$ to firms. At the accounting instant, bank equity is set to zero, so $F_t = L_t$. The next assumption clarifies the source of investment finance.

**Assumption 2.4 (External Finance for Investment)** Firms do not retain earnings and do not issue equity. All gross investment orders are financed *exclusively* by bank loans arranged through the funding book. At the accounting instant, new loans for the coming period are issued one-for-one with new funding:

$$L_{t+1} = F_{t+1}.$$

*Implications* Within the period, firms draw on funding-book credit as needed to cover investment outlays (subject to one-period gestation). Settlement flows run through the payments book and are cleared at the accounting instant. With $E_t^F = 0$ at close, firms carry no internal funds across the book close; in this sense, investment is financed externally through the funding book.

**Remark 2.3 (Capital Accumulation and Rollover Identities)** With one-period gestation, capital evolves according to

$$K_{t+1} = (1 - \delta)K_t + I_t.$$

Funding positions are rolled over at the accounting instant: Old instruments $(L_t, F_t)$ are redeemed and new instruments $(L_{t+1}, F_{t+1})$ are issued for the next period, with $L_{t+1} = F_{t+1}$ under zero bank equity at close. In the instant TFM, these operations appear as "redeem old $(L_t, F_t)$" and "issue new $(L_{t+1}, F_{t+1})$."

**Funding Identity and Implication** Since the bank closes with zero equity, $F_t = L_t$ at the accounting instant. Hence, there is no *reserves→lending* channel: reserves cap *payments capacity*, not credit supply.

In contrast to traditional banking models, the 100%-reserve architecture ensures that changes in money demand do not affect the supply of lending. To better understand the dynamics of this system, we introduce the following remark about the interaction between money demand and interest rates.

**Remark 2.4 (No Reserve-Lending Channel Under 100% Reserves)** In this architecture, reserves back deposits in the payments book, while loans are funded one-for-one by nonmonetary bank paper $F$ in the funding book. Hence, a decline in transactions money demand contracts deposits $D$ and, by identity, reserves $R$. Because lending is funded through the funding book rather than reserves, no reserves are "freed" for credit expansion. Instead, the funding rate $r_t^F$ moves through household portfolio reallocation: shifting from payments balances into funding paper tends to lower $r_t^F$ (and eventually the loan rate $r_t^L$), whereas shifts away from financial saving into consumption or outside assets tend to raise $r_t^F$.

**One-Period Rollover: A Concrete Flow** To fix ideas, consider a firm that repays principal $\Delta$ at the accounting instant. The funding book reduces loans by $\Delta$, and the bank simultaneously redeems funding liabilities by the same amount to keep $L = F$. The associated settlement flows pass through the payments book, so deposits and reserves contract in lockstep during the close. Every step preserves the two key identities—$D = R$ in the payments book and $L = F$ in the funding book—and none requires "using reserves for lending."

### 2.5.3 *Fixed Base Money and the Role of "Unused" Money*

In what follows, reserves are fiat money issued by the government in a fixed nominal quantity $M^S$, under which payments balances are settlement instruments rather than funding instruments.

Deposits $\mathcal{D}_t$ are fully reserved within the payments book, so $\mathcal{R}_t = \mathcal{D}_t$, and satisfy the settlement cap $\mathcal{D}_t \leq M^S$. Any base money not held as deposits is held outside the bank as currency or idle settlement balances, $\mathcal{C}_t := M^S - \mathcal{D}_t$. Credit intermediation occurs separately through the funding book: loans $L_t$ are financed one-for-one by non-monetary bank paper $F_t$, with $L_t = F_t$ at the accounting instant. Hence a decline in deposits does not "free reserves for lending": lending conditions adjust only through households' portfolio allocation into $F_t$. Table 2.3 makes this separation explicit as a two-book balance-sheet snapshot.

## 2.6 Risk Capital and the Funding Rate

Loans are funded one-for-one by marketable bank paper $F$, whereas payments capacity is constrained by the reserve stock. This section (a) clarifies which instruments can become illiquid under 100% reserves, and (b) derives the funding rate $r_t^F$.

**Table 2.3** Two-book balance sheet snapshot under a fixed nominal base money stock

| | Assets | Liabilities/equity |
|---|---|---|
| Households | Currency/idle settlement $\mathcal{C}_t$ | |
| | Deposits (payments) $\mathcal{D}_t$ | |
| | Funding claims (bank paper) $F_t$ | |
| Bank: payments book (custody) | Reserves $\mathcal{R}_t = \mathcal{D}_t$ | Deposits $\mathcal{D}_t$ |
| Bank: funding book (intermediation) | Loans to firms $L_t$ | Funding liabilities $F_t$ |
| | | Equity $E_t^B = 0$ |

*Note.* The nominal base money stock is fixed at $M^S$. At any time, it decomposes into currency/idle settlement balances and reserves: $M^S = \mathcal{C}_t + \mathcal{R}_t = \mathcal{C}_t + \mathcal{D}_t$. The payments book is fully reserved and never funds lending. Lending is financed separately by nonmonetary claims $F_t$, so $L_t = F_t$ (with $E_t^B = 0$ at the accounting instant)

Building on the monetary architecture developed in Sect. 2.5, we now impose the accompanying policy closure.

**Remark 2.5 (Maintained Policy Assumption)** The monetary base consists of fiat money issued by a monetary authority. The authority holds safe assets on its balance sheet as the counterpart to fiat-money liabilities. For simplicity, the safe asset is assumed to yield a zero interest rate. The quantity of fiat money, denoted $M^S$, is treated as an exogenous policy variable. Given posted within-period prices $P_t$, the associated real settlement capacity is

$$\bar{m}_t := \frac{M^S}{P_t}.$$

As noted in Chap. 1, Chicago economists—most notably Irving Fisher—advanced the *Chicago Plan* (100%-reserve deposits) in 1933, separating the payments function from credit intermediation.[9] While compelling as a stability blueprint, critics argued that if loans are illiquid, then deposits must be illiquid as well.[10]

**Remark 2.6 (What Actually "Loses Liquidity" Under 100% Reserves?)** With two legally separated books, demand deposits $D$ are fully backed by reserves $R$ and are never used for lending. Illiquidity can emerge in the funding market $F$ that finances loans $L$, not in $D$. A freeze in $F$ tightens credit supply (and raises loan and funding rates) but—absent leakage between books—cannot impair deposit convertibility.

The practicality of a 100%-money architecture hinges on institutional details, yet renewed interest reflects persistent concerns about run risk and settlement safety. A modern payment–funding split revives Fisher's idea in a workable form: fully reserved deposits handle payments, while marketable, loss-bearing claims fund lending. Securitized funding with retention can help align incentives and support secondary-market liquidity, making a safe payments core compatible with active intermediation.[11] More broadly, modern credit-market technologies—risk transfer,

[9] By the "Chicago Plan," we mean 100%-money proposals that segregate fully reserved, par-convertible deposits from market-priced credit; see Fisher (1935), Simons et al. (1933), and Benes and Kumhof (2012) (and updates) for *Dynamic Stochastic General Equilibrium* (DSGE) treatments; our focus is investment dynamics and stochastic stability under the same institutional split.

[10] That objection presumes deposits fund loans. Under strict segregation, illiquidity arises (if at all) in the funding market $F$ that pays for loans $L$, not in deposits $D$. Legal ring-fencing and nonrecourse make senior $F$ claims bankruptcy-remote from the payments book; see Chen et al. (2006).

[11] For background on fully reserved payments coexisting with market-priced credit, see Benes and Kumhof (2012), Cochrane (2014), and Pennacchi (2012). Retention as incentive alignment is standard (e.g., Keys et al. 2010). Our results do not hinge on particular funding instruments or market designs.

securitization, and improved disclosure/settlement infrastructure—can, when well designed and supervised, improve risk sharing and secondary-market liquidity; poorly designed versions can also amplify run risk.

### *2.6.1 Funding Rate*

**Assumption 2.5 (Market Funding with Risk Retention)** Each loan cohort is funded by tradable claims ("bank paper" $F$). The bank retains a fixed fraction $\theta \in (0, 1)$ of the first-loss tranche and sells senior claims broadly. Retention $\theta$ aligns incentives; senior claims are bankruptcy-remote from the payments book.

The funding rate embeds the base settlement yield $i_*$ plus a risk–liquidity premium compensating expected pool loss, priced systematic exposure, and rollover/trading illiquidity. Let $\pi_t^r$ denote this risk–liquidity premium.

**Remark 2.7 (Risk–Liquidity Premium)** In reduced form, $\pi_t$ is decreasing in retention $\theta$ and in market liquidity.

Here $\pi_t$ is distinct from the transactions-demand parameter $\lambda$ used in the money-demand block. Inside the stability corridor studied in Chap. 8, aggregate risk is abstracted from; accordingly, we treat $\pi_t^r = \bar{\pi}$ as constant (or, if desired, $\pi_t^r = \bar{\pi} + \kappa_r/\phi_t$ with $\phi_t$ capturing market liquidity).[12]

With the payments cap fixed at $M^S$ (equivalently $\bar{m}$ in real terms), the LM component adjusts with expenditure $C_t$. The funding rate is therefore

$$r_t^F = i_* + \bar{\pi} + \mu(C_t; \bar{m}),$$

where $\mu(C_t; \bar{m})$ captures the endogenous tightening of funding conditions required to clear payments as expenditure approaches the settlement cap. We assume $\mu$ is increasing in $C_t$ and decreasing in $\bar{m}$; see Chap. 4. The PM–LM results are robust to any monotone risk premium $\pi_t^r$ with bounded elasticity, so the microstructure of $F$ is inessential.[13]

#### 2.6.1.1 Relation to the Chicago Plan Literature

The institutional separation adopted here—fully reserved deposits for payments and credit funded by nonmonetary liabilities—coincides with the Chicago Plan

[12] $\kappa_r$ measures the liquidity wedge: the magnitude of the extra premium that appears when markets are illiquid (low $\phi_t$).

[13] With the payments–funding split and instant closures $D = R = 0$ and $E^B = 0$, we have $F = L$ at the accounting instant. Funding shocks move $r_t^F$ and credit quantities but do not impair deposit convertibility.

tradition (e.g., Benes and Kumhof 2012, Cochrane 2014, Pennacchi 2012). While that literature evaluates monetary reform in calibrated DSGE models, this book studies investment behavior and stochastic stability under the same architecture using a non-DSGE, evolutionary framework.

This chapter fixes the institutional and accounting environment within which the model operates. We turn in the next chapter to household consumption and money demand, completing the behavioral structure needed for the macroeconomic dynamics that follow.

# Appendix

## 2.A Stock-Flow Consistency and Accounting Closure

This section provides a formal stock–flow consistency check for the model. The consolidated transaction-flow matrix and the corresponding stock matrix verify that all sectoral flows add up and that balance sheets close exactly under the split-bank, 100 %-reserve architecture.

Table 2.4 reports the transaction-flow matrices under one-period rollover. Panel (a) records within-period flows, while panel (b) displays the balance-sheet configuration at the accounting instant. In both panels, the upper block lists flow transactions, and the lower block records instantaneous redemption and issuance. At the accounting instant, we impose the split-bank closure $E_t^B = 0$ and $F_t = L_t$.

The table illustrates the stock–flow consistent accounting discipline: all sectoral flow identities and budget constraints hold exactly, and stocks evolve as the accumulation of flows (see Godley & Cripps 1983 and Godley & Lavoie 2007, 2012).

## 2.B Funding-Side Failure Modes Under a Two-Book Architecture

This section maps potential funding-side disruptions to their transmission channels and clarifies why, under a legally segregated two-book architecture, such shocks cannot impair deposit convertibility. Table 2.5 summarizes key *funding-book* failure modes and their impact channels. Under a 100%-reserve two-book system, shocks affect the price and quantity of funding $(r^F, F, L)$, not the payments ledger $(D, R)$.

**Table 2.4** Consolidated matrices

(a) Consolidated transaction-flow matrix (real terms; SQSS)

| | Bank | Firms | Households | Sum |
|---|---|---|---|---|
| Wages ($\omega N$) | 0 | $-\omega N$ | $+\omega N$ | 0 |
| Dividends (div) | $-\text{div}_B$ | $-\text{div}_F$ | $+\text{div}_F + \text{div}_B$ | 0 |
| Interest on loans ($r^L L$) | $+r^L L$ | $-r^L L$ | 0 | 0 |
| Interest on deposits ($i_* D$) | $-i_* D$ | 0 | $+i_* D$ | 0 |
| Consumption ($C$) | 0 | $+C$ | $-C$ | 0 |
| Investment (gross) ($I$) | 0 | $+I$ | 0 | 0 |
| Depreciation ($\delta K$) | 0 | $-\delta K$ | 0 | 0 |
| Organizational cost ($c_b \delta K$) | 0 | $-c_b \delta K$ | 0 | 0 |
| New loans ($\Delta L$) | $-\Delta L$ | $+\Delta L$ | 0 | 0 |
| Loan repayment ($\Delta L^{\text{redeem}}$) | $+\Delta L^{\text{redeem}}$ | $-\Delta L^{\text{redeem}}$ | 0 | 0 |
| Change in deposits ($\Delta D$) | $+\Delta D$ | 0 | $-\Delta D$ | 0 |
| Column sum | 0 | 0 | 0 | **0** |

(b) Post-closing stocks at the accounting instant (real terms)

| | Bank (funding) | Firms | Households |
|---|---|---|---|
| Loans outstanding $L_t$ | $+L_t$ | $-L_t$ | 0 |
| Funding claims $F_t$ | $-F_t$ | 0 | $+F_t$ |
| Firm equity claims $E_t$ | 0 | $-E_t$ | $+E_t$ |
| Fixed capital $K_t$ | 0 | $+K_t$ | 0 |
| Column sum | $L_t - F_t$ | $K_t - L_t - E_t$ | $F_t + E_t$ |

*Note.* This table verifies stock–flow consistency algebraically. Signs: inflow +, outflow −. All entries are in real units. For within-period timing and accounting-instant closures, see Table 2.1. The household column records household net wealth $F_t + E_t = K_t$

**Table 2.5** Funding-side failure modes, impact channels, and mitigations under a 100%-reserve, two-book architecture

| Failure mode (funding) | Impact channel | Mitigation under strict two-book segregation |
|---|---|---|
| Market freeze/rollover risk in $F$ | Funding premium $r^F \uparrow$; issuance $F \downarrow$; credit $L \downarrow$ | Maturity matching in $F$; transparent pool reporting; liquidity backstop for *senior* $F$ (e.g., repo eligibility) at a penalty rate (price, not a guarantee) |
| Asset losses in the loan pool | Losses borne by $F$ holders (principal/mark-to-market) | First-loss retention $\theta > 0$; senior/sub tranching; statutory *no recourse* to the payments book |
| Legal/operational leakage between books | Attempted cross-pledge of payments assets or grab at reserves $R$ | Separate legal entities; bankruptcy-remote payments bank; no cross-default/pledge; segregated systems and operational controls |
| "Reputation put" (attempt to support $F$ using $R$) | Governance pressure to shift resources from payments to funding | Statutory prohibition; resolution protocol for the funding book; ex post accountability tools (e.g., clawbacks) |

*Note.* The payments book is ring-fenced: deposits are fully reserved ($D = R$ within the period; $D = R = 0$ at the accounting instant as a snapshot at close). Funding shocks move $r^F$ and funding quantities ($F$, $L$) but cannot impair deposit convertibility

## References

Benes, J., & Kumhof, M. (2012) The Chicago Plan revisited. IMF Working Paper WP/12/202, International Monetary Fund. https://doi.org/10.5089/9781475505528.001.

Chen, N. F., Kobayashi, T., & Sai, R. (2006). Rethinking "100% money": Challenges from new financial technology (in Japanese). CIRJE J-Series CIRJE-J-166, CIRJE, Faculty of Economics, University of Tokyo. https://ideas.repec.org/p/tky/jseres/2006cj166.html.

Cochrane, J. H. (2014). Toward a run-free financial system. In A. S. Blinder, A. W. Lo, & R. M. Solow (Eds.) *Across the great divide: New perspectives on the financial crisis* (pp. 197–249). Hoover Institution Press.

Fisher, I. (1935). *100% money: Designed to keep checking banks 100% liquid; to prevent inflation and deflation; largely to cure or prevent depressions; and to wipe out much of the national debt.* Adelphi Company. https://books.google.com/books?id=O0tZLbBNlBUC, revised edition.

Godley, W., & Cripps, F. (1983). *Macroeconomics.* Oxford University Press.

Godley, W., & Lavoie, M. (2007). *Monetary economics: An integrated approach to credit, money, income, production and wealth.* Palgrave Macmillan.

Godley, W., & Lavoie, M. (2012). *Monetary economics: An integrated approach to credit, money, income, production and wealth* (2nd edn.). Palgrave Macmillan.

Iwai, K. (1974). The firm in uncertain markets and its price, wage and employment adjustments. *The Review of Economic Studies, 41*(2), 257–268. https://doi.org/10.2307/2296393.

Iwai, K. (1981). *Disequilibrium dynamics: A theoretical analysis of inflation and unemployment.* Cowles Foundation Monograph No. 27. Yale University Press.

Iwai, K. (1987). *Hukinko Dogaku-no Riron, A theory of disequilibrium dynamics.* Iwanami Shoten.

Keys, B. J., Mukherjee. T., Seru, A., & Vig, V. (2010). Did securitization lead to lax screening? evidence from subprime loans. *Quarterly Journal of Economics, 125*(1), 307–362. https://doi.org/10.1162/qjec.2010.125.1.307.

Mehrling, P. (2011). *The New Lombard Street: How the Fed Became the Dealer of Last Resort.* Princeton University Press.

Pennacchi, G. (2012). Narrow banking. *Annual Review of Financial Economics, 4*, 141–159. https://doi.org/10.1146/annurev-financial-110311-101826.

Simons, H., Cox, G., Director, A., Douglas, P., Hart, A., Knight, F., Mints, L. W., Schultz, H. (1933). *Banking and currency reform. manuscript*, printed in Warren Samuels, ed, Research in the History of Economic Thought and Methodology, Archival Supplement 4.

Solow, R. M. (1998). *Monopolistic competition and macroeconomic theory.* Cambridge University Press.

# Chapter 3
# Consumption and Money Demand

*The fundamental psychological law, upon which we are entitled to depend with great confidence, is that men are disposed, as a rule and on the average, to increase their consumption as their income increases, but not by as much as the increase in their income.*

— John Maynard Keynes, *The General Theory,* Ch. 13

## 3.1 Introduction

The purpose of this chapter is to provide a microeconomic foundation for a consumption function that depends linearly on current income and the (predetermined) capital stock. This functional form plays a pivotal role throughout the book because, by construction, it is independent of the interest rate. This restriction greatly simplifies the macroeconomic dynamics and is consistent with a Keynesian emphasis on short-run income determination and expenditure behavior.

Our accounting discipline follows the stock–flow consistent approach associated with Godley and Cripps (1983), and further developed by Godley and Lavoie (2007, 2012). We adopt their transaction-flow matrix (TFM) logic and quadruple-entry consistency, but impose a different monetary closure: a 100%-reserve architecture with separate payments and funding books.

**Methodological Note**
Consistent with the modular Keynesian structure adopted throughout this book, consumption and money demand are treated here as behavioral responses to realized income and wealth rather than as components of a unified representative-agent optimization problem. Employment determination is developed separately through the interaction of product markets, wage setting, and monetary conditions in subsequent chapters.

In the intermediate run—the stationary quasi-steady state (SQSS)—the aggregate capital stock is treated as fixed at $K$. Investment displays inertia and adjusts only when rare, sentiment-driven disturbances ("mutants") cross critical thresholds,

 
I. Takahashi, *Microfoundations of Keynesian Economics*, Evolutionary Economics and Social Complexity Science 37, https://doi.org/10.1007/978-981-95-9390-3_3

triggering upward or downward revisions. Accordingly, within the SQSS, macroeconomic variables such as $Y_t$, $C_t$, $P_t$, and $r_t$ fluctuate conditional on a given $K$, and we suppress time subscripts on capital.

Within this environment, we posit that consumption depends only on current real income and on the predetermined capital stock:

$$C_t = c_y \frac{Y_t}{P_t} + c_w K, \tag{3.1}$$

where $c_y$ and $c_w$ denote the propensities to consume out of income and wealth. This two-component linear consumption function captures the idea that spending is driven both by contemporaneous resources and by accumulated wealth. Under the accounting closure of the model, household nominal wealth equals the value of installed capital, $\mathcal{W}_t = P_t K$, so the wealth term reduces to $c_w K$. A key implication is that, under replacement investment $I_t = \delta K$ in the SQSS, aggregate absorption $X_t \equiv C_t + I_t$ becomes proportional to $K$. If transactions money demand scales with this expenditure level, then under a fixed nominal money supply in a 100%-reserve regime (with $P_t$ taken as given within the period), the interest rate adjusts endogenously to clear the payments market (see Remark 4.1). Capital $K$ therefore becomes the central state variable governing the short- and intermediate-run system.

While a fully rigorous derivation of (3.1) from an infinite-horizon model with explicit portfolios is rarely available in closed form, several strands of theory and evidence support it as a disciplined reduced form. Overlapping generations models naturally generate wealth effects (e.g., Blanchard, 1985); liquidity constraints and precautionary saving rationalize dependence on current resources (e.g., Carroll, 2001); and aggregate evidence suggest consumption responds to transitory income and asset wealth (e.g., Campbell & Mankiw, 1989; Bank of Japan, 2016).

The model developed here builds on Otaki (2015, Chap. 2) and Bénassy (2007), extending their Keynesian foundations in an *Overlapping Generations* (OLG) framework to an environment with physical capital. This extension is essential for analyzing how capital accumulation shapes consumption, effective demand, and long-run macroeconomic dynamics. Throughout the main text, we assume households consume a single homogeneous good, which is inconsistent with the monopolistic-competition structure in production. Appendix 3.C generalizes the consumption block to differentiated goods for completeness.

The remainder of this chapter proceeds as follows. Section 3.2 develops the household optimization problem and derives the consumption and money-demand functions from the underlying preferences and budget constraint. Section 3.3 characterizes the properties of these demand functions, with particular emphasis on the role of the opportunity cost of holding money and the elasticity parameters governing substitution between consumption and liquidity services. Section 3.4 then aggregates individual behavior to the macroeconomic level and establishes the relationship between consumption, real balances, and aggregate expenditure within the model. Finally, Sect. 3.5 summarizes the key implications and prepares the ground for the analysis of short-run dynamics in the subsequent chapter.

**Notation and Conventions**

In the payments architecture developed in Chap. 2, transactions balances take the form of demand deposits $D$ held for settlement. In this chapter, however, we adopt the conventional macroeconomic notation $M$ for aggregate transactions money, so that $M$ corresponds to the deposits used for payments in that settlement framework.

The symbol $M$ is used in two related but distinct senses in the book. In the monetary analysis of Chaps. 3–8, $M$ denotes nominal settlement balances supplied through the payments system, with $M^d$ and $M^S$ representing transactions money demand and the exogenous money supply, respectively. In Chap. 6, by contrast, $M$ (and its variants $M_L$, $M_U$) denotes the number of steps used in the stochastic reachability arguments; this usage is purely technical and unrelated to the monetary variable.

## 3.2 Households and Consumption Behavior

Households transition through three life stages and make consumption and money-holding decisions subject to the institutional constraints of the monetary architecture. Since the two-component linear consumption function arises in a symmetric quasi-steady state (SQSS), we analyze household behavior in an SQSS in which the aggregate capital stock and employment are constant over time. For this reason, throughout this chapter, we suppress time and individual indices and work in a symmetric equilibrium.

To keep notation consistent with the model chapter, we denote the household-held funding instrument by $F$ ("bank paper"), i.e., a nonmonetary claim issued by the bank's funding book to finance loans $L$; this corresponds to what is often called a bank bond in simpler treatments.

### *3.2.1 Representative Firms in a Symmetric Equilibrium*

The economy comprises a large number of monopolistically competitive firms producing differentiated but substitutable goods. In a symmetric equilibrium, all firms behave identically, so we represent them by a single representative firm.

Output can be consumed or invested. Capital depreciates at rate $\delta$, and in the SQSS, we impose replacement investment,

$$I = \delta K,$$

so the aggregate capital stock remains constant.

Firms finance investment externally through the bank's funding book and distribute profits at the accounting instant, so that retained earnings and firm equity do not carry across periods. Profits are given by

$$\Pi^F = P(C + I) - \left(WN + (r + \delta)L^F\right)$$

and are paid out as dividends to households at period end.

**Assumption 3.1 (Funding Structure under the Split-Bank System)** We maintain the split-bank accounting structure established in Lemma 2.1. In particular, deposits are fully reserved within the payments book, while firm loans are financed one-for-one through non-monetary bank paper held by households. Accordingly, at the accounting instant,

$$L = F.$$

In what follows, the household portfolio consists of payments balances for transactions during the period and funding claims $F$ that finance the loan book. The distinction between within-period transactions balances and end-of-period portfolio positions is made precise in the money-demand block developed below.

### *3.2.2 Budget Constraint of Households*

The detailed, individual-level budget constraints are derived in Appendix 3.A. In the main text, we suppress individual and time indices and treat each cohort as a *representative household within its life stage*, as appropriate in a symmetric quasi-steady state (SQSS). Our objective is to derive the two-component Keynesian consumption rule (3.1), which links aggregate consumption to current income and the predetermined capital stock.

**Funding Instrument Notation** Throughout this chapter, the household saving instrument is the bank's *funding paper* $F$ (a nonmonetary, tradable one-period claim issued by the funding book). It finances private loans $L$ and is distinct from transactions balances held in the payments book.

At the aggregate level, households receive wage income $WN$, firm dividends $Div^F$, and (cohort-specific) rebated interest flows from the funding book. As we show in Lemma 4.3, firm dividends satisfy

$$Div^F = (r + \delta)E = \frac{1}{\alpha(\eta - 1)} PK,$$

where $E$ denotes the household equity claim prior to payout.

Since this chapter only considers households, the subscript H indicating households will be omitted except for consumer loans provided to dissaving households,

$L^H$. The bank settles funding liabilities at the end of each period. Retired households sell all their securities at the end of stage 2.

Using the Cobb–Douglas cost-share relation $\omega N + (r+\delta)K = (r+\delta)K/\alpha$, we obtain

$$Y = WN + rPK + \frac{r+\delta}{\alpha(\eta-1)} P \cdot K,$$

which matches the real income identity in Eq. (4.5.5).

In this chapter, we treat the bank as a pass-through and write the bank transfer term directly as:

- Capital-income transfer $Y^I \equiv r(L^F - F)$ to retirees
- Interest-assistence $rL^H$ to dissavers.[1]

A household in the model passes through three life stages: the *working stage (stage 1)*, the *retired stage (stage 2)*, and the *dissaving stage (stage 3)*. In each life stage $s \in \{1, 2, 3\}$, a household enters the period with cash-on-hand (maximum feasible normal expenditure within the stage) $\mathcal{W}^s$, allocates resources between consumption expenditure $PC^s$ and money balances $M^s$ (which facilitate transactions), and carries forward remaining wealth to the next stage.

In each stage, the household with financial resources $\mathcal{W}^s$ (the maximum amount that they can spend on consumption, i.e., "cash on hand" in Deaton's terminology (Deaton, 1991) consumes goods $PC^s$ and holds money $M^s$ to facilitate transactions, where superscript $s \in \{1, 2, 3\}$ denotes the respective stages of life. The act of holding money is necessary for facilitating consumption in the economy, but comes with the opportunity cost of foregoing interest income. Each stage's financial resources $\mathcal{W}^s$ are divided between (i) "total spending" $S^s$—which includes the opportunity cost of holding money, i.e., $PC^s + rM^s/(1+r)$ (see Bénassy 2007)—and (ii) financial wealth carried forward $R^s$.

To clarify the link between household wealth and the installed capital stock, Table 3.1 reports balance-sheet snapshots by sector and household life stage in the SQSS, both before and at the end-of-period accounting instant. Relative to Chap. 2, the household sector is refined by separating stages 1–2 (working/retired) from stage 3 (dissaving). The post-closing panel highlights the key closure used throughout the book: under the split-bank architecture and full payout at the accounting instant, aggregate household net worth equals the capital stock, $NW_H = K$. This identity provides the accounting bridge for the wealth term in the reduced-form consumption rule, allowing $c_w(\mathcal{W}/P)$ to be written as $c_w K$ in (3.1).

Stage 1 accumulates wealth; stage 2 consumes out of asset income and liquidates portfolios; stage 3 dissaves and closes with a zero terminal condition. This three-

[1] In this chapter, we assume a three-period OLG model, and the assumption of a steady-state economy means that the number of households in each stage is equal.

**Table 3.1** Sectoral balance sheets in quasi-steady state and at end-of-period clearing

(a) Balance-sheet snapshots by sector and household life stage in the SQSS (real units)

| Bank—payments | | Bank—funding | | Firms | | HH: stages 1 and 2 | | HH: stage 3 (dissaving) | |
|---|---|---|---|---|---|---|---|---|---|
| Assets | Liab./eq. | Assets | Liab./eq. | Assets | Liab./eq. | Assets | Liab./eq. | Assets | Liab./eq. |
| Reserves $R$ | Deposits $D$ | Loans to private sector $L^F+L^H$ | Funding $F$ | Deposits $D^F$ | Loans $L^F$ | Deposits $D^{(1,2)}$ | | Deposits $D^{(3)}$ | Loans $L^H$ |
| | | | Bank equity $E^B$ | Capital $K$ | Equity $E^F$ | Funding paper $F^{(1,2)}$ | Net worth $NW^{(1,2)}$ | | Net worth $NW^{(3)}$ |

*Notes*. All entries are reported in real units (nominal values divided by $P_t$). This panel is a *within-period* snapshot taken before the end-of-period book closing. Deposits and reserves are active in the payments book and move in lockstep during the period, so $D = R$. The funding book's asset is total private credit, $L \equiv L^F + L^H$, financed by nonmonetary bank paper $F$; before closing, bank equity $E^B$ need not be zero. System deposits decompose as $D = D^F + D^{(1,2)} + D^{(3)}$, where $D^F$ denotes firms' transactions balances, and $D^{(1,2)}$ and $D^{(3)}$ denote household deposits held by stages 1–2 and stage 3 households, respectively. Firm equity is the residual claim, $E^F = K - L^F$, prior to payout.

*Closing convention (used throughout the book)*. At the accounting instant, the payments book is swept, so $D = R = 0$; the bank rebates any residual spread so that funding-book equity closes ($E^B = 0$), implying $F = L$; and firms distribute profits so that the residual equity claim closes ($E^F = 0$).

(b) Balance sheets at the accounting instant (post-closing, real units)

| Bank—payments | | Bank—funding | | Firms | | HH: stages 1 and 2 | | HH: stage 3 (dissaving) | |
|---|---|---|---|---|---|---|---|---|---|
| Assets | Liab./eq. | Assets | Liab./eq. | Assets | Liab./eq. | Assets | Liab./eq. | Assets | Liab./eq. |
| Reserves $R$ | Deposits $D$ | Loans to firms $L^F$ | Funding $F$ | Capital $K$ | Loans $L^F$ | Funding paper $F^{(1,2)}$ | Net worth $NW^{(1,2)}$ | 0 | $L^H = 0$ |
| 0 | 0 | | $E^B = 0$ | | $E^F = 0$ | | | | $NW^{(3)} = 0$ |

*Baseline closing (accounting instant)*. At the end-of-period close, payments are swept, so $D = R = 0$. The funding book closes with zero equity, $E^B = 0$, so funding matches the loan book, $F = L$. Firms distribute retained earnings and profits at the instant, so the residual equity claim closes, $E^F = 0$. Installed capital is carried at replacement cost ($q^{\text{acc}} = 1$).

*Stage-3 clearing*. Dissaving households settle consumer loans at the close, so $L^H = 0$ and $NW^{(3)} = 0$ in the post-closing snapshot

*Consolidated wealth identity*. Under these closures, aggregate household net worth equals the installed capital stock: $NW_H = K$.

stage structure generates an aggregate consumption rule with linear dependence on current income and the predetermined capital stock.

In each stage, we decompose resources into current absorption $S^s$ and carry-over wealth $R^s$. Regarding the bank profits, i.e., $\Pi^B = r(L - F)$, we make the following assumptions:

**Remark 3.1 (Bank Transfers as Reallocated Interest Flows)** In the baseline closure, the bank is a zero-net-worth intermediary whose net cash flow is rebated within the household sector. Accordingly, the banking block does not create net resources; it only *reallocates* interest flows across life stages. In particular, we decompose the net transfer from the funding book as

$$\Pi^B \equiv r(L - F) = \underbrace{r(L^F - F)}_{=: Y^I \text{ (capital-income transfer to retirees)}} + \underbrace{rL^H}_{\text{(interest-assistance to dissavers)}} .$$

Thus, introducing $\Pi^B$ adds no new income beyond the loan–funding spread; it merely specifies how that spread is distributed across cohorts.[2]

### 3.2.2.1 Stage 1: Working Household

A household starts stage 1 with no assets or debts. The household works $N$ hours, earning labor income $W \cdot N$ (note that $N$ here can either refer to the hours worked or be used as a variable to indicate the total employment in the economy). The household allocates its income between consumption and saving, building financial assets to secure their livelihood after retirement. The household purchases funding paper $F^1$ and acquires corporate equities $E^1$.

The budget constraint for the working household is

$$\begin{aligned} \mathcal{W}^1 &= W \cdot N = PC^1 + M^1 + F^1 + E^1 \\ &= \underbrace{\left(PC^1 + \frac{r}{1+r}M^1\right)}_{S^1} + \underbrace{\left(F^1 + E^1 + \frac{1}{1+r}M^1\right)}_{R^1}. \end{aligned} \tag{3.2}$$

$WN$ in Eq. (3.2) is the funds available in stage 1, which is decomposed into total spending for the current period, $S^1$, on the right side, and funds they carry over to the next period, $R^1$. The amount of money carried over to the end of the period $M^1$ can be divided into the discounted present value evaluated at the beginning of the period $M^1/(1 + r)$ and the interest lost during the current period $rM^1/(1 + r)$. The latter can be considered as the opportunity cost of holding money, that is, the

[2] In the baseline closure, we abstract from intermediation spreads and assume a single interest rate applies to both loans and funding claims, ${r^L}_t = {r^F}_t \equiv r_t$. The bank, therefore, acts as a zero-net-worth intermediary and any net cash flow is rebated within the household sector.

transaction-motivated cost of holding money. Since it is not spent, it is added to the funds carried over to the next period.

#### 3.2.2.2 Stage 2: Retired household

In stage 2, the household relies on assets carried over from stage 1: funding paper $F^1$, corporate equities $E^1$, and money holdings $M^1$. The retired household consumes using interest income from its funding paper $rF^1$, dividend income from corporate equities $Div^F = (r + \delta)E^F$ ($E^F = E^1$ since $E^2 = 0$), and redistributed capital income $Y^I = r(L^F - F)$. By the end of stage 2, all funding paper and equity are liquidated, so $F^2 = E^2 = 0$. Hence, the household enters stage 3 holding only money balances,

$$M^2 = F^1 + E^1 + M^1.$$

Thus, the budget constraint in stage 2 is

$$\begin{aligned}
\mathcal{W}^2 &= (1+r)F^1 + (1+r+\delta)E^1 + r(L^F - F^1) + M^1 \\
&= PC^2 + M^2 \\
&= \underbrace{\left(PC^2 + \frac{r}{1+r}M^2\right)}_{S^2} + \underbrace{\left(\frac{1}{1+r}M^2\right)}_{R^2}.
\end{aligned}$$

#### 3.2.2.3 Stage 3: Dissaving Household

In stage 3, the household is in the process of dissaving, having already sold assets except for money. Dissaving households also receive a transfer income $rL^H$ from the bank. They use their money holdings $M^2$ to finance consumption and to facilitate transactions. At the beginning of stage 3, they still hold some money, but they are no longer earning income from funding paper or equities.

Under perfect foresight and no bequest motive, the dissaving household exhausts net wealth by the end of stage 3. This means that the value of money holdings and liabilities should balance out. To achieve this, the dissaving household borrows through a consumer loan $L^H$ arranged via the fund book, ensuring that $M^3 = (1 + r)L^H$. Thus, the budget constraint for the dissaving household is

$$\begin{aligned}
\mathcal{W}^3 = M^2 + rL^H &= PC^3 + M^3 - L^H \\
&= \underbrace{\left(PC^3 + \frac{r}{1+r}M^3\right)}_{S^3} + \underbrace{\left(-L^H + \frac{1}{1+r}M^3\right)}_{R^3}. \qquad (3.3)
\end{aligned}$$

At the end of stage 3, the terminal condition is $R^3 = 0$, i.e.,

$$R^3 = -L^H + \frac{1}{1+r} M^3 = 0 \Longleftrightarrow M^3 = (1+r)L^H. \tag{3.4}$$

Finally, interest paid by dissaving households is treated here as an intra-household transfer within the private sector rather than as net value added and is therefore excluded from the aggregate income concept used in this chapter. The interest income on loans of investment funds to companies is included in GDP and national income because it compensates for the creation of capital stock that generates long-term wealth.

## 3.3 Derivation of the Two-Component Linear Consumption Function

As a preparation for deriving the two-component linear consumption function, we proceed in two steps. First, we show that in a SQSS equilibrium, the total real net wealth of households coincides with the aggregate capital stock $K$. Second, using the three-stage OLG structure, we demonstrate that the consumption expenditure of dissaving households equals exactly one half of aggregate wealth, that is, $K/2$. The subsequent subsection then collapses the three-stage model into an equivalent two-stage representation and derives the consumption function of the form (3.1).

### *3.3.1 Consumption Demand of Dissaving Households*

The following lemma establishes that, in a SQSS equilibrium, the total net wealth of households equals the aggregate capital stock.

**Lemma 3.1** *In a symmetric quasi-steady-state equilibrium, aggregate household real net wealth equals the installed capital stock:*

$$\frac{\mathcal{W}}{P} = K.$$

***Proof*** Aggregate household nominal net wealth is

$$\mathcal{W} = M + F + E - L^H.$$

At the accounting instant, the payments book is cleared and the residual equity claims are closed. By Lemma 2.1, household net worth is therefore carried by funding claims:

$$\mathcal{W} = F - L^H.$$

The same lemma gives

$$F = L^F + L^H.$$

Under replacement-cost valuation, firm loans finance the installed capital stock, so

$$L^F = PK.$$

Hence

$$\mathcal{W} = PK,$$

and therefore

$$\frac{\mathcal{W}}{P} = K.$$

□

The next lemma shows that rational terminal-stage behavior implies a simple and interest-rate independent dissaving expenditure in the SQSS.

**Lemma 3.2** *In a symmetric quasi-steady state equilibrium, stage 3 (dissaving) households consume*

$$C^3 = \frac{K}{2}.$$

***Proof*** Substituting the terminal condition (3.4), $M^3 = (1+r)L^H$, into the stage 3 budget constraint (3.3) yields

$$\mathcal{W}^3 = M^2 + rL^H = PC^3 + M^3 - L^H = PC^3 + rL^H.$$

Canceling $rL^H$ from both sides gives

$$PC^3 = M^2.$$

Under the liquidation assumption in stage 2, all interest-bearing assets are sold at the end of stage 2, so beginning-of-stage 3 money holdings satisfy

$$M^2 = M^1 + F^1 + E^1 = \mathcal{W}^2.$$

In the stationary symmetric equilibrium, stages 1 and 2 have equal measure and identical saving behavior, implying

$$\mathcal{W}^1 = \mathcal{W}^2 = \frac{\mathcal{W}}{2},$$

where $\mathcal{W}$ denotes aggregate nominal household net wealth. By Lemma 3.1, $\mathcal{W}/P = K$; hence,

$$C^3 = \frac{M^2}{P} = \frac{\mathcal{W}^2}{P} = \frac{1}{2} \cdot \frac{\mathcal{W}}{P} = \frac{K}{2}.$$

□

Lemma 3.2 relies on perfect foresight, in particular on agents knowing the interest-rate path and their planning horizon (i.e., no mortality risk). Absent perfect foresight, the terminal condition (3.4) would fail to hold, and consumption would depend on $r$. Under perfect foresight, however, interest-rate changes affect both sides of the stage 3 budget constraint symmetrically and therefore do not alter the optimal level of $C^3$.

This result also hinges on the assumption that the three life stages are of equal length. Under this assumption, one third of the population is dissaving at any point in time. Because all accumulated saving is liquidated into money at the start of stage 3, dissaving households collectively hold one half of aggregate wealth in the stationary three-stage structure. In this three-stage structure, stages 1 and 2 carry wealth forward, while stage 3 liquidates and spends it, so the terminal cohort holds the entire stock of accumulated saving. In other words, in the stationary three-stage OLG, stage 3 is the "spending phase," so it holds the pooled liquidation of prior savings. The result $C_3 = K/2$ is therefore a composition identity, not a behavioral elasticity result.

Appendix 3.B shows that if the final stage is shorter, dissaving households hold less wealth and account for a smaller population share, with both effects reducing $c_w$. While we do not pursue a full generalization here, the present structure captures the core mechanism linking life cycle behavior to the wealth–consumption ratio.

## 3.4 Connection to Aggregate Consumption

Lemma 3.2 links individual life cycle behavior to aggregate consumption dynamics. In this section, we aggregate households in stages 1 and 2 into a single *active generation*. Consequently, in the SQSS, aggregate consumption $C$ consists of expenditures by two cohorts: the active generation (stages 1–2) and the dissaving generation (stage 3). As shown in Lemma 3.2, the dissaving cohort contributes a fixed amount $C^3 = K/2$, reflecting terminal wealth exhaustion and asset liquidation.

Let $V_t^a$ denote the subutility of active households and $V_{t+1}^d$ that of dissaving households born in period $t$. Define $H_t := M_t/P_t$ as average intraperiod real money services (distinct from the period-end money stock, which is swept to zero; cf. Remark 3.2). Subutility in each stage takes the CES form

$$V(C^s, H^s) = \left[ (C^s)^{\frac{\beta-1}{\beta}} + \left(\frac{H^s}{\lambda}\right)^{\frac{\beta-1}{\beta}} \right]^{\frac{\beta}{\beta-1}}, \qquad s = a, d,$$

where $\beta > 0$ governs the elasticity of substitution between consumption and money services, and $\lambda > 0$ is the transactions-demand coefficient.

Let $\vartheta \in (0, 1)$ denote the welfare weight on the active (working) stage. The lifetime utility of a household born in period $t$ is

$$U_t = \vartheta \log V_t^a + (1 - \vartheta) \log V_{t+1}^d, \qquad \vartheta \in (0, 1),$$

where stage-$s$ within-period utility is given by

$$V_t^s = V(C_t^s, H_t^s), \qquad s \in \{a, d\}.$$

**Remark 3.2 (Backward Induction and the Two-Stage Reduction)** The household's optimization problem is solved by backward induction. First, dissaving households (stage 3) choose $(C^3, M^3)$ to maximize $V_{t+1}^d$ subject to available resources $\mathcal{W}^3 = M^2 + rL^H$, yielding an indirect utility function that depends only on inherited wealth. Second, households maximize lifetime utility $U_t$ by allocating resources across the active stages (stages 1 and 2). This recursive structure collapses the original three-stage problem into a tractable two-stage formulation.

**Lemma 3.3 (Keynesian Aggregation with a Dissaving Cohort)** *In a symmetric quasi-steady-state equilibrium, aggregate consumption satisfies*

$$C = c_y \frac{Y}{P} + \frac{K}{2}, \qquad 0 < c_y < 1.$$

***Proof*** Aggregating the budget constraints of stage 1 and stage 2 households yields

$$\begin{aligned} \mathcal{W}^1 + \mathcal{W}^2 &= WN + (1+r)F^1 + (1+r+\delta)E^1 + r(L^F - F^1) + M^1 \\ &= PC^1 + PC^2 + M^2 + M^1 + F^1 + E^1. \end{aligned}$$

Subtracting $M^1 + F^1 + E^1$ from both sides and using the stage 3 identity $PC^3 = M^2$ (from the stage 3 budget constraint and (3.4)) gives

$$WN + rL^F + (r+\delta)E = P(C^1 + C^2 + C^3).$$

The left-hand side equals national income $Y$; hence,

$$Y = P(C^1 + C^2 + C^3).$$

We now collapse the three-stage problem into a two-stage one. Let $C^w = C^1 + C^2$ denote active-stage consumption. The two-stage log allocation delivers a constant share rule; the first-order condition implies $C^w = \vartheta(Y/P)$. Thus, the optimal active-stage expenditure share is constant, and we denote it by $c_y := \vartheta \in (0, 1)$, so that $C^w = c_y(Y/P)$. Since $C^3 = K/2$ by Lemma 3.2, aggregate consumption is

$$C = C^w + C^3 = c_y\frac{Y}{P} + \frac{K}{2}.$$

□

In the three-stage benchmark, $c_w = 1/2$. Appendix 3.B generalizes this to unequal stage lengths.

**Corollary 3.4 (Two-Component Linear Consumption (Reduced Form))** *Under the maintained assumptions, aggregate consumption can be written in reduced form as*

$$C = c_y\frac{Y}{P} + c_w K, \qquad 0 < c_y < 1,\ c_w > 0,$$

*where $c_y$ and $c_w$ are treated as short-run constants. In the benchmark three-stage OLG environment above, $c_w = 1/2$ (Lemma 3.3).*

**Corollary 3.5 (SQSS Aggregation)** *In the SQSS block ($K$ fixed, $I = \delta K$),*

$$\frac{Y}{P} = \frac{c_w}{1 - c_y}K \equiv cK, \qquad X \equiv C + I = (c + \delta)K.$$

**Corollary 3.6 (Interest-Rate Independence)** *Within the SQSS horizon, $c_y$ and $c_w$ are invariant to the short-term interest rate. Hence, consumption depends on $Y/P$ and $K$ only, and the interest rate affects equilibrium exclusively through the money market.*

## 3.5 Money Demand Function

### *3.5.1 Household Money Demand*

Households hold transactions balances and interest-bearing assets. Holding one additional unit of money today entails the opportunity cost of not investing it in a one-period interest-bearing asset. The resulting foregone net return is $r$ next period, whose present value is

$$\tilde{r} := \frac{r}{1 + r}.$$

In real terms, defining $H \equiv M/P$, the intratemporal budget constraint is

$$C + \tilde{r}\, H = y,$$

where $y$ denotes real resources available for consumption and liquidity services (effective intratemporal expenditure). Preferences are represented by a constant-elasticity-of-substitution (CES) money-in-the-utility (MIU) function:

$$V(C, \tfrac{M}{P}) = \left[ C^{\frac{\beta-1}{\beta}} + \left( \frac{M}{\lambda P} \right)^{\frac{\beta-1}{\beta}} \right]^{\frac{\beta}{\beta-1}}, \qquad \beta > 0,\ \beta \neq 1,$$

where $\beta$ is the elasticity of substitution between consumption and real balances, and $\lambda > 0$ scales real balances into transaction services. A higher $\lambda$ means that a given level of real balances provides fewer transaction services.[3]

Maximizing $V(C, H)$ subject to $C + \tilde{r}\, H = y$ yields the intratemporal first-order condition $V_H / V_C = \tilde{r}$, implying the standard MIU ratio

$$\frac{H}{C} = \lambda^{1-\beta}\, \tilde{r}^{-\beta} \qquad \Longrightarrow \qquad \frac{M}{P} = \lambda^{1-\beta} \left( \frac{r}{1+r} \right)^{-\beta} C.$$

In the Cobb–Douglas limit $\beta \to 1$, the demand system becomes log-linear in $\tilde{r}$, with $H/C \propto \tilde{r}^{-1}$. Equivalently, in levels using the intratemporal budget constraint,

$$C = \frac{y}{1 + (\lambda \tilde{r})^{1-\beta}}, \qquad \frac{M}{P} = \frac{y\, \lambda^{1-\beta} \tilde{r}^{-\beta}}{1 + (\lambda \tilde{r})^{1-\beta}}.$$

### 3.5.2 Firm Money Demand

Firms also hold transactions balances to smooth timing mismatches between outlays and receipts. Holding money carries the same opportunity cost $\tilde{r} = r/(1 + r)$, as it displaces interest-bearing assets. For simplicity, assume firms share the same liquidity-services parameter $\lambda$ as households. Then, by the same reduced-form intratemporal liquidity trade-off,

$$\frac{M^F}{P} = \lambda^{1-\beta} \left( \frac{r}{1+r} \right)^{-\beta} C^F,$$

where $C^F$ denotes firm outlays that must be financed through working balances.

[3] Equivalently, the implied semi-elasticity satisfies $\partial \ln(H/C)/\partial \ln \tilde{r} = -\beta$. When $\beta < 1$, consumption and money are short-run complements.

### 3.5.3 Aggregate Money Demand

Before introducing the LM relation, we consolidate firms' working balances into aggregate transactions demand.

**Remark 3.3 (Consolidation of Firms' Working Balances)** Let $\phi_m \in [0, 1]$ denote the share of consumer-goods sales that requires firms to hold working balances. With aggregate household consumption $C$, firm transactions outlays satisfy $C^F = \phi_m C$. With a mild abuse of notation, we continue to write the rescaled parameter as $\lambda$. Aggregating household and firm money demands yields

$$\frac{M^d}{P} = \lambda^{1-\beta}\tilde{r}^{-\beta}(C + C^F) = \lambda^{1-\beta}\tilde{r}^{-\beta}(1 + \phi_m)C.$$

For notational economy, we normalize the liquidity-services scale so that $\lambda^{1-\beta}$ absorbs the factor $(1 + \phi_m)$. With this convention, aggregate transactions money demand takes the compact form.

**Lemma 3.7 (Aggregate Transactions Money Demand)** *Under the baseline intratemporal trade-off and the consolidation of firms' working balances described above, aggregate real money demand satisfies*

$$\frac{M^d}{P} = \lambda^{1-\beta} C \left(\frac{r}{1+r}\right)^{-\beta}.$$

This expression serves as the baseline specification of money demand for the analysis that follows. Together with the labor-market and pricing relations developed in the next chapter, it forms the PM–LM structure that determines the equilibrium levels of capital accumulation and the interest rate in the broader macroeconomic system.

## Appendix

## 3.A Budget Constraints

This appendix provides a detailed derivation of the intertemporal budget constraints for a representative household in the three-stage overlapping generations (OLG) model used in the main text. A household progresses through three life stages: working (stage 1), retired (stage 2), and dissaving (stage 3). Throughout, we maintain the assumptions of a symmetric steady state and perfect foresight.

**Notation**

Aggregate funding paper, equity, and money holdings at time $t$ are denoted by $F_t$, $E_t$, and $M_t$, respectively, and aggregate household consumer loans by $L_t^H$. A stock variable with superscript $s \in \{1, 2, 3\}$ and subscript $\tau$ denotes the *beginning-of-period balance* for a household who is in life stage $s$ during period $\tau$. For example, $F_{j,\tau}^2$ denotes the fund paper held by retired household $j$ at the beginning of period $\tau$.

Let $H_t^s$ denote the set of households in stage $s$ at time $t$.[4] Then total outstanding funding paper and consumer loans satisfy

$$F_t := \sum_{j \in H_t^1} F_{j,t}^1 + \sum_{j \in H_t^2} F_{j,t}^2, \qquad L_t^H := \sum_{j \in H_t^3} L_{j,t}^H.$$

Let $L_t^F$ denote total business loans outstanding and $L_t \equiv L_t^F + L_t^H$. The bank redistributes profits $\Pi_t^B = r_t(L_t - F_t)$ through transfers to stage 2 and stage 3 households. Specifically, each retired household receives redistributed capital income

$$Y_t^I := \frac{r_t(L_t^F - F_t)}{\bar{N}},$$

and each dissaving household receives an interest-assistance transfer

$$T_t^H := \frac{r_t L_t^H}{\bar{N}},$$

where $\bar{N}$ is the (constant) size of each cohort. Note that $Y_t^I + T_t^H = \Pi_t^B / \bar{N}$, so total bank profits are fully rebated across stage 2 and stage 3 households. Accordingly, per-household consumer loans satisfy $L_{j,t}^H = L_t^H / \bar{N}$ in the symmetric steady state.

**Liquidation at the End of Stage 2**

Retired households liquidate all interest-bearing assets by the end of the period, so they carry only money into stage 3, so $F_{j,t+2}^2 = E_{j,t+2}^2 = 0$ at the start of stage 3, and thus all remaining wealth is carried as money $M_{j,t+1}^2$.

Dissaving households enter stage 3 carrying their liquid wealth $M_{j,t+1}^2$. During this final stage, they may borrow $L_{j,t+2}^H$ to smooth consumption. We impose a no-bequest terminal condition: households fully run down their net asset position by the end of stage 3, so terminal wealth is normalized to zero at the end of life.

**Intertemporal Budget Constraints**

For a household $j$ entering stage 1 in period $t$, the stage-by-stage budget constraints are

[4] Under the steady-state assumption, cohort sizes are constant over time.

$$\begin{aligned}\mathcal{W}^1_{j,t} &= W_t N_{j,t} = P_t C^1_{j,t} + M^1_{j,t} + F^1_{j,t} + E^1_{j,t} \\ &= \underbrace{\left(P_t C^1_{j,t} + \frac{r_t}{1+r_t} M^1_{j,t}\right)}_{S^1_{j,t}} + \underbrace{\left(F^1_{j,t} + E^1_{j,t} + \frac{1}{1+r_t} M^1_{j,t}\right)}_{R^1_{j,t}}, \end{aligned} \tag{3.5a}$$

$$\begin{aligned}\mathcal{W}^2_{j,t+1} &= (1+r_{t+1})F^1_{j,t} + (1+r_{t+1}+\delta)E^1_{j,t} + Y^I_{t+1} + M^1_{j,t} \\ &= P_{t+1}C^2_{j,t+1} + M^2_{j,t+1} \\ &= \underbrace{\left(P_{t+1}C^2_{j,t+1} + \frac{r_{t+1}}{1+r_{t+1}} M^2_{j,t+1}\right)}_{S^2_{j,t+1}} + \underbrace{\left(\frac{1}{1+r_{t+1}} M^2_{j,t+1}\right)}_{R^2_{j,t+1}}, \end{aligned} \tag{3.5b}$$

$$\begin{aligned}\mathcal{W}^3_{j,t+2} &= M^2_{j,t+1} + T^H_{t+2} = P_{t+2}C^3_{j,t+2} + M^3_{j,t+2} - L^H_{j,t+2} \\ &= \underbrace{\left(P_{t+2}C^3_{j,t+2} + \frac{r_{t+2}}{1+r_{t+2}} M^3_{j,t+2}\right)}_{S^3_{j,t+2}} + \underbrace{\left(-L^H_{j,t+2} + \frac{1}{1+r_{t+2}} M^3_{j,t+2}\right)}_{R^3_{j,t+2}}, \end{aligned}$$

$$\mathcal{W}^4_{j,t+3} = (1+r_{t+2})\, R^3_{j,t+2} = 0. \tag{3.5c}$$

The terminal condition (3.5c), under perfect foresight, ensures that dissaving households deplete wealth exactly at the end of stage 3. $R^3_{j,t+2}$ is carried from the beginning of period $t+2$ to the end of that period; it accrues at the period-$(t+2)$ interest rate $r_{t+2}$. The second term on the right-hand side of Eq. (3.5b) is the return on real capital. It exceeds $(1 + r_{t+1} + \delta)$ because capital investment commands a risk premium, so its expected growth rate is scaled up by the factor $1/(\alpha(\eta - 1))$. This premium reflects the compensation required for bearing profit/markup risk in imperfectly competitive production (details in Sect. 5.2.4, see also Eq. (5.2.3) in Chap. 5).

Equations (3.5a)–(3.5c) constitute the complete lifetime budget framework. The notation parallels the main text; this appendix simply records the fully indexed version for reference.

## 3.B Propensity to Consume Out of Income and Wealth: $c_y$ and $c_w$

This appendix provides a simple OLG derivation showing how life-cycle lengths map into the propensities to consume out of income and wealth, $(c_y, c_w)$. Following Modigliani's life cycle hypothesis (LCH) (Modigliani, 1986), we use a stripped-down OLG setup with a stationary population, zero interest, no taxes, and no bequests. The goal is intuition rather than calibration.

### 3.B.1 Model Setup

Assume:

1. Individuals earn constant labor income $\bar{y}$ for $N$ years and none thereafter.
2. They consume a constant amount over a lifespan of length $L$, with no bequests and zero interest.
3. The population is stationary and uniformly distributed across ages.

Let $M := L - N$ denote retirement length.

### 3.B.2 Consumption Smoothing and Wealth

Lifetime resources satisfy $\bar{y}\,N = c\,L$, so constant consumption is

$$c = \frac{\bar{y}\,N}{L}.$$

Saving during working years is $\bar{y} - c = \bar{y}\,(1 - N/L) = \bar{y}\,M/L$, so wealth at retirement is

$$\bar{w} = (\bar{y} - c)\,N = \bar{y}\,\frac{MN}{L}.$$

During retirement, the individual dissaves uniformly over $M$ years, so $c = \bar{w}/M$, consistent with the expression above.

### 3.B.3 Aggregates and Propensities

Let $n$ denote the constant number of individuals per unit age. Then total labor income (aggregate income) is

$$Y = nN\bar{y}.$$

Aggregate wealth $\mathcal{W}$ equals the area under the cohort-wealth triangle:

$$\mathcal{W} = \frac{nL}{2}\,\bar{w} = \frac{nL}{2}\,\bar{y}\,\frac{MN}{L} = n\bar{y}\,\frac{MN}{2}.$$

Aggregate consumption of dissaving (retired) households equals the number of retirees times per-capita consumption:

$$C^d = nM \cdot c = nM \cdot \bar{y}\,\frac{N}{L}.$$

Hence, the (average, and hence also marginal) propensity to consume out of wealth in this stylized Modigliani-triangle economy is

$$c_w := \frac{C^d}{\mathcal{W}} = \frac{nM\,\bar{y}\,N/L}{n\bar{y}\,MN/2} = \frac{2}{L}.$$

Similarly, consumption of working households is $C^w = nN \cdot c$, so

$$c_y := \frac{C^w}{Y} = \frac{nN\,(\bar{y}N/L)}{nN\bar{y}} = \frac{N}{L}.$$

### *3.B.4 Interpretation*

The coefficients $c_y = N/L$ and $c_w = 2/L$ summarize how the fraction of life spent working and dissaving shapes aggregate demand in the Modigliani triangle. This appendix is intended as heuristic support for the sign and magnitude of the wealth term in the main text; changing stage lengths or cohort shares would change $c_w$ quantitatively without altering the qualitative mechanism. In particular, a longer lifespan $L$ mechanically lowers both propensities, while a longer working life $N$ raises $c_y$ by increasing the fraction of life spent earning income.

## 3.C Differentiated Goods and Household Demand

This appendix extends the single-good consumption block of the main text to an economy with a continuum of differentiated varieties produced under monopolistic competition. The Dixit–Stiglitz CES aggregator delivers the standard price index and variety demands. We then combine this with the money-in-the-utility (MIU) intratemporal trade-off used in the main text. The results in the main text do not rely on this structure, but embedding the analysis in a standard Dixit-Stiglitz/MIU environment strengthens the model's internal consistency and provides an additional microfoundation. We adopt the normalization conventions summarized in Sect. 5.2.2.

### *3.C.1 Dixit–Stiglitz Aggregator and Demand*

A differentiated consumption variety is indexed by $i \in [0, 1]$. Aggregate consumption is the Dixit–Stiglitz composite

$$C_t := \left(\int_0^1 c_t(i)^{\frac{\eta-1}{\eta}}\,di\right)^{\frac{\eta}{\eta-1}}, \qquad \eta > 1, \tag{3.6}$$

where $\eta$ is the elasticity of substitution across varieties. The corresponding CES price index is

$$P_t := \left( \int_0^1 p_t(i)^{1-\eta} \, di \right)^{\frac{1}{1-\eta}}. \tag{3.7}$$

Cost minimization for a given composite $C_t$ implies the standard variety-level demand system

$$c_t(i) = \left( \frac{p_t(i)}{P_t} \right)^{-\eta} C_t, \qquad i \in [0, 1]. \tag{3.8}$$

### 3.C.2 Money in Utility and the User Cost of Money

Let real payments balances be $H_t := M_t / P_t$. Denote by $i_t^B$ the nominal yield on the representative one-period interest-bearing claim (e.g., bank funding paper) and by $i_t^M$ the nominal yield on payments balances. The *user cost of money* is

$$\tilde{r}_t := \frac{i_t^B - i_t^M}{1 + i_t^B}.$$

In the 100%-reserve baseline of the main text, we set $i_t^M = 0$, so that $\tilde{r}_t = i_t^B/(1 + i_t^B)$ (and $\tilde{r}_t \simeq i_t^B$ for small rates).

The intratemporal allocation between consumption and real balances is governed by the real effective-expenditure budget

$$C_t + \tilde{r}_t H_t = y_t, \qquad y_t := \frac{P_t C_t + \tilde{r}_t M_t}{P_t}.$$

**Lemma 3.8 (CES–MIU Block)** *Let $H \equiv M/P$ denote real transactions balances. Preferences over $(C, H)$ are represented by the CES MIU aggregator*

$$V(C, H) = \left[ C^{\rho} + (H/\lambda)^{\rho} \right]^{1/\rho}, \qquad \rho := \frac{\beta - 1}{\beta}, \qquad \beta > 0, \ \beta \neq 1,$$

*where $\beta$ is the elasticity of substitution between consumption and real balances, and $\lambda > 0$ scales real balances into transaction services. Let the opportunity cost of holding money be the user cost*

$$\tilde{r} := \frac{i^B - i^M}{1 + i^B},$$

*and in the 100%-reserve baseline, set* $i^M = 0$, *so* $\tilde{r} = i^B/(1 + i^B)$. *Given real effective expenditure* $y$ *and budget* $C + \tilde{r}\,H = y$, *the intratemporal optimum satisfies*

$$\frac{H}{C} = \lambda^{1-\beta}\,\tilde{r}^{-\beta}.$$

*Equivalently,*

$$H = \lambda^{1-\beta}\tilde{r}^{-\beta}C, \qquad C = \frac{y}{1 + (\lambda\tilde{r})^{1-\beta}}, \qquad H = \frac{y\,\lambda^{1-\beta}\tilde{r}^{-\beta}}{1 + (\lambda\tilde{r})^{1-\beta}}.$$

*Moreover, the marginal utility of consumption is*

$$V_C(C, H) = \left[C^{\rho} + (H/\lambda)^{\rho}\right]^{\frac{1}{\rho}-1} C^{\rho-1},$$

*and at the intratemporal optimum, it can be written as a function of* $\tilde{r}$ *alone:*

$$V_C(\tilde{r}) = \left[1 + (\lambda\tilde{r})^{1-\beta}\right]^{-\frac{1}{1-\beta}}.$$

*Finally, if the intertemporal margin features a one-period nominal claim with gross return* $1 + i_t^B$ *and (gross) inflation* $1 + \pi_{t+1}$, *the Euler equation is*

$$V_C(\tilde{r}_t) = \delta\,\frac{1 + i_t^B}{1 + \pi_{t+1}}\,V_C(\tilde{r}_{t+1}), \qquad \tilde{r}_t = \frac{i_t^B}{1 + i_t^B}.$$

### 3.C.3 Remark on Symmetry and Smooth Interior Demand

When all firms face identical costs and choose prices optimally under monopolistic competition, symmetry implies a common equilibrium price $p_t(i) = P_t$ for all $i \in [0, 1]$, and variety demands satisfy $c_t(i) = C_t$. A finite elasticity $\eta < \infty$ ensures smooth, interior substitution across varieties and prevents knife-edge reallocation of demand in response to arbitrarily small price differentials.

**Bridge to the Single-Good Block in the Main Text**

The main text treats consumption as a single composite good $C_t$ with price level $P_t$. The differentiated-goods structure above provides a microfoundation for this representation: $C_t$ is the Dixit–Stiglitz consumption index (3.6), and $P_t$ is the associated unit-expenditure (price) index (3.7). Since optimal within-period allocation across varieties delivers the standard demand system (3.8), households' intertemporal choice problem can be written directly in terms of the composite pair $(C_t, P_t)$. Consequently, the MIU intratemporal money-demand condition in Lemma 3.8 applies without modification once real balances are defined as $H_t = M_t/P_t$ and effective expenditure is measured in composite units.

## References

Bank of Japan. (2016). Wealth effects on private consumption. Outlook for Economic Activity and Prices (April 2016), Box 5, URL https://www.boj.or.jp/en/mopo/outlook/box/data/1604BOX5a.pdf. Accessed 30 Sept 2025.

Bénassy, J. P. (2007). IS–LM and the multiplier: A dynamic general equilibrium model. *Economics Letters, 96*(2), 189–195. https://doi.org/10.1016/j.econlet.2006.12.028.

Bénassy, J. P. (2007). Ricardian equivalence and the intertemporal Keynesian multiplier. *Economics Letters, 94*(1), 118–123. https://doi.org/10.1016/j.econlet.2006.08.010.

Blanchard, O. J. (1985). Debt, deficits, and finite horizons. *Journal of Political Economy, 93*(2), 223–247. https://doi.org/10.1086/261297.

Campbell, J. Y., & Mankiw, N. G. (1989). Consumption, income, and interest rates: Reinterpreting the time series evidence. *NBER Macroeconomics Annual, 4*, 185–216. https://doi.org/10.2307/3584973.

Carroll, C. D. (2001). A theory of the consumption function, with and without liquidity constraints. *Journal of Economic Perspectives, 15*(3), 23–45. https://doi.org/10.1257/jep.15.3.23.

Deaton, A. (1991). Saving and liquidity constraints. *Econometrica, 59*(5), 1221–1248. https://doi.org/10.2307/2938366.

Godley, W., & Cripps, F. (1983). *Macroeconomics*. Oxford University Press.

Godley, W., & Lavoie, M. (2007). *Monetary economics: An integrated approach to credit, money, income, production and wealth*. Palgrave Macmillan.

Godley, W., & Lavoie, M. (2012). *Monetary economics: An integrated approach to credit, money, income, production and wealth* (2nd ed.). Palgrave Macmillan.

Modigliani, F. (1986). Life cycle, individual thrift, and the wealth of nations. *Science, 234*(4777), 704–712. https://doi.org/10.1126/science.234.4777.704.

Otaki, M. (2015). *Keynesian economics and price theory: Re-orientation of a theory of monetary economy* (Vol. 7). Springer.

# Chapter 4
# Short-Run Dynamics: Demand, Finance, and Markups (PM–LM)

*I now feel that if I were writing the book again I should begin by setting forth my theory on the assumption that short-period expectations were always fulfilled; and then have a subsequent chapter showing what difference it makes when short-period expectations are disappointed.*

—John Maynard Keynes, *Collected Writings,* Vol. XIV, p. 181

## 4.1 Introduction

The behavioral relations derived in the preceding chapter—linking consumption, saving, and the demand for money—provide the microfoundation for the present analysis of short-period macroeconomic adjustment. Having specified how households allocate income between spending and real balance holdings, we now turn to the interaction between the goods market—pricing and markups—and the payments–funding system. In this setting, investment is financed first through bank funding, while income and money flows subsequently adjust through the Keynesian multiplier.

The resulting pricing-markup (PM)–liquidity-money (LM) framework connects the real and financial sides of the economy under a 100%-reserve payments regime introduced in Chap. 2 and serves as the analytical stage for the Wicksell–Fisher cumulative process. This chapter develops a short-run PM–LM framework in which investment is financed first: banks meet marginal outlays by issuing nonmonetary liabilities ("funding paper") rather than creating payments money.

We work with a stripped-down environment in which the nominal money stock $M^S$ (hereafter denoted $M$) and nominal wage $W$, the two-component propensity to consume $(c_y, c_w)$, and long-horizon investment sentiment, summarized by a fixed *marginal efficiency of investment* (MEI) schedule are taken as given. With long-term expectations fixed, net investment equals replacement by assumption, so the installed capital stock $K$ is effectively predetermined over the horizon of interest. Against this backdrop, we separate *short-run impact analysis* from *short-period SQSS equilibrium*:



I. Takahashi, *Microfoundations of Keynesian Economics*, Evolutionary Economics and Social Complexity Science 37, https://doi.org/10.1007/978-981-95-9390-3_4

**Impact Analysis (One Period, "Intraperiod Rigidity")** At the outset of the period, $P_t$ and $r_t$ are preset, since households and firms require them to form demands. Employment $N_t$ is quasi-fixed; firms meet unexpected demand mainly by varying work intensity and utilization rather than headcount. Thus, an innovation in effective demand (driven by short-run expectations) primarily moves transactions, velocity, and profits, rather than posted prices or interest rates. This reflects the Keynesian view that over very short horizons, money wages and the prevailing employment configuration anchor the price level, while quantities absorb shocks through intensity and usage rather than immediate factor reallocation.

**SQSS Equilibrium (Resetting of Prices and Interest)** When the same disturbance is understood to persist, firms and banks re-set price and interest policies so that the goods market (via the Pricing–Markup locus PM) and the money market (via the Liquidity–Money locus LM) clear for the given $K$. Here we solve simultaneously for $(r, \omega)$ and establish existence and local stability, as well as how $(r, \omega)$—and the induced factor demand $N^d(K)$—depend on $K$. Throughout, we classify the SQSS as a short-period equilibrium in the Keynesian sense: akin to a medium-run reference point in modern usage, because $K$ and long-term expectations are fixed.[1]

Section 4.2 traces the *impact* of demand shifts when prices and interest rates are sticky within the period and loan-market spreads adjust only sluggishly. This provides the PM–LM benchmark under a 100%-reserve payments system and sets up the multiplier process that feeds into the Wicksell–Fisher cumulative dynamics. Sections 4.3 and 4.4 derive the PM and LM relations that summarize short-run equilibrium conditions in the goods and money markets. Section 4.5 investigates the intermediate-run symmetric stationary equilibrium in which employment, the price level, the interest rate, and effective demand are endogenously determined by the level of a single exogenous variable: the capital stock. This result establishes the foundation for the stochastic macroeconomic dynamics studied in later chapters.

## 4.2 Impact Step (SQSS-Anchored, with Utilization Margin)

We analyze the impact step starting from an SQSS in which prices, wages, and administered rates are posted. A revision of long-run expectations shifts planned investment, which banks finance ex ante at posted terms. Within the period, prices and rates are fixed, employment and capital are quasi-fixed, and income adjusts ex post through the multiplier, subject to payments feasibility.

[1] Keynes's "short period" fixes $K$ and long-term expectations, while prices, output, and interest rates may adjust. We avoid the term "long run," since capital accumulation and revisions to long-term expectations are excluded.

**Protocol 4.1 (One-Period Protocol (Short Run))**

**(P1) Gestation as clock** An investment chosen in $t$ becomes productive only in $t+1$:

$$K_{t+1} = (1-\delta)K_t + I_t.$$

**(P2) Start-of-period commitments** Given $(M, W, K_t, \bar{N}_t)$, solve PM–LM to pin $(r_t, \omega_t)$ and set the price once for the period,

$$P_t = \frac{W}{\omega_t}, \qquad \text{with } r_t \text{ and } P_t \text{ fixed within } t.$$

**(P3) Quasi-fixed employment** Employment used in production is preset:

$$N_t = \bar{N}_t,$$

but utilization/work intensity can adjust within $t$; a convenient reduced form is

$$Q_t = \bar{Q}_t + \kappa_q\big(X_t - \bar{Q}_t\big), \qquad \bar{Q}_t = F(K_t, \bar{N}_t), \quad X_t \equiv C_t + I_t^K, \quad \kappa_q \in [0, 1],$$

where $I_t^K$ denotes gross fixed capital formation.[2]

**(P4) Intraperiod multiplier completion** With $(r_t, P_t)$ fixed and $N_t$ quasi-fixed, an absorption shock (e.g., $\Delta I_t$) is accommodated intraperiod by utilization adjustment. Given $C_t = c_y Y_t^r + c_w K_t$, the shock propagates through the income–expenditure loop until the cycle's reduced-form outcome is reached:[3]

$$\Delta Y_t^r = \Delta Q_t = \frac{\kappa_q}{1-\kappa_q c_y}\Delta I_t, \qquad 0 < c_y < 1. \tag{4.2.1}$$

The coefficient

$$m_I(\kappa_q) \equiv \frac{\kappa_q}{1-\kappa_q c_y} \tag{4.2.2}$$

is the intraperiod investment multiplier, capturing the degree of supply flexibility through utilization. Ephemeral inventories and netting absorb any residual imbalance within $t$; repricing and factor adjustment are deferred to $t+1$.

**(P5) Expectation freeze (short run)** The short-run beliefs used to determine PM–LM and $C_t$ $(c_y, c_w, \lambda, \beta$, and the baseline long-horizon view) are held constant within $t$; they may update only when the next period begins.

---

[2] When $X_t$ falls short of $\bar{Q}_t$, $Q_t = C_t + I_t^K + \widetilde{W}_t$, where $\widetilde{W}_t > 0$ denotes unintended inventory absorption or perishable oversupply that does not carry over.

[3] Microlink: with $Q = K^{\alpha}(eN)^{1-\alpha}$, where $e$ denotes labor effort (efficiency per worker), and for small intensity changes, $\kappa_e \approx (1-\alpha)\,\partial \ln e / \partial \ln X$.

**Table 4.1** Finance leg (100%-reserve payments; loans funded by nonmoney liabilities)

| | Households (H) | Firms (F) | Banks (B) | Sum |
|---|---|---|---|---|
| Funding paper/time liabilities (F) | $+\Delta F$ | 0 | $-\Delta F$ | 0 |
| Loans to firms (L) | 0 | $-\Delta L$ | $+\Delta L$ | 0 |
| Demand deposits (D) | $-\Delta F$ | $+\Delta L$ | $\Delta F - \Delta L$ | 0 |
| Row sums | 0 | 0 | 0 | |

Notes: Row sign convention (Godley–Lavoie style): + for asset acquisition by the sector (or reduction of a liability), − for liability issuance (or reduction of an asset). Demand deposits are a *bank liability*. If the bank matches funding one-for-one, $\Delta F = \Delta L$, then the bank's net change in demand-deposit liabilities is zero and aggregate $D$ is unchanged: $\Delta D = (-\Delta F) + (\Delta L) = 0$

Consider an unanticipated change in investment as a local perturbation. This point is emphasized by Keynes[4] and motivates the accounting representation below. The flow of funds can be summarized as follows:

**Funding Leg (Before Outlays)** Banks issue nonmoney funding liabilities to the public and use the proceeds to make loans.

- Banks issue $\Delta F_t$ (funding paper) to households. Households pay with existing demand deposits: $\Delta D_t^H = -\Delta F_t$.
- Banks extend loans $\Delta L_t = \Delta I_t$ and credit firms' deposits: $\Delta D_t^F = +\Delta L_t$.

If funding is matched one-for-one, $\Delta F_t = \Delta L_t$, then total demand deposits are unchanged:

$$\Delta D_t = \Delta D_t^H + \Delta D_t^F = (-\Delta F_t) + \Delta L_t = 0.$$

The adjustment margin is the funding spread, not $M$. Table 4.1 summarizes the investment-driven transaction flows, highlighting that the quantity of payments money is unchanged.

**Profits and the Intraperiod Boost** Let $\pi_t$ denote gross nominal profits (before interest on loans) and let $u_t$ be utilization (or intensity). With $P_t$ and $W_t$ fixed within the period,

$$\pi_t = P_t\, Q_t - W_t\, N_t,$$

and the investment shock raises $Q_t$ via the multiplier and $u_t$ on the supply side, so $\Delta\pi_t > 0$ even at fixed $P_t$. Interest payments $r^L$ are paid later; the gross margin rises now, validating the entrants' move.

[4] Planned investment—i.e., investment ex-ante—may have to secure its "financial provision"... There has, therefore, to be a technique to bridge this gap between the time when the decision to invest is taken and the time when the correlative investment and saving actually occur (Keynes, 1937, p. 246).

**Next Period Price Adjustment and the Wicksell–Fisher Upswing** Under conditions of short-run wage rigidity, the wage remains constant in period $t+1$, while prices begin to adjust to demand pressures according to the function

$$P_{t+1} = P_t \left[ 1 + \kappa_p \left( \frac{Q_t - \bar{Q}}{\bar{Q}} \right) \right], \qquad W_{t+1} = W_t.$$

Then the real wage falls, the markup widens, and unexpected profits lift further in $t+1$. This initiates the familiar cumulative process: higher profitability → more planned investment → continued credit funding (via funding paper/time liabilities) → further demand pressure—until spreads rise enough (or wages reset, or capacity binds) to arrest the upswing.[5]

When entrants expand investment,[6] banks fund the additional loans by issuing nonmonetary liabilities (time deposits or funding paper) to the public. Hence, the money stock $M$ does not change[7], no sterilization is required, and the adjustment occurs via loan-market spreads (the cost of bank funding). This preserves stock–flow consistency while keeping the debt channel transparent in the Fisherian sense: expansions raise leverage first; whether subsequent sales validate the finance determines whether the cumulative process continues or stalls. Adjustment occurs via loan-market spreads rather than quantities.

The short-run analysis proceeds with the assumption of fixed price level $P$, interest rate $r$, capital stock $K$, and a quasi-fixed employment level $N$ within the relevant period. The PM–LM closures then determine the short-run allocation in a neighborhood of SQSS. This isolates the income channel before we "turn on" price-setting and money-market responses.

We follow the standard treatment that temporary excess demand is absorbed by labor hoarding and higher capital utilization (e.g., Oi 1962; Burnside et al. 1993; Greenwood et al. 1988; Shapiro 1993; Basu 1996) and therefore omit an explicit utilization block.

The observation that the SQSS anchoring alongside baseline investment persistence leads to a shock that is a pure demand displacement indicates the necessity of two conditions for the short-run multiplier to operate effectively.

---

5 Fisher emphasized that leverage is the prime mover in expansions and the amplifying mechanism in contractions; see Fisher (1932), Fisher (1933). Our 100%-bank-finance closure makes this debt channel transparent without varying the stock of payments money.

6 We model a perturbed change in investment ($\Delta I \neq \delta K$) as an entry of mutants whose investment differs from the replacement level (see Chap. 5).

7 This financing logic is related to Keynes's "finance" motive (see (Keynes 1973, p. 208)) and resonates with portfolio-balance critiques of strict money-stock targeting associated with Tobin (1963).

**Assumption 4.1 (Two Ingredients for the Short-Run Multiplier)**

1. **Intraperiod commitment** At the start of period $t$, the price level $P_t$ (hence $\omega_t = W/P_t$) and posted rates $(r_t^F, r_t^L)$ are fixed at their SQSS reference (no repricing or resetting within $t$).[8]
2. **Quasi-fixed factors with utilization margin** Employment $N_t$ and installed capital $K_t$ are preset, but firms can raise work intensity $(e)$ and capacity utilization on impact. Thus, at fixed $P_t$, short-run supply can expand up to a capacity ceiling $\bar{Q}(K_t, N_t)$.

For completeness—and to avoid ambiguity about what adjusts within the period—we collect four remarks on (1) inventories, (2) LM as a feasibility condition, (3) velocity as the payments-side margin, and (4) comparison to the *Investment–Saving/ Liquidity Preference–Money Supply (IS–LM) framework*.

**Remark (Why No Inventory Term $\widetilde{W}_t$)** With intraperiod adjustment of utilization and work intensity and no carryover, corners appear as *rationing at the ceiling* $Q_{\max} = Q(K, e_{\max} N)$ rather than "negative inventory." Thus, $\widetilde{W}_t$ is not modeled; excess demand beyond $Q_{\max}$ shows up as queues/lost sales.

**Remark (LM as a Feasibility Check (Payments Capacity))** Within the period, the money/settlement block serves as a *feasibility constraint*, not an optimizing margin. Let real settlement capacity be $\overline{m} \equiv M/P_t$, and let required real balances for payments be $L(Y, r^F; \nu)$, where $\nu$ summarizes institutional "velocity" (netting, timing, intraday credit). The short-run feasibility condition is

$$L\big(Y, r^F; \nu\big) \;\le\; \overline{m}.$$

Given $(\bar{m}, r^F, \nu)$, define the payments-feasible activity set $\mathcal{Y}_{\mathrm{LM}} = \{Y : L(Y, r^F; \nu) \le \overline{m}\}$ and the implied ceiling $Y_{\max}^{\mathrm{LM}}$ by $L\big(Y_{\max}^{\mathrm{LM}}, r^F; \nu\big) = \overline{m}$. In the Impact Step, $Y$ must lie in $\mathcal{Y}_{\mathrm{LM}}$; the PM relation is defined only on this feasible set.

**Remark (Velocity as the Payments-Side Margin)** We parameterize institutional velocity by $\nu$ so that $L_Y > 0$ and $L_\nu < 0$ (higher $\nu$ lowers required balances for a given $Y$). Small, temporary increases in $\nu$ reflect operational adjustments (netting, batching, payment timing, intraday credit lines), which relax the LM feasibility and raise $Y_{\max}^{\mathrm{LM}}$. In the short run, $(r^F, \overline{m})$ are fixed and $\nu$ can move only modestly; hence,

[8] The within-period rigidity of the posted commodity price and lending rate may be interpreted as a commitment device—commitment via posted terms—since if sellers or lenders could revise terms after observing demand, buyers, or borrowers might strategically delay transactions; see Sobel and Takahashi (1983). A related but distinct microfoundation for price rigidity is provided by Negishi (1979), where informational asymmetry gives rise to a kinked subjective demand curve, rendering firms reluctant to adjust prices across periods. While that mechanism operates at the intertemporal margin, the present framework emphasizes intra-period rigidity arising from bargaining and posting frictions.

we treat $\nu$ as a *shifter* of the feasibility set rather than a control variable. At the SQSS reference, $\nu = \bar{\nu}$ and $L(Y^*, r^{F*}; \bar{\nu}) = \overline{m}$.

**Remark (Comparison to IS–LM framework)** Our short-run closure is IS–LM-like in its multiplier logic but differs in the finance step and the interest-rate closure: administered rates $(r^L, r^F)$ and the spread $s = r^L - r^F$ are fixed within the period, banks bridge *ex-ante* investment at posted terms, and income adjusts *ex post* under a payments-capacity (PM) rather than a money-market (LM) closure.

Having isolated the impact response (with a possible capacity kink), we now open the closures that will determine the short-run point around the anchor: PM (price/markup, mapping user cost into $MC$) and LM (payments/portfolio, pinning down the opportunity cost $r^F$). The interaction of the demand displacement with PM–LM yields the short-run equilibrium.

## 4.3 The Price–Markup (PM) Relation

We derive the PM locus from (i) cost minimization under Cobb–Douglas technology and (ii) markup pricing under monopolistic competition. Throughout this subsection, interpret variables as *within-period values at the SQSS reset* (capital $K$ fixed, and posted $(W, r)$ treated as given).

**Technology and Factor Prices**
A representative firm produces

$$q = k^{\alpha} n^{1-\alpha}, \qquad \alpha \in (0, 1) \tag{4.3.1}$$

and takes as given the real wage $\omega \equiv W/P$ and the user cost of installed capital $r + \delta$ (absorbing any constant, sluggish spread into $\delta$ when convenient).

**Unit Cost**
Let $c(q; \omega, r + \delta)$ denote the minimum real cost of producing $q$:

$$c(q; \omega, r + \delta) = \min_{k,n} \ \omega n + (r + \delta)k \quad \text{s.t.} \quad q = k^{\alpha} n^{1-\alpha}.$$

Cost minimization yields the standard Cobb–Douglas unit cost (real marginal cost):

$$mc(\omega, r + \delta) = \frac{\omega^{1-\alpha}(r + \delta)^{\alpha}}{(1 - \alpha)^{1-\alpha}\alpha^{\alpha}}.$$

**Markup Pricing**
Under Dixit–Stiglitz monopolistic competition with elasticity $\eta > 1$, the optimal price is a constant markup $\mathfrak{m}$ over nominal marginal cost:

$$P = \mu P \cdot mc(\omega, r+\delta), \qquad \mathfrak{m} \equiv \frac{\eta}{\eta - 1}.$$

Canceling $P$ implies the PM condition

$$mc(\omega, r+\delta) = \frac{1}{\mathfrak{m}} = \frac{\eta - 1}{\eta}. \tag{4.3.2}$$

Substituting the expression for $mc$ gives the explicit PM locus in $(r, \omega)$:

$$\left(\frac{\omega}{1-\alpha}\right)^{1-\alpha}\left(\frac{r+\delta}{\alpha}\right)^{\alpha} = \frac{\eta - 1}{\eta}.$$

**PM as a Wage–Interest Relation**
Solving (4.3.2) for $\omega$ yields

$$\omega_{\mathrm{PM}}(r) = \theta_{\mathrm{PM}}\,(r+\delta)^{-\frac{\alpha}{1-\alpha}}, \qquad \theta_{\mathrm{PM}} = \left[\frac{(1-\alpha)^{1-\alpha}\alpha^{\alpha}}{\mathfrak{m}}\right]^{\frac{1}{1-\alpha}}. \tag{4.3.3}$$

Hence, $\omega'_{\mathrm{PM}}(r) < 0$: the PM curve is strictly decreasing in $(r, \omega)$-space.

**Lemma 4.1 (Shifters of the PM Curve)** *The* PM *locus shifts only with technology and market-power parameters:*

*(i) Higher competition (larger $\eta$, smaller markup $\mathfrak{m}$) raises $(\eta - 1)/\eta$ and shifts* PM up *(higher $\omega$ at any $r$).*
*(ii) Changes in $\alpha$ or $\delta$ rotate/shift* PM *through the exponents and the $(r + \delta)$ term.*
*(iii) Demand and monetary variables $(M, K, \lambda, \beta, c_y, c_w)$ do* not *enter* PM*; they belong to* LM.

**Definition 4.1 (Pricing–Markup (PM) Curve)** The *pricing–markup (PM) curve* is the locus in the $(r, \omega)$-plane defined by

$$\left(\frac{\omega}{1-\alpha}\right)^{1-\alpha}\left(\frac{r+\delta}{\alpha}\right)^{\alpha} = \frac{\eta - 1}{\eta},$$

equivalently (4.3.3). It summarizes cost minimization and markup pricing under monopolistic competition at the SQSS reset.

## 4.4 The Liquidity–Money (LM) Relation

Having pinned down the pricing side via PM, we now turn to the asset side: the LM relation.

**Money Demand Scales with Consumption Rather Than Income**
In deriving the LM relation, we follow an empirical and behavioral tradition that short-run money demand is more tightly linked to consumption expenditures than to contemporaneous income. The key intuition is consumption smoothing. Households plan transactions around desired spending flows, not around transitory fluctuations in income. In the short run, consumption adjusts smoothly relative to income, while saving absorbs much of the high-frequency variation in $Y$. This empirical regularity—that consumption is substantially smoother than income in the short run—is well documented by Campbell and Mankiw (1989).

This logic aligns naturally with money-in-utility formulations: transactions balances are demanded to facilitate purchases, not to mirror income receipts. Accordingly, we scale transactions money demand with $C$ rather than $Y/P$. In the SQSS block of the model, where consumption itself is proportional to the capital stock, this choice yields a parsimonious LM relation while remaining consistent with both empirical evidence and the microfoundation developed in Chap. 3.

In an SQSS equilibrium,

$$Y = C = c_y + c_w K, \Longrightarrow C = c K, \quad \text{where } c = c_w/(1 - c_y).$$

**Scope (Transactions)** Lemma 3.7 gives the *period-average* demand for settlement balances,

$$\frac{M^d}{P} = \lambda^{1-\beta} C \left(\frac{r}{1+r}\right)^{-\beta}.$$

Under our settlement convention, the payments system supplies a fixed nominal capacity $M$ (the reserve cap), so market clearing requires $M^d = M$:[9]

$$\frac{M}{P} = \lambda^{1-\beta} c K \left(\frac{r}{1+r}\right)^{-\beta}. \qquad \text{(LM relation)} \qquad (4.4.1)$$

**Remark 4.1 (Slack Payments Capacity Implies a Lower Clearing Interest Rate)** In the 100% reserve regime, the fixed nominal stock of settlement balances $M$ represents the system's payments capacity. If this capacity exceeds the level required by current transactions, the opportunity cost of holding money must fall for the payments market to clear. Hence, a larger settlement capacity implies a lower clearing funding rate $r_t$, which raises money demand toward full utilization of settlement balances.

[9] Interpreting $M$ as system capacity, the equality $M^d = M$ means the reserve cap is calibrated to cover period-average settlement needs (or, equivalently, that any prudential peak cushion is folded into $\lambda^{1-\beta}$).

### 4.4.1 Quantity Rule and Rate Rule

Under a *quantity rule*, the nominal payments capacity $M$ is fixed and the opportunity cost $r$ adjusts to clear the money market. This yields a downward-sloping LM locus in the $(r, \omega)$-plane.

Under a *rate rule*, the payments authority administers a target rate $\bar{r}$ and supplies whatever quantity of reserves is required to support settlement at that rate (see Kaldor, 1985; Moore, 1988). In this case, the LM locus is horizontal:

$$r = \bar{r}.$$

The money stock then adjusts endogenously:

$$M(\omega; \bar{r}) = \frac{W\, c\, \lambda^{1-\beta} K}{\omega} \left(\frac{\bar{r}}{1+\bar{r}}\right)^{-\beta}.$$

In what follows, we focus on the *quantity rule*. Solving (4.4.1) for the real wage $\omega \equiv W/P$ yields the LM locus:

$$\omega_{\text{LM}}(r) = \theta_{\text{LM}} \left(\frac{r}{1+r}\right)^{-\beta} K, \qquad \theta_{\text{LM}} \equiv \frac{W}{M} \lambda^{1-\beta} c. \tag{4.4.2}$$

*Monotonicity.* Let $f(r) \equiv \left(\frac{r}{1+r}\right)^{-\beta}$ with $\beta > 0$. Then

$$\frac{f'(r)}{f(r)} = -\frac{\beta}{r(1+r)} < 0,$$

so $\omega'_{\text{LM}}(r) < 0$. Hence, the LM curve in $(r, \omega)$-space is *strictly decreasing*.

We now study how this stable equilibrium changes with the capital stock $K$.

## 4.5 Symmetric Quasi-Steady-State (SQSS) Equilibrium

In this section, we define the *Symmetric Quasi-Steady State (SQSS)* as the short-period stationary equilibrium in which the real capital stock $K$ and the long-term state of expectations are held fixed. Prices and the interest rate adjust to clear the goods market and the money market, while investment is just sufficient to maintain the existing capital stock, $I = \delta K$. The SQSS thus corresponds to Keynes's *short period*: output, the real wage, and the rate of interest may vary, but productive capacity itself is predetermined.

The SQSS plays a dual role in the analysis. First, it is the reference point toward which nominal and financial variables converge through short-run adjustment. Second, it provides the anchor from which medium-run movements in the capital stock are initiated once profitability conditions induce net investment. In this sense, the SQSS is not a long-run steady state with capital accumulation, but a *quasi-steady anchor*: a stationary attractor for nominal variables conditional on $K$.

### 4.5.1 Equilibrium in the $(r, \omega)$-Plane

Given a fixed capital stock $K$, the short-period equilibrium is determined by the interaction of two loci:

- The *pricing–markup (PM) locus*, which gives the real wage $\omega \equiv W/P$ consistent with cost minimization and optimal price setting as a decreasing function of the user cost of capital $(r + \delta)$.
- The *liquidity–money (LM) locus*, which gives the real wage consistent with money-market equilibrium. A higher real wage raises transactions demand through its effect on income, while the interest rate $r$ governs the opportunity cost of holding money.

Formally, the SQSS equilibrium is characterized by

$$\omega_{\mathrm{PM}}(r) = \omega_{\mathrm{LM}}(r; K), \tag{4.5.1}$$

where $\omega'_{\mathrm{PM}}(r) < 0$ and $\omega'_{\mathrm{LM}}(r; K) < 0$. Each solution $(r^*, \omega^*)$ to (4.5.1) represents a stationary short-period equilibrium conditional on $K$. The general price level is then determined by nominal wage setting as

$$P = \frac{W}{\omega^*}. \tag{4.5.2}$$

Multiple intersections of the PM and LM loci may exist, depending on parameter values. Only those that are locally stable under the short-run adjustment dynamics are economically relevant.

### 4.5.2 Local Stability of the SQSS Equilibrium

To assess local stability, we specify the adjustment mechanisms governing prices and the interest rate. *Prices adjust to restore goods-market balance, while the interest rate adjusts to restore money-market balance.*

**Goods-Market Adjustment (Relative to PM)**
Let $\omega = \mathrm{PM}(r)$ denote the real wage implied by firms' pricing and cost-minimization conditions. If the actual configuration satisfies $\omega > \mathrm{PM}(r)$, firms' posted prices imply unit costs that are inconsistent with clearing the goods market

at prevailing demand: *effective demand exceeds supply at posted prices*. Prices therefore rise ($P \uparrow$), reducing the real wage $\omega = W/P$, so $\dot{\omega} < 0$. Conversely, if $\omega < \mathrm{PM}(r)$, there is effective excess supply at posted prices; prices fall ($P \downarrow$) and the real wage rises, so $\dot{\omega} > 0$.

**Money-Market Adjustment (Relative to LM)**
Let $\omega = \mathrm{LM}(r; K)$ denote the real wage consistent with money-market equilibrium. If $\omega > \mathrm{LM}(r; K)$, transactions demand for money exceeds the available settlement capacity, placing upward pressure on the interest rate ($\dot{r} > 0$). If $\omega < \mathrm{LM}(r; K)$, there is excess supply of money services and the interest rate falls ($\dot{r} < 0$).

These mechanisms generate the local dynamic system

$$\dot{\omega} = -\kappa_\omega\big(\omega - \mathrm{PM}(r)\big), \qquad \dot{r} = -\kappa_r\big(\omega - \mathrm{LM}(r; K)\big), \tag{4.5.3}$$

where $\kappa_\omega, \kappa_r > 0$ are adjustment speeds.

These adjustment directions are summarized graphically in the phase diagrams reported at the end of the chapter (Fig. 4.1), where we compare the exponent-matching case $\beta = \alpha/(1-\alpha)$ with the general case $\beta < \alpha/(1-\alpha)$ and show that local stability does not hinge on the equality restriction.

### 4.5.3 *Stability Condition*

Linearizing (4.5.3) around a stationary point $(r^*, \omega^*)$ yields the Jacobian

$$J = \begin{pmatrix} -\kappa_\omega & \kappa_\omega \mathrm{PM}'(r^*) \\ -\kappa_r & \kappa_r \mathrm{LM}'(r^*; K)) \end{pmatrix}.$$

**Lemma 4.2 (Local Stability of the PM–LM Equilibrium)** *Suppose* $\mathrm{PM}'(r^*) < 0$ *and* $\mathrm{LM}'(r^*; K) < 0$. *The SQSS equilibrium* $(r^*, \omega^*)$ *is locally asymptotically stable if and only if*

$$\big|\mathrm{LM}'(r^*; K)\big| > \big|\mathrm{PM}'(r^*)\big|.$$

*That is, the LM locus must be steeper (in absolute value) than the PM locus at the intersection.*

***Proof (Sketch)*** Local stability requires $\det(J) > 0$ and $\mathrm{Tr}(J) < 0$. When both loci slope downward, $\det(J) > 0$ is equivalent to $\mathrm{PM}'(r^*) > \mathrm{LM}'(r^*; K)$, or $\big|\mathrm{LM}'(r^*; K)\big| > \big|\mathrm{PM}'(r^*)\big|$. The trace condition is automatically satisfied for $\mathrm{LM}'(r^*; K) < 0$ and positive adjustment speeds. □

### 4.5.4 Income, Costs, and Profits at the SQSS

The PM–LM equilibrium conditions determine prices, interest rates, and the scale of activity at the SQSS without explicit reference to national-income aggregates. Nevertheless, for purposes of economic interpretation and completeness, it is useful to record how real output and income are decomposed once the SQSS is established.

With prices posted and constant, real national income is defined as net output valued at posted prices:

$$\frac{Y}{P} \equiv Q - \delta K,$$

where $Q$ denotes gross real output and $\delta K$ the replacement of depreciated capital. In the SQSS, net output is fully absorbed, so that $Y/P = C$.

The linear homogeneity of the production function implies that gross output can be written as the markup-weighted sum of factor payments:

$$Q = \mathfrak{m}\big(\omega N + (r+\delta)K\big), \tag{4.5.4}$$

where $\omega$ is the real wage and $r$ the real interest rate. Subtracting depreciation yields the corresponding decomposition of real national income:

$$\frac{Y}{P} = \omega N + rK + \Pi, \qquad \Pi \equiv (\mathfrak{m}-1)\big(\omega N + (r+\delta)K\big), \tag{4.5.5}$$

where $\Pi$ represents gross real profits. The total cost is given by $\omega N + (r+\delta)K$, and average cost equals $1/\mathfrak{m}$ in units of output.

An immediate implication of the constant elasticity of substitution (CES) markup condition is the following lemma, which expresses firm dividends as a simple function of $r+\delta$, $\alpha$, and $\eta$.

**Lemma 4.3 (Dividends Under CES Markups)** *Suppose the final-good sector is monopolistically competitive with CES demand elasticity $\eta > 1$, so that optimal pricing implies the constant markup*

$$\mathfrak{m} := \frac{P}{MC} = \frac{\eta}{\eta - 1}.$$

*Let the real user cost of capital be $v \equiv r+\delta$, and assume cost minimization delivers a constant capital-cost share $\alpha \in (0,1)$, so that*

$$vK = \alpha\, MC \cdot Y,$$

*where Y denotes aggregate output in real terms. Then operating profits (and hence firm dividends in the SQSS) satisfy*

$$Div^F = \Pi^F = \frac{v}{\alpha(\eta-1)}K = \frac{r+\delta}{\alpha(\eta-1)}K. \tag{4.5.6}$$

*Equivalently, if firm equity E is defined by the reduced-form payout identity* $Div^F = (r+\delta)E$, *then*

$$\frac{E}{K} = \frac{1}{\alpha(\eta-1)}. \tag{4.5.7}$$

***Proof (Deferred)*** See Appendix 4.A, Proof of Lemma 4.3. □

At the SQSS, factor proportions and prices are pinned down by the equilibrium conditions, so that labor income, capital income, and profits are all proportional to the scale of the economy. Writing $\kappa \equiv K/N$, the real national income can be expressed equivalently as a function of the capital stock alone:

$$\frac{Y}{P} = y(K),$$

with the proportionality reflecting the stationary reproduction of the economy at posted prices and fixed markups.

These accounting relations play no role in determining the SQSS itself. Their purpose is solely to clarify the magnitudes and income shares associated with the equilibrium configuration that the PM–LM system selects.

### *4.5.5 Keynesian Interpretation*

The economic logic of the stability condition reflects Keynes's view that the interest rate is governed by *liquidity preference*, not by the marginal efficiency of capital. In this framework, changes in $r$ primarily redistribute income between capital and labor rather than equilibrating saving and investment.

When effective demand falls, prices decline and the real wage rises, reducing transactions demand for money and putting downward pressure on $r$. If money demand is weakly interest-elastic, a large fall in $r$ is required to clear the money market. Because capital income is proportional to $\alpha Y$, such a sharp fall in $r$ strongly compresses aggregate income when the capital share is large, reinforcing the contraction in demand. In this case, the PM locus is steeper than the LM locus and the equilibrium is locally unstable.

Stability instead requires that money demand be sufficiently interest-elastic:

$$\beta > \frac{\alpha}{1-\alpha}. \tag{4.5.8}$$

Under this condition, a modest change in $r$ suffices to restore money-market equilibrium. The redistribution toward wages dominates the loss of capital income, and aggregate demand stabilizes. The LM locus is then steeper than the PM locus, ensuring local stability of the SQSS.

## 4.6 Comparative Statics of the Equilibrium with Respect to $K$

Given a locally stable intersection $(r^*, \omega^*)$ for a fixed $K$, we ask how the equilibrium interest rate responds to changes in the capital stock. Throughout, we treat $K$ as a parameter that shifts the LM locus, while the PM locus is unaffected.

On the stable branch, the equilibrium condition is

$$\omega_{\mathrm{PM}}(r) = \omega_{\mathrm{LM}}(r; K), \qquad \text{equivalently} \qquad F(r, K) := \omega_{\mathrm{LM}}(r; K) - \omega_{\mathrm{PM}}(r) = 0. \tag{4.6.1}$$

Since $\partial_K F = \partial_K \omega_{\mathrm{LM}}(r; K) > 0$, implicit differentiation yields

$$\frac{dr^*}{dK} = -\frac{\partial_K F}{\partial_r F} = -\frac{\partial_K \omega_{\mathrm{LM}}(r^*; K)}{\mathrm{LM}'(r^*; K) - \mathrm{PM}'(r^*)}. \tag{4.6.2}$$

On the locally stable branch, the slope ordering $|\mathrm{LM}'(r^* ; K)| > |\mathrm{PM}'(r^*)|$ implies $\mathrm{LM}'(r^* ; K) - \mathrm{PM}'(r^*) < 0$ (when both loci are downward sloping). Therefore, $\partial_K \omega_{\mathrm{LM}} > 0$ and Eq. (4.6.2) gives

$$\frac{dr^*}{dK} > 0. \tag{4.6.3}$$

**Remark 4.2 (Regularity of the Stable Branch)** Throughout the stochastic analysis, we restrict attention to a compact state space $S = [1, \bar{K}]$ lying strictly below the tangency level $K_{\mathrm{crit}}$ (defined later in Eq. (4.6.6)). On this domain, the PM and LM schedules intersect uniquely and transversally, so the implicit function theorem implies that the equilibrium interest rate $r = r(K)$ is $C^1$ on $S$. Consequently, any object defined by composition, is continuous on $S$.

**Low- and High-Rate Branches Under Exponent Matching**

Under the exponent-matching specification (Appendix 4.D), the equilibrium condition can be written as

$$\frac{\theta_{\mathrm{LM}} K}{\theta_{\mathrm{PM}}} h(r) = 1, \qquad h(r) = \left[\frac{(r+\delta)(1+r)}{r}\right]^{\beta}, \tag{4.6.4}$$

where $h(r) \to \infty$ as $r \downarrow 0$ or $r \uparrow \infty$ and $h$ has a unique minimum at $r = \sqrt{\delta}$. Hence, for $K < K_{\mathrm{crit}}$, there are two roots $0 < r_- < \sqrt{\delta} < r_+$.

The lower-rate root $r_-$ is locally stable and satisfies

$$\frac{dr_-}{dK} = -\frac{h(r_-)}{K\,h'(r_-)} > 0, \qquad \text{since } h > 0 \text{ and } h'(r) < 0 \text{ for } r < \sqrt{\delta}.$$

By contrast, the upper-rate root $r_+ > \sqrt{\delta}$ is unstable; since $h'(r_+) > 0$, it follows that $dr_+/dK < 0$.

**Economic Interpretation** An increase in $K$ shifts the LM locus upward in $(r, \omega)$-space (higher real income and transactions raise money demand at given $r$). Stability requires the LM to be steeper (in absolute value) than PM, so the stable intersection slides *rightward* along PM as $K$ increases, implying a higher equilibrium interest rate. On the stable branch, this also entails a lower real wage along PM, $\omega_{\mathrm{PM}}(r)$, and correspondingly, a larger demand for labor $N^d$; formal expressions appear in Appendix 4.E.

For explicit expressions and the associated thresholds ($K_{\mathrm{crit}}$, $\theta_{\mathrm{LM,crit}}$, $M_{\min}$), see Appendix 4.D (e.g., Eqs. (4.D3)–(4.D11)). For derivative signs and comparative statics of $\omega(K)$, $\kappa(r(K))$, $N^d(K)$, $Q(K)$, and $M^d(K)/P$, see Appendix 4.E.

### 4.6.1 Explicit Characterization Under Exponent Matching

To obtain the closed-form expression for $K(r)$ and the characterization $r^* = \sqrt{\delta}$, we establish the *exponent-matching assumption* by setting our baseline parameter as $\beta = \alpha/(1-\alpha)$.[10]

**Assumption 4.2 (Exponent Matching)** Let $\gamma := \alpha/(1-\alpha) > 0$ and set $\beta = \gamma$ in the SQSS PM–LM system

$$\omega = \theta_{\mathrm{PM}}(r+\delta)^{-\gamma}, \qquad \omega = \theta_{\mathrm{LM}}\left(\tfrac{r}{1+r}\right)^{-\beta} K.$$

with $\theta_{\mathrm{PM}} > 0$ and $\theta_{\mathrm{LM}} > 0$ as defined in Appendix 4.D. Equating the two gives the equilibrium condition

$$\frac{\theta_{\mathrm{LM}} K}{\theta_{\mathrm{PM}}}\,h(r) = 1, \qquad h(r) = \left[\frac{(r+\delta)(1+r)}{r}\right]^{\beta}, \tag{4.6.5}$$

which is Eq. (4.D3) in Appendix 4.D.

[10] The chosen parameter values for $\alpha$ and the resulting interest elasticity $\gamma$ are within a plausible economic range. Specifically, with $\alpha \in (0, 1)$, $\gamma$ is defined as $\gamma = \alpha/(1-\alpha) \in (0, \infty)$. Typical values, such as $\alpha = 1/3$ (yielding $\gamma = 1/2$) and $\alpha = 0.4$ (yielding $\gamma \approx 2/3$), represent reasonable magnitudes for the money-demand interest elasticity in a constant elasticity of substitution (CES) money-in-the-utility-function (MIU) framework, suggesting the calibration is economically sound.

**Existence, Multiplicity, and Selection** The function $h(r)$ is U-shaped with a unique minimum at $r = \sqrt{\delta}$ and $h_{\min} = [(1 + \sqrt{\delta})^2]^{\beta}$ (Eq. (4.D3)). Consequently, there exists a *tangency (degenerate) equilibrium* at $r = \sqrt{\delta}$ if and only if

$$\frac{\theta_{\text{LM}} K}{\theta_{\text{PM}}} h_{\min} = 1 \quad \Longleftrightarrow \quad K = K_{\text{crit}} := \frac{\theta_{\text{PM}}}{\theta_{\text{LM}}} (1 + \sqrt{\delta})^{-2\beta}; \tag{4.6.6}$$

see Appendix 4.D. For $K < K_{\text{crit}}$ (*subtangency*), Eq. (4.6.5) admits two solutions $0 < r_-(K) < \sqrt{\delta} < r_+(K)$. For $K > K_{\text{crit}}$ (*supertangency*), there is no solution.

**Stability Classification** At any intersection, local stability requires $|\text{LM}'(r^* ; K) | > |\text{PM}'(r^*)|$ (Lemma 4.2). Under exponent matching, this holds on the *low-rate branch* $r_-(K) \in (0, \sqrt{\delta})$ and fails on the *high-rate branch* $r_+(K) > \sqrt{\delta}$; hence, $r_-(K)$ is locally asymptotically stable, whereas $r_+(K)$ is unstable (Appendix 4.D, Proposition "Existence and stability").

**Calibration Thresholds and Money Requirement** Using $\theta_{\text{LM}} = (W/M)\, c\, \lambda^{1-\beta}$, the tangency condition Eq. (4.6.6) is equivalent to a minimum nominal money supply

$$M_{\min} = \frac{W\, c\, \lambda^{1-\beta}}{\theta_{\text{PM}}} K \left[(1 + \sqrt{\delta})^2\right]^{\beta}, \tag{4.6.7}$$

as in Eq. (4.D11). For given $(K, \theta_{\text{PM}})$, intersections exist if and only if $M \geq M_{\min}$; when $M > M_{\min}$, the stable equilibrium satisfies $r^* < \sqrt{\delta}$.

**Proposition 4.4 (Closed-Form Equilibrium Interest Rate Under Exponent Matching)** *Let*

$$\textit{PM: } \omega_{\text{PM}}(r) = \theta_{\text{PM}}(r + \delta)^{-\gamma}, \qquad \textit{LM: } \omega_{\text{LM}}(r;\ K) = \theta_{\text{LM}} K \left(\frac{r}{1 + r}\right)^{-\beta},$$

*with* $\gamma := \frac{\alpha}{1-\alpha} > 0$, $\beta > 0$, *and*

$$\theta_{\text{LM}} := \frac{W}{M} c\, \lambda^{1-\beta} > 0.$$

*Impose the* exponent-matching assumption $\beta = \gamma$ *and define*

$$s(K) := \left(\frac{\theta_{\text{PM}}}{\theta_{\text{LM}} K}\right)^{1/\beta} = \left(\frac{1}{\sigma_0 K}\right)^{1/\beta}, \qquad \sigma_0 := \theta_{\text{LM}}/\theta_{\text{PM}}. \tag{4.6.8}$$

*Then the PM–LM equilibrium condition* $\omega_{\text{PM}}(r) = \omega_{\text{LM}}(r)$ *reduces to the quadratic equation*

$$r^2 + \big(1 + \delta - s(K)\big) r + \delta = 0,$$

*which admits two positive solutions*

$$r_{\pm}(K) = \frac{s(K) - (1+\delta) \pm \sqrt{\big(s(K) - (1+\delta)\big)^2 - 4\delta}}{2}, \qquad 0 < r_- < r_+. \tag{4.6.9}$$

**Proposition 4.5 (Equilibrium Selection and Stability Under Exponent Matching)** *Under the assumptions of Proposition 4.4, define the equilibrium interest-rate function as*

$$r(K) := r_-(K) = \frac{s(K) - (1+\delta) - \sqrt{\big(s(K) - (1+\delta)\big)^2 - 4\delta}}{2}, \; 0 < r_- < \sqrt{\delta}. \tag{4.6.10}$$

*Then, provided that*

$$K < \frac{\theta_{\mathrm{PM}}}{\theta_{\mathrm{LM}}} (1 + \sqrt{\delta})^{-2\beta},$$

*the PM–LM system admits a unique locally asymptotically stable equilibrium interest rate $r(K)$, while the upper branch $r_+(K)$ is unstable.*

***Proof*** Define the scale-free ratio

$$h(r) := (r+\delta)^{\beta} \Big(\frac{1+r}{r}\Big)^{\beta}.$$

A direct calculation yields

$$\frac{d}{dr} \log h(r) = \beta \, \frac{r^2 - \delta}{r(1+r)(r+\delta)},$$

so $h$ attains a unique global minimum at $r = \sqrt{\delta}$, with $h(r) \to \infty$ as $r \downarrow 0$ or $r \to \infty$. Thus, for sufficiently small $K$, the equation

$$\omega_{\mathrm{LM}}(r;\, K) = \omega_{\mathrm{PM}}(r) \quad \Longleftrightarrow \quad \frac{\theta_{\mathrm{LM}} K}{\theta_{\mathrm{PM}}} h(r) = 1$$

admits two solutions $0 < r_- < \sqrt{\delta} < r_+$.

Local stability follows from the slope condition $|\omega'_{\mathrm{LM}}(r^*;\, K)| > |\omega'_{\mathrm{PM}}(r^*)|$, which is equivalent to $r^* < \sqrt{\delta}$. Hence, the lower branch $r_-(K)$ is locally asymptotically stable, while the upper branch $r_+(K)$ is unstable. □

**Corollary 4.1 (Closed-Form $\omega(K)$ and $P(K)$)** *Under exponent matching, the equilibrium real wage and price level are given by*

$$\omega(K) = \theta_{\mathrm{PM}}\big(r(K) + \delta\big)^{-\beta}, \qquad P(K) = \frac{W}{\omega(K)},$$

*where* $r(K) = r_{-}(K)$.

**Corollary 4.6 (Tangency and Calibration at $r = \sqrt{\delta}$)** *Maintain* $\beta = \gamma = \alpha/(1-\alpha)$ *and the definitions*

$$\omega_{\mathrm{LM}}(r, K) = \theta_{\mathrm{LM}} K \Big(\frac{r}{1+r}\Big)^{-\beta}, \quad \omega_{\mathrm{PM}}(r) = \theta_{\mathrm{PM}}(r+\delta)^{-\beta}, \quad h(r) = (r+\delta)^{\beta}\Big(\frac{1+r}{r}\Big)^{\beta}.$$

*Then $h$ attains its global minimum at* $r = \sqrt{\delta}$ *with*

$$h_{\min} = h(\sqrt{\delta}) = \big[(1+\sqrt{\delta})^2\big]^{\beta}.$$

*The PM–LM system has a* tangent (degenerate) equilibrium *iff*

$$\frac{\theta_{\mathrm{LM}} K}{\theta_{\mathrm{PM}}}\, h_{\min} = 1 \iff r^{\dagger} = \sqrt{\delta}, \;\; \omega^{\dagger} = \omega_{\mathrm{PM}}(\sqrt{\delta}) = \omega_{\mathrm{LM}}(\sqrt{\delta}).$$

*Equivalently, the tangency delivers the following* ***critical thresholds****:*

$$K_{\mathrm{crit}} = \frac{\theta_{\mathrm{PM}}}{\theta_{\mathrm{LM}}}(1+\sqrt{\delta})^{-2\beta} \qquad \theta_{\mathrm{LM,crit}} = \frac{\theta_{\mathrm{PM}}}{K}(1+\sqrt{\delta})^{-2\beta}.$$

*Hence:*

- *If* $K < K_{crit}$, *there are two intersections:* $0 < r_{-} < \sqrt{\delta} < r_{+}$. *In this case, the lower-rate root* $r_{-}$ *is locally stable, while the higher-rate root* $r_{+}$ *is unstable.*
- *If* $K = K_{\mathrm{crit}}$, *there is a unique degenerate tangent equilibrium at* $r = \sqrt{\delta}$.
- *If* $K > K_{\mathrm{crit}}$, *there is* no *intersection.*

*Using* $\theta_{\mathrm{LM}} = \frac{W}{M} c\,\lambda^{1-\beta}$, *the tangency condition is also equivalent to a minimum nominal money supply:*

$$M_{\min} = \frac{W c\,\lambda^{1-\beta}}{\theta_{\mathrm{LM,crit}}} = \frac{W c\,\lambda^{1-\beta}}{\theta_{\mathrm{PM}}} K\big[(1+\sqrt{\delta})^2\big]^{\beta}.$$

*Thus, for the given* $(K, \theta_{\mathrm{PM}})$, *intersections exist iff* $M \geq M_{\min}$; *when* $M > M_{\min}$, *the lower-rate equilibrium satisfies* $r^* < \sqrt{\delta}$ *and is locally asymptotically stable.*

### 4.6.2 Robustness to Parameter Perturbations

The explicit characterization above relies on the exponent-matching restriction $\beta = \gamma = \alpha/(1-\alpha)$. We now show that the qualitative properties of the stable SQSS equilibrium are locally robust to small deviations from this restriction.

Let $\beta = \gamma + \varepsilon$ with $|\varepsilon| \ll 1$. Writing the PM and LM loci as

$$\omega_{\text{PM}}(r) = \theta_{\text{PM}}(r+\delta)^{-\gamma}, \qquad \omega_{\text{LM}}(r;K) = \theta_{\text{LM}} K\left(\tfrac{r}{1+r}\right)^{-\beta},$$

the equilibrium condition $\omega_{\text{PM}}(r) = \omega_{\text{LM}}(r;K)$ can be expressed as

$$K(r) = A_\omega\,[f(r)]^{-\gamma}\left(\tfrac{r}{1+r}\right)^{\varepsilon}, \qquad \log K = \log A_\omega - \gamma \log f(r) + \varepsilon \log\left(\tfrac{r}{1+r}\right), \tag{4.6.11}$$

with $A_\omega = \theta_{\text{PM}}/\theta_{\text{LM}}$ and $f(r) = (r+\delta)\,(1+r)/r$ (see Appendix 4.E).

At $\varepsilon = 0$, $K(r)$ attains its unique minimum at $r = \sqrt{\delta}$; this is the tangency point that separates the low-rate and high-rate branches (Appendix 4.D). The first-order perturbation of the first-order condition yields the first-order shift of the minimizer:

$$\delta r^* \approx -\frac{\varepsilon}{\gamma}\left.\frac{\partial_r \log\left(\frac{r}{1+r}\right)}{\partial_{rr}\log f(r)}\right|_{r=\sqrt{\delta}}, \tag{4.6.12}$$

so for sufficiently small $|\varepsilon|$ the unique minimizer persists near $\sqrt{\delta}$ and the U-shaped profile of $K(r)$ is unchanged.

**Proposition 4.1 (Local Robustness of Branch Structure and Signs)** *Fix $K$ and suppose the PM and LM loci are smooth with $\gamma > 0$. There exists $\bar{\varepsilon} > 0$ such that for all $|\varepsilon| < \bar{\varepsilon}$: (i) the equilibrium condition admits two branches for $K$ below the (perturbed) tangency level; (ii) the low-rate branch $r_-(K)$ remains locally asymptotically stable (Lemma 4.2); and (iii) on this branch, the comparative-static sign*

$$\frac{dr_-}{dK} > 0$$

*continues to hold. Consequently, the qualitative statements established under exponent matching—existence/multiplicity, selection of the low-rate solution, and the positive response of $r^*$ to $K$—are locally robust to small departures $\beta \neq \gamma$.*

**Economic Interpretation** Small departures from exponent matching tilt the LM schedule relative to PM but do not eliminate the basic geometry: a U-shaped $K(r)$ with a unique minimum near $r = \sqrt{\delta}$ and a stable low-rate branch. Hence, the mechanism by which higher $K$ raises money demand, steep LM dominates PM, and the stable intersection slides rightward (remitting $dr^*/dK > 0$) remains intact. Formal details for Eqs. (4.6.11)–(4.6.12) are provided in Appendix 4.E.

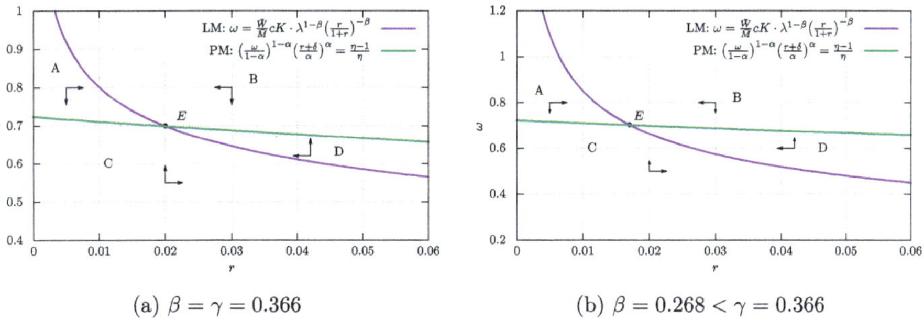


**Fig. 4.1** PM–LM phase diagrams under two calibrations.
In Panel (**a**), the exponent-matching baseline yields the textbook geometry with a unique locally stable intersection on the low-rate branch ($r < \sqrt{\delta}$). Panel (**b**) shows that local stability does not require $\beta > \gamma$: when $|\text{LM}'| > |\text{PM}'|$ holds (e.g., at sufficiently low $r$), the intersection is still locally stable even if $\beta < \gamma$. This corroborates the robustness result that the stability criterion is geometric $\big(|\text{LM}'| > |\text{PM}'|\big)$, with $\beta > \gamma$ and $r < \sqrt{\delta}$ a convenient sufficient pair

Figure 4.1 contrasts the geometry of the PM–LM system under the benchmark case $\beta = \gamma$ and a perturbed case $\beta < \gamma$,[11] illustrating how the slope relation $|\text{LM}'| > |\text{PM}'|$ governs local stability across parameter configurations.

The baseline calibration $\beta = \gamma = 0.366$ in panel (a) aligns with the closed-form characterization and delivers a uniquely stable intersection on the low-rate branch ($r < \delta$). Panel (b) illustrates that the sufficient pair ($\beta = \gamma$, $r < \sqrt{\delta}$) is not *necessary*: local stability is governed by the *slope inequality* $|\text{LM}'| > |\text{PM}'|$. For $\beta = 0.268 < \gamma = 0.366$, the LM locus remains steeper at low $r$, so the intersection is still locally stable. This contrast clarifies that our preferred parameterization in (a) is didactically clean, while (b) demonstrates robustness beyond the exponent-matching benchmark.

**Economic Summary** Under exponent matching, the SQSS admits either zero, one (tangent), or two candidate interest-rate equilibria. The economically relevant (stable) branch is the low-rate solution $r_-(K)$, which increases with capital, implies a declining real wage along the PM locus, and supports higher output and money demand; closed-form expressions for $N^d(K)$, $Q(K)$, and $M^d(K)/P$ are reported in Appendix 4.D.

## 4.7 Summary of SQSS Equilibrium

In the symmetric quasi-steady state (SQSS), the capital stock $K$ and long-term expectations are held fixed, while prices and the interest rate adjust to clear the goods market and the money market. The equilibrium is determined by the intersection of

[11] The depreciation rate is set at $\delta = 0.2$.

the pricing–markup (PM) locus, which reflects optimal pricing by firms, and the liquidity–money (LM) locus, which captures money-market equilibrium.

An intersection of these loci yields a stationary pair $(r^*, \omega^*)$, with the general price level determined by nominal wage setting. However, not every intersection is economically relevant: local stability requires that the LM curve be steeper in absolute value than the PM curve. This ensures that short-run adjustments in prices and interest rates converge back to the equilibrium.

Once stability is imposed, the SQSS equilibrium is unique. Comparative statics show that, along the stable branch, the equilibrium interest rate increases monotonically with the capital stock. This relationship holds both under the exponent-matching specification—where closed-form solutions are available—and under small deviations from that specification, indicating local robustness of the equilibrium structure.

Under exponent matching, the PM–LM system admits a low-rate stable equilibrium and a high-rate unstable equilibrium when the nominal money supply exceeds a critical threshold; tangency between the two loci is attained at $r = \sqrt{\delta}$. The stable branch exhibits $r^*(K) \in (0, \sqrt{\delta})$, increasing in $K$, associated with a decreasing real wage, higher output, and higher transactions money demand. These closed-form expressions and calibration thresholds are provided in Appendices 4.D and 4.E.

Figure 4.2 illustrates how the equilibrium interest rate and real wage vary with the capital stock along the stable SQSS branch. The curve $r^*(K)$ shows how the equilibrium interest rate increases with the capital stock on the low-rate branch $(r < \sqrt{\delta})$, where $\frac{dr^*}{dK} > 0$. The corresponding $\omega^*(K)$ curve illustrates how the equilibrium real wage declines as capital accumulates. Each pair $\left(r^*(K), \omega^*(K)\right)$ represents the stationary $(r^*, \omega^*)$ outcome at the symmetric quasi-steady state

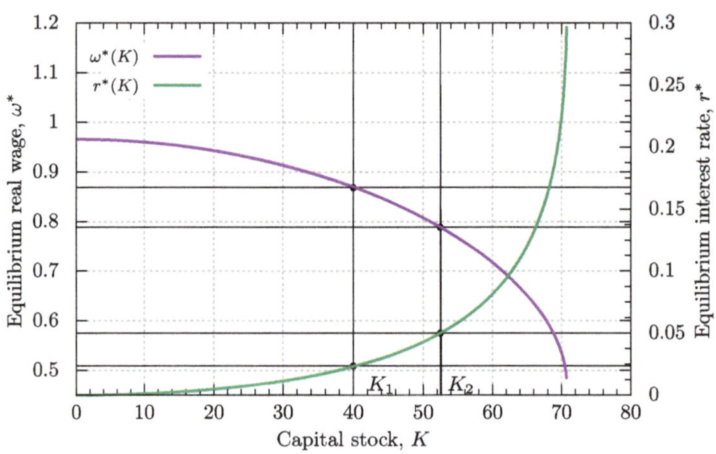


**Fig. 4.2** Equilibrium interest rate and real wage as functions of capital stock.
The stable SQSS branch lies in the region $r < \sqrt{\delta}$, where $r'(K) > 0$. As capital increases, the equilibrium interest rate rises moderately while the real wage declines, tracing a smooth locus of short-run equilibria. The economy moves along this branch as investment adjusts the capital stock over time

(SQSS) for the given $K$. As investment gradually alters the capital stock, the economy moves from one SQSS position to another along this branch, tracing the evolution of short-run equilibria.

This completes the characterization of the SQSS as the short-period anchor of the macroeconomic system. In the next section, we study how profit-driven investment decisions cause the capital stock to evolve over time, generating dynamic transitions between successive SQSS positions.

### 4.7.1 Capital Adjustment Around the SQSS

The SQSS describes the short-period equilibrium for a given capital stock $K$, but it does not determine $K$ itself. Firms in this economy make investment decisions based on profitability conditions, comparing the expected rate of return on capital with the cost of finance. When the incentive to invest is positive, the capital stock increases; when it is negative, the capital stock contracts. Consequently, the economy does not remain at a fixed SQSS position over time. Instead, it transitions across a continuum of SQSS equilibria, each associated with a different value of $K$.

The key implication is that the SQSS is not a terminal state but a moving anchor. At each instant, the short-run adjustment of prices and the interest rate drives the system to the SQSS consistent with the current $K$. Then, capital accumulation responds to profitability, moving the economy to a new SQSS position. This two-tier adjustment mechanism—fast nominal adjustment followed by slower real adjustment—is the defining characteristic of the macroeconomic dynamics developed in this chapter.

To formalize this mechanism, we specify how firms evaluate investment returns and how these returns determine the evolution of the capital stock.

**Transition to Investment Dynamics**

The analysis in this chapter has established the symmetric quasi-steady state (SQSS) as the short-period equilibrium position to which nominal variables converge for a given capital stock $K$. However, the SQSS is not a terminal resting point of the macroeconomy. Firms continuously reassess the profitability of their capital holdings relative to the prevailing interest rate $r^*(K)$; when profitable opportunities exist, investment activity shifts the capital stock away from its current level.

In this way, the SQSS serves as a moving anchor rather than a fixed endpoint: prices and the interest rate adjust rapidly within each period to restore the SQSS corresponding to the current $K$, while the capital stock evolves discretely over time as firms expand or contract their scale of production. The economy thus traverses a sequence of SQSS positions, with each position representing a short-run equilibrium conditional on the inherited capital stock.

The mechanism by which firms evaluate returns, form expectations, and alter their capital stocks is developed formally in the subsequent chapter. There we introduce the evolutionary investment process—driven by mutation and imitation—

and characterize the dynamic law of motion for $K$. This mechanism determines whether the economy converges to a steady configuration, fluctuates persistently, or undergoes cumulative expansion or contraction. The SQSS characterized in the present chapter provides the foundation upon which those dynamic forces act.

# Appendix

## 4.A Proof of Lemma 4.3 (Dividends Under CES Markups)

***Proof*** Under CES demand, the markup condition implies $P = \mathfrak{m}\, MC$ with $\mathfrak{m} = \eta/(\eta - 1)$. Operating profits are therefore

$$\Pi^F := PY - MC \cdot Y = (\mathfrak{m} - 1)\, MC \cdot Y.$$

Since $\mathfrak{m} - 1 = 1/(\eta - 1)$, this becomes

$$\Pi^F = \frac{1}{\eta - 1} MC \cdot Y.$$

Using the capital-cost share identity $vK = \alpha\, MC \cdot Y$, we obtain $MC \cdot Y = (vK)/\alpha$ and hence

$$\Pi^F = \frac{1}{\eta - 1} \cdot \frac{vK}{\alpha} = \frac{v}{\alpha(\eta - 1)} K = \frac{r + \delta}{\alpha(\eta - 1)} K.$$

In the SQSS, dividends equal operating profits, $Div^F = \Pi^F$, establishing (4.5.6). Finally, if $Div^F = (r + \delta)E$, then dividing (4.5.6) by $(r + \delta)$ yields (4.5.7). □

## 4.B Derivation of First- and Second-Order Conditions

The demand function of a representative firm is given by the equation:

$$x = X \cdot \left(\frac{p}{P}\right)^{-\eta} = a\, p^{-\eta}, \tag{4.B1}$$

where $a := P^{-\eta} X$ denotes the aggregate demand intensity. The sales volume, denoted by $y$, is determined by the minimum of supply ($q$) and demand ($x$), expressed as $y = \min\{x, q\}$. Consequently, the optimal price and the sales volume are

$$p^* = \left(\frac{a}{q}\right)^{1/\eta}, \quad \text{and} \quad y = q.$$

Taking the capital stock as predetermined, the firm chooses labor and price to maximize static profit.

$$py - wn = \left(\frac{XP^{\eta}}{q}\right)^{1/\eta} \cdot q - wn \tag{4.B2}$$

$$= X^{1/\eta} P \left(k^{\alpha} n^{1-\alpha}\right)^{\frac{\eta-1}{\eta}} - wn \tag{4.B3}$$

$$= X^{1/\eta} P \cdot k^{\frac{\alpha(\eta-1)}{\eta}} \cdot n^{\frac{(1-\alpha)(\eta-1)}{\eta}} - wn. \tag{4.B4}$$

Differentiating Eq. (4.B4) with respect to $n$ and $k$, we have

$$\frac{\partial \pi}{\partial n} = \frac{(1-\alpha)(\eta-1)}{\eta}\left(X^{1/\eta} P\right) \cdot k^{\frac{\alpha(\eta-1)}{\eta}} n^{\frac{(1-\alpha)(\eta-1)}{\eta}-1} - w = 0, \tag{4.B5}$$

$$\frac{\partial \pi}{\partial k} = \frac{\alpha(\eta-1)}{\eta}\left(X^{1/\eta} P\right) \cdot k^{\frac{\alpha(\eta-1)}{\eta}-1} n^{\frac{(1-\alpha)(\eta-1)}{\eta}} - p(r+\delta) = 0. \tag{4.B6}$$

Under symmetry, all firms charge the same price $p = P$. Let $A = \frac{(1-\alpha)(\eta-1)}{\eta}\left(X^{1/\eta} P\right)$ and $\bar{\eta} = \frac{\eta-1}{\eta}$. The second-order derivatives are

$$\frac{\partial^2 \pi}{\partial n^2} = A\big((1-\alpha)\bar{\eta} - 1\big) k^{\alpha\bar{\eta}} n^{(1-\alpha)\bar{\eta}-2}, \tag{4.B7}$$

$$\frac{\partial^2 \pi}{\partial n \partial k} = A\big(\alpha\bar{\eta}\big) k^{\alpha\bar{\eta}-1} n^{(1-\alpha)\bar{\eta}-1}, \tag{4.B8}$$

$$\frac{\partial^2 \pi}{\partial k^2} = A(\alpha\bar{\eta} - 1) k^{\alpha\bar{\eta}-2} n^{(1-\alpha)\bar{\eta}}. \tag{4.B9}$$

The Hessian matrix of second derivatives is negative definite if $\partial^2\pi/\partial n^2 < 0$ and $\det(H) > 0$. Since $(1-\alpha)\bar{\eta} < 1$, we have $\partial^2\pi/\partial n^2 < 0$. Moreover, direct calculation shows that $\det(H) > 0$. Hence, the second-order conditions for a maximum are satisfied.

The optimal pricing and symmetry condition implies that $X = Q = k^{\alpha} n^{1-\alpha}$ and $w = W$; Eq. (4.B5) and Eq. (4.B6) can be written as

$$\frac{\eta-1}{\eta} MP_L = \frac{\eta-1}{\eta}(1-\alpha) k^{\alpha} n^{-\alpha} = \frac{W}{P} = \omega, \tag{4.B10}$$

$$\frac{\eta-1}{\eta} MP_K = \frac{\eta-1}{\eta} \alpha k^{\alpha-1} n^{1-\alpha} = (r+\delta). \tag{4.B11}$$

Dividing Eq. (4.B10) by Eq. (4.B11), we get capital-labor ratio $\kappa$:

$$\kappa = \frac{\alpha}{1-\alpha} \cdot \frac{\omega}{r+\delta}. \tag{4.B12}$$

Equations (4.B10) and (4.B11) are the necessary conditions for cost minimization. Since $W/MP_L = MC$, Eq. (4.B10) implies that

$$\frac{P - MC}{P} = \frac{1}{\eta} \quad \Rightarrow \quad P = \mathfrak{m}\, MC, \qquad \mathfrak{m} \equiv \frac{\eta}{\eta - 1}, \tag{4.B13}$$

which is the Lerner condition and serves as the starting point for deriving the PM equation below.

## 4.C Derivation of the PM Equation

**Unit Cost** Given factor prices $(\omega, r+\delta)$, cost minimization for a given output level $Q$ yields the familiar Cobb–Douglas real unit cost

$$c(\omega, r+\delta) = \frac{\omega^{1-\alpha}(r+\delta)^{\alpha}}{(1-\alpha)^{1-\alpha}\alpha^{\alpha}}. \tag{4.C1}$$

**PM Equation (Symmetric Short-Run Evaluation)** We note that AC=MC for linearly homogeneous production function. Thus, it follows from Eqs. (4.B13) and (4.C1) that

$$P = \mathfrak{m}\, P \times c(\omega, r+\delta).$$

Dividing by $P$ and rearranging yields the stationary relation between $\omega$ and $r+\delta$,

$$\left(\frac{\omega}{1-\alpha}\right)^{1-\alpha}\left(\frac{r+\delta}{\alpha}\right)^{\alpha} = \frac{\eta - 1}{\eta}. \tag{4.C2}$$

This is the PM equation used in the text. In reduced form, allowing for a loan–funding spread $s \equiv r^L - r^F$ (constant or sluggish), replace $r$ by $r^F + s$ wherever it appears in $r + \delta$.

## 4.D Closed-Form Objects as Functions of *K* (Exponent Matching)

**Standing Pieces** Under symmetry and optimal pricing,

$$Q = K^{\alpha}N^{1-\alpha}, \qquad \frac{\eta - 1}{\eta} MP_L = \omega, \qquad \frac{\eta - 1}{\eta} MP_K = r + \delta.$$

With SQSS replacement $I = \delta K$, the two-component consumption function implies

$$C = cK, \qquad X = (c+\delta)K.$$

**PM and LM** In $(r, \omega)$-space,

$$\text{PM}: \quad \omega = \theta_{\text{PM}}(r+\delta)^{-\gamma}, \qquad \theta_{\text{PM}} = (1-\alpha)\left(\tfrac{\eta-1}{\eta}\right)^{\frac{1}{1-\alpha}} \alpha^{\frac{\alpha}{1-\alpha}}, \tag{4.D1}$$

$$\text{LM}: \quad \omega = \theta_{\text{LM}} K \left(\tfrac{r}{1+r}\right)^{-\beta}, \qquad \theta_{\text{LM}} = \frac{W}{M} c\,\lambda^{1-\beta}. \tag{4.D2}$$

Adopt the *exponent-matching assumption* ($\beta = \gamma = \dfrac{\alpha}{1-\alpha}$).

**Determination of $r(K)$ and $\omega(K)$** Define

$$h(r) := \left[\frac{(r+\delta)(1+r)}{r}\right]^{\beta}, \qquad H(r;K) := \frac{\theta_{\text{LM}} K}{\theta_{\text{PM}}} h(r).$$

The equilibrium condition (PM = LM) is

$$H(r;K) = 1. \tag{4.D3}$$

Since $\beta > 0$, minimizing $h$ is equivalent to minimizing $\log h(r) = \beta[\log(r+\delta) + \log(1+r) - \log r]$, whose derivative is proportional to $(r^2 - \delta)$; thus, $h(r)$ has a unique minimum at $r = \sqrt{\delta}$, with the negative for $0 < r < \sqrt{\delta}$ and positive derivative for $r > \sqrt{\delta}$.

Therefore, (4.D3) admits two solutions for $K < K_{\text{crit}}$, with the stable branch given by $r(K) \in (0, \sqrt{\delta})$. Then, define

$$\omega(K) = \theta_{\text{PM}}\,(r(K)+\delta)^{-\beta}. \tag{4.D4}$$

**Factor Demands, Money Demand, and Output as Functions of $K$** From cost minimization,

$$\kappa(r) := \frac{K}{N} = \frac{\alpha}{1-\alpha} \cdot \frac{\omega}{r+\delta}.$$

Using $\omega = \theta_{\text{PM}}(r+\delta)^{-\beta}$ and $\beta = \frac{\alpha}{1-\alpha}$,

$$\kappa(r) = \frac{\alpha}{1-\alpha}\,\theta_{\text{PM}}\,(r+\delta)^{-\frac{1}{1-\alpha}}, \tag{4.D5}$$

which is a decreasing function of $r$. Using $Q/K = \kappa(r)^{-(1-\alpha)}$ and $X = (c+\delta)K$, the capital stock required to support SQSS output satisfies

$$\frac{X}{K^d} = \frac{(c+\delta)K}{K^d} = \Big(\frac{1}{\kappa(r(K))}\Big)^{1-\alpha} \quad \Rightarrow \quad K^d(K) = \kappa\big(r(K)\big)^{1-\alpha}\,(c+\delta)\,K. \tag{4.D6}$$

The unit cost–minimizing condition $\kappa = K/N = \kappa(r(K))$ also implies that

$$N^d(K) = \frac{K}{\kappa(r(K))} = \frac{1-\alpha}{\alpha}\,\frac{K}{\theta_{\mathrm{PM}}}\,(r(K)+\delta)^{\frac{1}{1-\alpha}} \tag{4.D7}$$

on the stable branch $r(K)$.

Output:

$$Q(K) = K^{\alpha}\big(N^d(K)\big)^{1-\alpha} = \frac{K}{\kappa(r(K))^{\,1-\alpha}} = \Big(\frac{1-\alpha}{\alpha}\Big)^{1-\alpha}\theta_{\mathrm{PM}}^{-(1-\alpha)}\,K\,\big(r(K)+\delta\big) \tag{4.D8}$$

Transactions money demand:

$$\frac{M^d(K)}{P} = \lambda^{1-\beta}\Big(\frac{r(K)}{1+r(K)}\Big)^{-\beta} C(K) = \lambda^{1-\beta}\,c\,K\,\Big(\frac{r(K)}{1+r(K)}\Big)^{-\beta} \tag{4.D9}$$

Under 100% reserves with nominal ($M$) given, Eq. (4.D3) *is exactly* the money-market clearing condition that pins down ($r(K)$).

**Existence/Tangency Thresholds (Calibration)** At tangency, where $r = \sqrt{\delta}$, $H(\sqrt{\delta};\,K) = 1$,

$$K_{\mathrm{crit}} = \frac{\theta_{\mathrm{PM}}}{\theta_{\mathrm{LM}}}\,(1+\sqrt{\delta})^{-2\beta}, \qquad \theta_{\mathrm{LM,crit}} = \frac{\theta_{\mathrm{PM}}}{K}\,(1+\sqrt{\delta})^{-2\beta}. \tag{4.D10}$$

Equivalently, with $\theta_{\mathrm{LM}} = \dfrac{W}{M}c\lambda^{1-\beta}$,

$$M_{\min} = \frac{W\,c\,\lambda^{1-\beta}}{\theta_{\mathrm{PM}}}\,K\,\big[(1+\sqrt{\delta})^2\big]^{\beta}. \tag{4.D11}$$

These thresholds characterize the domain of existence of the stable branch.

## 4.E Comparative Statics at the PM–LM Intersection

We present comparative statics results (key derivatives and signs) for the stable SQSS equilibrium, based on the exponent-matching assumption ($\beta = \gamma = \alpha/(1-\alpha)$). As was stated in Sect. 4.D, the **stable branch** ($r(K) \in (0, \sqrt{\delta})$) is defined implicitly by

$$\frac{\theta_{\text{LM}} K}{\theta_{\text{PM}}}\, h(r) = 1, \qquad h(r) = \left[\frac{(r+\delta)(1+r)}{r}\right]^{\beta},$$

where $h'(r) < 0$ on the stable branch $r \in (0, \sqrt{\delta})$.

**Implicit Derivative for** $r(K)$ From $\theta_{\text{LM}}\, K h(r)/\theta_{\text{PM}} = 1$

$$\frac{dr}{dK} = -\frac{h(r)}{K\,h'(r)} > 0 \quad \text{on the stable branch} \tag{4.E1}$$

since $(h > 0)$ and $(h' < 0)$ for $(r < \sqrt{\delta})$.

**Derived Objects**
*Real wage along PM*

$$\omega(K) = \theta_{\text{PM}}(r+\delta)^{-\beta}$$

$$\frac{d\omega}{dK} = -\frac{\beta}{r+\delta}\,\omega\,\frac{dr}{dK} < 0$$

*Capital-labor ratio*

$$\kappa(r) = \frac{\alpha}{1-\alpha}\,\theta_{\text{PM}}(r+\delta)^{-\frac{1}{1-\alpha}}$$

$$\frac{d\kappa}{dK} = \kappa'(r)\frac{dr}{dK} = -\frac{1}{1-\alpha}\,\frac{\kappa}{r+\delta}\,\frac{dr}{dK} < 0 \tag{4.E2}$$

*Labor demand*

$$N^d(K) = K/\kappa(r(K))$$

$$\frac{dN^d}{dK} = \frac{1}{\kappa}\Big[1 + \frac{K}{1-\alpha}\,\frac{1}{r+\delta}\,\frac{dr}{dK}\Big] > \frac{1}{\kappa} > 0 \tag{4.E3}$$

*Output*

$$Q(K) = \left(\tfrac{1-\alpha}{\alpha}\right)^{1-\alpha}\theta_{\text{PM}}^{-(1-\alpha)}\,K\,(r+\delta) \;\equiv C_Q\,K\,(r+\delta)$$

$$\frac{dQ}{dK} = C_Q\big[(r+\delta) + K\,\frac{dr}{dK}\big] > 0$$

**Consumption, Income, Expenditure (SQSS Closure)**

$$C(K) = cK, \quad \frac{Y(K)}{P} = C(K) = cK, \quad X(K) = (c+\delta)K$$

$$\frac{dC}{dK} = c, \quad \frac{d(Y/P)}{dK} = c, \quad \frac{dX}{dK} = c + \delta$$

**Transactions Money Demand**

$$\frac{M^d}{P}(K) = \lambda^{1-\beta}\, c\, K\left(\frac{r}{1+r}\right)^{-\beta}$$

$$\frac{d}{dK}\left(\frac{M^d}{P}\right) = \lambda^{1-\beta} c\left(\frac{r}{1+r}\right)^{-\beta}\left[1 - \frac{\beta K}{r(1+r)}\frac{dr}{dK}\right]$$

The sign is positive provided the direct scale effect dominates the interest-rate effect; for sufficiently strong responses of $r(K)$, the derivative may approach zero.

## References


Basu, S. (1996). Procyclical productivity: Increasing returns or cyclical utilization? *Quarterly Journal of Economics, 111*(3), 719–751. https://doi.org/10.2307/2946670.

Burnside, C., Eichenbaum, M., & Rebelo, S. (1993). Labor hoarding and the business cycle. *Journal of Political Economy, 101*(2), 245–273. https://doi.org/10.1086/261875.

Campbell, J. Y., & Mankiw, N. G. (1989). Consumption, income, and interest rates: Reinterpreting the time series evidence. *NBER Macroeconomics Annual 4*, 185–216. https://doi.org/10.2307/3584973.

Fisher, I. (1932). *Booms and depressions*. Adelphi.

Fisher, I. (1933). The debt-deflation theory of great depressions. *Econometrica, 1*(4), 337–357. https://doi.org/10.2307/1907327.

Greenwood, J., Hercowitz, Z., & Huffman, G. W. (1988). Investment, capacity utilization, and the real business cycle. *American Economic Review, 78*(3), 402–417.

Kaldor, N. (1985). *The scourge of monetarism*. Oxford University Press.

Keynes, J. M. (1937). Alternative theories of the rate of interest. *The Economic Journal, 47*(186), 241–252.

Keynes, J. M. (1973). *The general theory and after: Part II, The Collected Writings of John Maynard Keynes, vol 14*. Macmillan, London.

Moore, B. J. (1988). *Horizontalists and verticalists: The macroeconomics of credit money*. Cambridge University Press.

Negishi, T. (1979). *Microeconomic foundations of Keynesian macroeconomics*. North-Holland Pub. Co.

Oi, W. Y. (1962). Labor as a quasi-fixed factor. *Journal of Political Economy, 70*(6), 538–555. https://doi.org/10.1086/258715.

Shapiro, M. D. (1993). Cyclical productivity and the workweek of capital. *American Economic Review, 83*(2), 229–233. Papers and Proceedings.

Sobel, J., & Takahashi, I. (1983). A multistage model of bargaining. *Review of Economic Studies, 50*(3), 411–426. https://doi.org/10.2307/2297673

Tobin, J. (1963). Commercial banks as creators of "money". In D. Carson (Ed.), *Banking and monetary studies* (pp. 408–419). Richard D. Irwin.

# Chapter 5
# Investment Under Uncertainty: Confidence, Thresholds, and Coordination

*The theory can be summed up by saying that, given the psychology of the public, the level of output and employment as a whole depends on the amount of investment.*

— John Maynard Keynes, *The General Theory* (1937)

## 5.1 Introduction

In the preceding chapters, we analyzed the symmetric quasi-steady-state (SQSS) economy—a short-period equilibrium in the Keynesian sense—in which each stationary level of the capital stock $K$ determines a consistent macroeconomic environment. Given $K$, aggregate investment is pinned down at $I = \delta K$, while consumption follows from the fixed propensities $c_y$ and $c_w$. Under fixed money supply and nominal wages, the resulting aggregate expenditure $C$ then determines money demand, the interest rate $r$, and the price level $P$. Thus, each level of $K$ corresponds to a unique SQSS equilibrium, conditional on fixed expectations and institutional parameters.

The present chapter turns to the dynamic forces that shift $K$ itself. Specifically, it asks how firms make investment decisions when confronted with shocks to confidence or demand, and how these decisions interact across firms to shift the economy from one equilibrium environment to another. To answer the research question, we develop a model in which firms occasionally deviate from the incumbent rule of simply replacing depreciation, $I = \delta K$. These deviators—called *mutants*—may fade out or, if they surpass a critical threshold, trigger a cascade of imitation that tips the system into a new equilibrium. Repeated tipping events generate evolutionary dynamics, echoing the equilibrium-selection frameworks of Kandori et al. (1993) and Young (1993), while remaining grounded in Keynesian macroeconomics.

The subsequent chapter builds on this investment mechanism to construct a fully stochastic evolutionary law of motion for the aggregate capital stock and to characterize the resulting long-run dynamics.

 
I. Takahashi, *Microfoundations of Keynesian Economics*, Evolutionary Economics and Social Complexity Science 37, https://doi.org/10.1007/978-981-95-9390-3_5

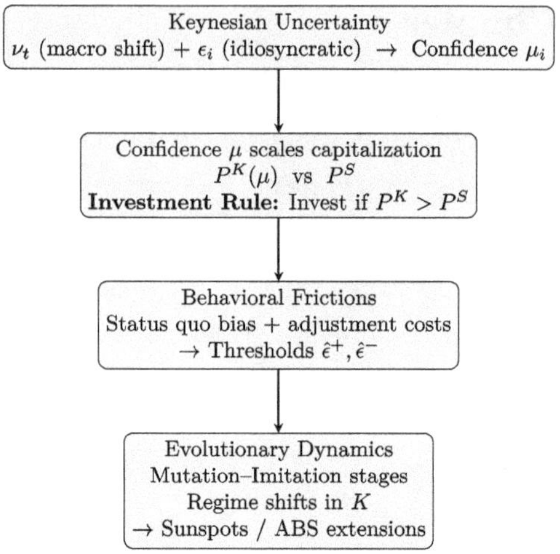


**Fig. 5.1** Chapter roadmap. The mutant–incumbent framework formalizes Keynes's intuition: Confidence-driven actions, once imitated, can destabilize equilibrium even without fundamental shocks. This schematic summarizes the logical flow of the chapter

A central feature of real capital is its fixity: Once installed, it cannot be freely withdrawn or redeployed. Neoclassical theory largely abstracts from this constraint, treating capital as if it were perfectly malleable. In contrast, our analysis acknowledges the importance of fixity, demonstrating how the irreversibility of investment decisions generates frictions, thresholds, and coordination effects that shape macroeconomic dynamics.

Wc will show that the economy evolves through repeated tipping events driven by the entry and imitation of such mutants. Figure 5.1 summarizes the chapter's logical flow.

### *5.1.1 Mechanism Overview: Complementarity, Externality, and Utilization*

Investment plays a central role in macroeconomic dynamics due to its dual role. On the one hand, investment stimulates aggregate demand through the Keynesian multiplier effect, generating income and employment in the short run. On the other hand, it expands the productive capacity of the economy, influencing future supply. This dual character makes investment a key driver of both business cycle fluctuations and long-run growth.

In what follows, the term "firm" should be understood in a stylized sense, referring not to an atomistic individual enterprise but to a representative production unit or sector composed of many firms. This interpretation avoids assigning macroeconomic significance to the investment of a single small firm, while also abstracting from detailed inter-industry substitution effects. The strategic complementarity considered here operates through aggregate demand: when one production unit expands investment, the resulting income and expenditure spillovers raise the demand conditions faced by other units.

**Strategic Complementarity and Equilibrium Selection**

The duality of investments creates strategic complementarity: When one firm invests, it raises aggregate demand and thereby the returns to investment for other firms. Such positive feedback loops can sustain multiple self-fulfilling equilibria (Cooper & John, 1988).

The challenge of selecting among multiple equilibria has been central in both game theory and macroeconomics. Game theorists pursued refinements such as trembling-hand perfection (Selten, 1975), purification (Harsanyi, 1975), and strategic information transmission (Crawford & Sobel, 1982). Behavioral approaches, such as *gift-exchange models* (Fehr et al., 1993), highlighted fairness norms that prevent markets from clearing at competitive levels. Evolutionary perspectives focused on *mutation and learning* (Kandori et al., 1993; Young, 1993). Macroeconomists, in turn, introduced extrinsic coordination devices, leading to the theory of *sunspot equilibria* (Azariadis, 1981; Cass & Shell, 1983), whereas macrolabor studies stressed *hysteresis effects* (Blanchard & Summers, 1986).

One way to formalize how strategic complementarities operate under bounded rationality is to model investment behavior as an *evolutionary process of mutants and incumbents*. Kandori et al. (1993) famously illustrate this mechanism with the analogy of competing operating systems: Windows and macOS represent alternative strategies in a coordination game. Even if macOS—standing here for a *Pareto-superior outcome* such as full employment—offers superior features, the prevalence and familiarity of Windows—the *risk-dominant outcome* of unemployment—can sustain its dominance over time. This example highlights how *evolutionary dynamics favor stability and robustness* rather than optimality.

In our investment setting, this evolutionary perspective is sharpened by the presence of *behavioral frictions*. Costs associated with overcoming status quo bias or organizational inertia make deviation from the incumbent strategy costly, effectively rendering the decision set discontinuous. These frictions create *threshold effects*: Only when a sufficient number of firms ("mutants") act together can they surpass the adjustment cost and tip the system toward a new equilibrium. Such thresholds serve as *focal points* that coordinate expectations, ensuring that once the critical mass is reached, the transition to the new stationary level of capital becomes self-reinforcing.

**Externalities and Utilization**

Strategic complementarity, when combined with state-dependent utilization, introduces a critical asymmetry through the externalities of capacity expansion. The spillover effects of investment create state dependence:

- When capacity is scarce and utilization is high, new investment generates both private profits and public benefits.

- When capacity is abundant and much of it lies idle, the private incentive to invest collapses—even if the social return remains positive.

Thus, the capital utilization rate becomes a critical state variable. It determines whether strategic complementarity amplifies investment or dampens it. This interaction between utilization and private incentives underlies the thresholds that govern tipping dynamics.

Taken together, these insights suggest that a realistic investment theory must incorporate two key features:

- *Positive feedback* from strategic complementarity, through which each firm's decision affects aggregate demand and thereby the incentives of others.
- *Frictions and state-dependent externalities* that generate persistence and coordination failures. In particular, demand spillovers depend on utilization: With scarce capital, private and public returns align; with idle capacity, private incentives collapse even though the public benefit is large.

### 5.1.2 Evolutionary Selection: Mutants and Thresholds

In coordination games, *focal points* guide expectations. Here, the incumbent rule—replacing depreciation with $I = \delta K$—serves as the default focal point. As Keynes (1937, p. 214) noted we tend to regard the present as a more reliable guide to the future than past experience justifies, reinforcing the status quo as a coordination device.

Yet, as Keynes (1936, Ch. 22) stressed, expectations are "based on shifting and unreliable evidence" and subject to sudden revisions. Motivated by such shifts, mutants disrupt this stability by introducing an alternative focal point:

- Optimistic mutants double investment when confidence is high.
- Pessimistic mutants halt investment when confidence collapses.[1]

Whether the economy tips depends on whether the mutant share crosses the relevant threshold. Above this critical mass, imitation cascades and the economy tips to a new SQSS. Below it, mutants fade and the incumbent rule prevails.

The mutation–imitation structure is closely related to the stochastic evolutionary dynamics of Kandori et al. (1993), in which rare mutations and payoff-based

[1] This three-point adjustment rule should be interpreted as a replacement-scale discretization rather than as a literal restriction on technologically feasible capital choices. For example, if a firm operates 100 machines and one depreciates each period ($\delta = 0.01$), then net investment choices $\{-K, 0, +K\}$ correspond to moving from 100 installed units to 99, 100, or 101. The assumption captures gradual lumpy adjustment at the replacement margin. Accordingly, the mutant mechanism introduced below should be interpreted as altering replacement and expansion intensity at the margin, rather than implying an instantaneous doubling or elimination of the aggregate capital stock. Because capital is durable and adjusts incrementally, even substantial shifts in investment sentiment generate only limited changes in installed capital within a single period.

selection determine long-run stochastic stability. Unlike informational cascade models (Chamley, 2004), the present framework does not rely on belief updating or sequential information aggregation; instead, macroeconomic dynamics emerge from repeated payoff-driven selection in a finite strategy space.

Because the analysis in this chapter employs a relatively rich notation system, Table 5.4 in Appendix 5.A provides a consolidated reference table of symbols, definitions, and units used throughout the chapter.

### *5.1.3 Roadmap of the Chapter*

The chapter proceeds as follows:

1. Section 5.2 introduces uncertainty in Keynesian economics, distinguishing it from measurable risk, and develops a confidence index as a scaling factor for capital valuation.
2. Section 5.3 analyzes behavioral frictions—status quo bias and adjustment costs—that create investment thresholds.
3. Section 5.4 integrates these elements into a two-stage evolutionary model, showing how utilization and mutant shares interact to generate regime shifts.
4. Section 5.5 compares this framework with sunspot equilibria and agent-based simulations, highlighting how our model balances tractability and realism.

## 5.2 Investment Decision Rule Under Uncertainty

This section develops the Keynesian investment decision rule under uncertainty in three steps; progressing from uncertainty to confidence, followed by capitalization, and finally leading to the investment rule. First, we define a *confidence index* that translates subjective beliefs into a measurable parameter, combining aggregate sentiment and idiosyncratic shocks. Second, we show how this index modifies the capitalization rate, thereby scaling the present value of depreciating capital assets. Finally, we derive the investment decision rule itself, demonstrating how higher confidence raises the marginal efficiency of capital, while lower confidence suppresses it. Together, these steps provide a foundation for understanding how fluctuations in sentiment drive investment behavior in our framework.

### *5.2.1 Macroshift and Confidence Index*

The previous section emphasizes a crucial role of focal points in coordinating investment decisions among individual firms. Each firm has a different view on the future prospectives due to idiosyncratic reasons. Those dispersed views will be summarized into a unified collective view through the interaction among firms. This

**Table 5.1** Core notation (investment theory I: stocks, flows, shocks)

| Symbol | Meaning |
|---|---|
| $K$ | Installed capital stock (incumbent baseline) |
| $\delta$ | Depreciation/replacement ratio |
| $I_i$ | Gross investment of firm $i$ |
| $z_i = I_i - \delta K_i$ | Net investment relative to replacement |
| $\nu$ | Aggregate macroshift (sentiment shock) |
| $\epsilon_i$ | Idiosyncratic shock for firm $i$ |
| $\mu_i = \nu + \epsilon_i$ | Confidence index of firm $i$ |
| $\zeta$ | Excess replacement ratio, $(K^d - K)/(\delta K)$ |
| $\epsilon^+$, $\epsilon^-$ | Share of optimistic and pessimistic mutants |

subsection shows how an individual firm forms such views—confidence index—based on macroenvironment and idiosyncratic shocks.

In Keynes's usage, *uncertainty* does not mean the measurable variance of known probability distributions. It denotes *fundamental ignorance*: Future outcomes cannot be objectively predicted, and expectations must instead rely on subjective inference. Such judgments rest on layered assertions—some grounded in evidence, others speculative or intuitive. The degree of uncertainty reflects how weak or indirect the chains of reasoning are, and hence, how fragile the resulting judgment may be. A high degree of uncertainty corresponds to low *confidence* in the investment outlook (Keynes, 2013, Chapter 6). Table 5.1 lists notations that appear in this subsection.

We measure confidence by the index $\mu$, with its reciprocal $\varsigma = 1/\mu$ as an *anxiety index*. A higher $\varsigma$ (lower $\mu$) means more pessimism and a higher effective discount rate.

We model the macroshift as an exogenous term $\nu_t$, capturing period-$t$ sentiment about improvement or deterioration around the stationary capital stock $K$. Formally, $\nu_t$ is i.i.d. (uniform) across periods. Unlike indicators such as the Purchasing Managers' Index (PMI)—which reflect managers' assessments of orders and production—$\nu_t$ is better interpreted as a sentiment shock: a groundless, mood-like deviation from the current state rather than a forecast based on fundamentals. In empirical terms, one might think of it as the first difference of a sentiment index, representing shifts in confidence rather than its absolute level.[2]

## Industry Layer

Between aggregate sentiment and firm-level perceptions lies an intermediate layer: industry operating conditions. Even for a given aggregate capital stock, industries differ in their degree of utilization and tightness. When most industries operate with substantial slack, optimistic sentiment generates only limited private gains, and

[2] Although $\nu_t$ is i.i.d., the effective macroeconomic environment generated through its interaction with the state variable $K_t$ may exhibit substantial apparent persistence around the stationary region. Because transition probabilities and economic responsiveness depend on the prevailing capital stock, similar values of $K_t$ tend to produce similar aggregate conditions over time even under independently distributed shocks. For brevity, we refer to this phenomenon as a "virtual random walk."

deviations from the incumbent investment rule tend to remain isolated. By contrast, when utilization is broadly high, even modest optimism can trigger sizable payoff differentials and widespread imitation. In this sense, aggregate sentiment shocks act through the cross-industry distribution of operating conditions, which is implicitly embedded in the transition kernel governing the evolution of the capital stock.

**Firm Layer**

Each firm $i$ forms a confidence measure

$$\mu_{it} = \nu_t + \epsilon_{it},$$

where $\epsilon_{it}$ is firm-specific perceptual noise. Thus $\mu_{it}$ represents a confidence-adjusted *perception* of the macroshift. For readability, we continue to call $\mu_{it}$ "confidence," while $\nu_t$ denotes the underlying (observable) macroimprovement/deterioration.

Two confidence thresholds, $\mu_H$ (optimism) and $\mu_L$ (pessimism), reflect both expectations and the adjustment costs of changing investment:

- If $\mu_i > \mu_H$ in an *expanding* state ($\nu > 0$), the firm becomes an *optimistic mutant*, doubling its investment relative to incumbents.
- If $\mu_i < \mu_L$ in a *contracting* state ($\nu < 0$), it becomes a *pessimistic mutant*, halting new investment.

The share of mutants in each state, denoted $\epsilon^+$ and $\epsilon^-$, depends on the distribution of $\mu_i$ relative to these thresholds.

In the *imitation stage*, after the current excess replacement ratio $\zeta$ and tentative plans are revealed, firms revise their decision. This revision depends on aggregate thresholds $\hat{\epsilon}^+(\zeta)$ and $\hat{\epsilon}^-(\zeta)$. If the mutant share exceeds the relevant threshold, imitation triggers a regime shift; otherwise, incumbents persist.

This structure separates idiosyncratic confidence shocks (driving $\epsilon^\pm$) from coordination thresholds (driving $\hat{\epsilon}^\pm$), while embedding both Keynesian confidence effects and utilization-dependent inertia. The formal definitions and payoff structure appear in the next section.

The coordination thresholds are another manifestation of the fixity of the capital stock. The capital-adjustment-cost specification developed in the subsequent sections, i.e., $\phi(z)$, follows the spirit of Uzawa (1968)'s reinterpretation of Penrose's theory of firm growth, in which *real capital is fixed in form once installed.* Capital goods are specific and location-bound, so expanding or contracting the operational scale requires reallocating physical assets and managerial capacity under constraints. These constraints generate convex adjustment costs that rise steeply as the magnitude of the change increases, and may become effectively prohibitive at a finite capacity limit. Such frictions imply a *finite inaction range*: Unless the risk-adjusted present value of marginal returns clears a positive threshold above replacement cost, no change in capital stock occurs.

In contrast, the neoclassical model's "jelly" capital is perfectly divisible, homogeneous, and instantly reconfigurable for any productive use. In that setting, the cost of an infinitesimal change in the capital stock is zero, so the inaction range collapses: Even the smallest deviation between the marginal product of capital and its user cost triggers an adjustment. This smooth, continuous investment behavior tends to drive the model rapidly toward full-capacity or full-employment equilibria, leaving no room for persistent slack.

By adopting a Penrose–Uzawa-type $\phi(z)$, augmented here with behavioral frictions such as status quo bias and asymmetric irreversibility, we retain the realistic feature that investment is *lumpy and threshold driven*, in the spirit of $(S, s)$ adjustment models (Doms & Dunne, 1998; Caballero & Engel, 1999). The combination of physical fixity and behavioral costs makes capital adjustment discontinuous and allows underemployment equilibria to persist in the face of moderate changes in profitability or confidence. This contrasts sharply with the neoclassical jelly-capital benchmark, in which the investment threshold is effectively infinitesimal.

### 5.2.2 *Valuation and Normalization*

**Capital Valuation: $P^K$**

By incorporating confidence index built in the previous subsection, this subsection develops the capitalization rate, which will be instrumental in deriving the Keynesian investment decision rule. Table 5.2 is a list of notations referenced in the rest of section.

According to Keynes, the act of purchasing a capital asset amounts to acquiring *a right to a sequence of prospective returns*—namely, the net earnings from selling the output of the asset, after deducting the costs of production, over the course of its life (Keynes, 1936, p. 135). He termed this sequence of anticipated returns $Q_{t+1}, Q_{t+2}, \ldots$ the *prospective yields* of the investment. The market value of a

**Table 5.2** Extended notation (investment theory II: prices, frictions, thresholds)

| Symbol | Meaning |
|---|---|
| $P^K$ | Profit-normalized capital valuation |
| $P^K(\mu)$ | Confidence-adjusted capital valuation |
| $\widehat{P}^K(\mu)$ | Risk-adjusted capital valuation |
| $P^S$ | Supply price of capital (commitment) |
| $\varphi = P^S/P^K$ | Cost-to-return ratio (investment wedge) |
| $q$ | Tobin's $q$ |
| $\widehat{q}$ | Risk-adjusted Tobin's $q$ |
| $\phi(z)$ | Adjustment cost (Penrose–Uzawa type) |
| $c_b$ | Behavioral wedge (status quo/organizational bias) |
| $\hat{\epsilon}^+(\varsigma)$ | Expansion threshold |
| $\hat{\epsilon}^-(\varsigma)$ | Contraction threshold |
| $K^d$ | Demand-required capital (see Eq. (5.4.3)) |
| $K^{\text{op}}$ | Operating capital (short side, see Eq. (5.D.1)) |

capital asset, $P^K$, is formed by discounting this stream of yields. By contrast, the *supply price of capital*, $P^S$, represents the minimum imputed value that induces the owner to install and replace a unit of capital. It reflects more than just the cost of physical investment goods—it incorporates the incentive necessary to keep the capital stock intact. Keynes defined the *marginal efficiency of capital (MEC)* as the rate of return that equates $P^K$ with $P^S$.[3] We emphasize that this analysis presupposes the fixity of capital, in contrast to the neoclassical notion of liquid "jelly capital."[4]

Keynes's marginal efficiency of capital (MEC), developed in Chapter 11 of *The General Theory*, denotes the expected rate of return that equates the discounted stream of prospective yields with the supply price (replacement cost) of capital goods (Keynes, 1936). Yet, as Lerner (1944) and Uzawa (1969) emphasized, this term is somewhat misleading: The concept refers not to the productivity of existing capital but to the inducement to undertake new projects. For this reason, it is more precise to speak of the marginal efficiency of investment (MEI)—a terminology that will be adopted throughout this book.[5]

Figure 5.2 illustrates the series of yield streams and cash payments associated with the loans, based on the following assumption:

**Assumption 5.1 (Price Concession to Investing Firms)** It is assumed that, although construction of capital begins at the start of period $t$, contractors grant investing firms a concession by deferring payment during the gestation period. The cash outlay is therefore due only at the end of $t + 1$, coinciding with the point at which the capital becomes productive and gross profits begin to accrue. Consequently, both the demand price of capital $P^K$ (the present value of gross profits) and the supply price $P^S$ (the contractual commitment) are evaluated at $t + 1$.[6]

Let a unit of capital yield $(1 - \delta)^{\tau-1}$ in constant dollars at each period $\tau = 1, 2, \ldots$, with geometric depreciation rate $\delta$. Under perfect foresight, the capitalization factor is

$$C^K = \sum_{\tau=1}^{\infty} \frac{(1-\delta)^{\tau-1}}{(1+r)^{\tau}} = \frac{1}{r+\delta}.$$

[3] MEC and Fisher's "rate of return over cost" are the same as the internal rate of return (IRR) concept—Keynes acknowledged Fisher's priority; see Fisher (1930, Ch. 5) and Keynes (1936, Ch. 11).

[4] See the Introduction, and Uzawa (1969), for a fuller discussion of this distinction.

[5] Uzawa (1969) gives a formal definition of the marginal efficiency of investment (MEI) in an intertemporal setting, distinguishing it from the productivity of the existing capital stock.

[6] *Remark*: This "price concession" should be understood as a timing convention rather than an additional frictional cost. It reflects the assumption that contractors allow investing firms to defer payment until the installation is complete, thereby aligning the timing of both the supply price $P^S$ and the demand price $P^K$ at the start of production in $t + 1$. The concession does not introduce any extra adjustment cost beyond those already modeled.

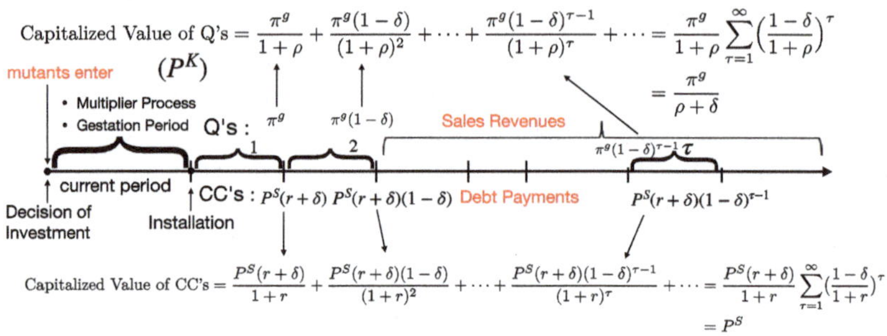


**Fig. 5.2** Value of capital assets and supply price.
The timeline illustrates two flows associated with an installed capital asset: (i) the operating yields ($Q$) accruing to the owner and (ii) the contractual cash payments to lenders ($CC$), in the sense of Minsky (1975). Discounting the yield stream gives the market value $P^K$, while discounting the replacement outlays determines the supply price $P^S$. Under uncertainty, a confidence factor $\mu$ influences the *subjective discount rate* $\rho$, whereas $P^S$ uses the interest rate as the discount rate

Under uncertainty, firms discount prospective yields at a subjective rate $\rho$, giving

$$C_\rho^K = \frac{1}{\rho + \delta}.$$

Let

$$P^K = C_\rho^K \pi^g = \frac{1}{\rho + \delta}\left(1 + \frac{1}{\alpha(\eta - 1)}\right)(r + \delta) \tag{5.2.1}$$

denote the *profit-normalized capital valuation*. This object provides a confidence-free benchmark, corresponding to $\mu = 1$.

Uncertainty weakens confidence in future profits and capital durability. We capture this by a confidence index $\mu \in (0, \infty)$, which scales the valuation of prospective yields proportionally.[7] The resulting *confidence-adjusted demand price of capital* is

$$P^K(\mu) = \frac{\mu}{\hat{\rho} + \delta}\left(1 + \frac{1}{\alpha(\eta - 1)}\right)(r + \delta), \tag{5.2.2}$$

where $\hat{\rho}$ denotes the perceived internal discount rate applied to future operating profits. Optimism ($\mu > 1$) raises, while caution ($\mu < 1$) lowers, the valuation of capital.

[7] The level of anxiety $1/\mu = \varsigma$ scales up both the discount and depreciation rates. Rewriting the capitalization series yields

$$C^K(\mu) = \frac{1}{1 + \varsigma r} + \frac{1 - \varsigma\delta}{(1 + \varsigma r)^2} + \cdots = \frac{1}{\varsigma(r + \delta)} = \frac{\mu}{r + \delta}.$$

In the benchmark SQSS, firms face stationary financing conditions and posted prices. The perceived long-run discount rate therefore coincides with the prevailing user cost of capital, so that $\hat{\rho} = r$. Under this identification,

$$P^K(\mu) = \mu\left(1 + \frac{1}{\alpha(\eta - 1)}\right). \tag{5.2.3}$$

**Tobin's $q$**

We work in Tobin-$q$ units so that all values are measured relative to replacement cost:

$$q := \frac{P^K}{P^S}.$$

Since the model is expressed in real terms, the supply (replacement) price of capital is normalized to unity, $P^S \equiv 1$.

**MEI versus D-MEI**

Keynes's marginal efficiency of capital (MEI) is defined as the discount rate that equates the present value of a stream of prospective yields from a capital asset to its supply price. In its original formulation, MEI is best suited to long-lived or non-depreciating assets, where depreciation can be treated implicitly.

In the present framework, capital depreciates at a constant rate $\delta$, and investment decisions concern gross investment relative to replacement. Accordingly, the relevant comparison is between the total user cost of capital, $r + \delta$, and the return required to justify installing and maintaining capital. We therefore work with the *depreciation-adjusted marginal efficiency of investment* (D-MEI), defined as

$$\text{D-MEI} \equiv \hat{\rho} + \delta.$$

Here $\hat{\rho}$ denotes the firm's *perceived internal discount rate* applied to prospective operating profits. It summarizes the firm's assessment of the long-run normal return required to justify committing resources to capital installation, given uncertainty, risk, and limited confidence in future conditions. Importantly, $\hat{\rho}$ is not a policy-controlled interest rate, nor does it represent a short-term financial cost of funds. Rather, it is a subjective benchmark that firms use to evaluate the profitability of investment projects relative to replacement cost.

In the symmetric quasi-steady state (SQSS), posted prices, utilization, and financing conditions are stationary. In this environment, arbitrage between financial returns and operating profits implies that the perceived long-run return norm coincides with the prevailing user cost of capital. We therefore set $\hat{\rho} = r$ in the benchmark SQSS. Confidence shocks do not alter this benchmark return; instead, they affect investment decisions through valuation by scaling the demand price of capital. As a result, the D-MEI remains fixed at $r + \delta$ in the SQSS, while Tobin's $q$ fluctuates with confidence.

### 5.2.3 Confidence, D-MEI, and the Investment Rule

Using (5.2.2), investment is induced whenever the confidence-adjusted valuation of capital exceeds replacement cost:

$$\text{Invest if} \quad q = \frac{\mu}{\hat{\rho}+\delta}\left(1+\frac{1}{\alpha(\eta-1)}\right)(r+\delta) > 1. \tag{5.2.4}$$

Equivalently,

$$\mu\left(1+\frac{1}{\alpha(\eta-1)}\right)(r+\delta) > \hat{\rho}+\delta.$$

In the SQSS, $\hat{\rho} = r$, so the depreciation-adjusted marginal efficiency of investment is fixed at $r+\delta$. Confidence shocks therefore do not alter the required rate of return. Instead, they operate entirely through valuation: A rise in confidence raises $P^K$, and hence, Tobin's $q$, while D-MEI remains unchanged.

**Remark 5.1 (Valuation Versus Required Returns)** In the stationary environment considered here, D-MEI is pinned down by the user cost of capital and does not respond to confidence. Investment is therefore triggered not by a rise in the required rate of return, as in the textbook comparison between MEI and the market interest rate, but by valuation crossing the replacement-cost threshold. Confidence governs investment through the demand price of capital, while D-MEI serves as a fixed benchmark reflecting the cost of maintaining capital in the SQSS.

### 5.2.4 Risk Adjustment and Stationarity

Under monopolistic competition, positive markups imply $1+\frac{1}{\alpha(\eta-1)} > 1$. Absent further adjustment, (5.2.4), would therefore imply $q > 1$ even at $\mu = 1$, leading to perpetual net investment and violating stationarity.

We address this by introducing a risk premium proportional to $1/(\alpha(\eta-1))$, capturing dividend requirements and expected bankruptcy costs associated with expansion. With this adjustment, the *risk-adjusted demand price of capital* becomes

$$\widehat{P}^K(\mu) = \mu. \tag{5.2.5}$$

Under the normalization $P^S = 1$, the associated risk-adjusted Tobin's $q$ satisfies $\widehat{q} = \widehat{P}^K = \mu$, so that the investment rule simplifies to

$$\text{Invest if} \quad \widehat{q} = \mu > 1.$$

This normalization nets out the structural effect of market power and restores indifference at the benchmark $\mu = 1$. Net investment occurs only when confidence is sufficiently high to push valuation above replacement cost.

Investment nevertheless responds discontinuously to confidence because firms face behavioral and organizational frictions. We turn next to these frictions and the resulting threshold effects in capital accumulation.

## 5.3 Behavioral Frictions

This section turns from the drivers of investment to the frictions that restrain it. We focus on two sources of inertia: *status quo bias*, the psychological tendency to remain at the current operational scale, and *investment adjustment costs*, the managerial and organizational expenses involved in altering the capital stock. Together, these forces generate a lump-sum deviation cost whenever firms depart from the replacement strategy, thereby shaping the thresholds that govern imitation (see Fig. 5.3). By embedding these frictions in our framework, we show why investment adjustments are discontinuous and how capital accumulation dynamics are bounded by both behavioral and organizational constraints.

### 5.3.1 *Status Quo Bias*

Status quo bias, as documented by Samuelson and Zeckhauser (1988), is the tendency to prefer the current state even when alternatives may yield higher expected payoffs. In investment decisions, this bias manifests as a reluctance to alter the existing capital scale. A closely related—but conceptually distinct—mechanism is *loss aversion* (Kahneman & Tversky, 1982; Kahneman et al., 1986): Decision-makers weigh potential losses more heavily than equivalent gains, producing an

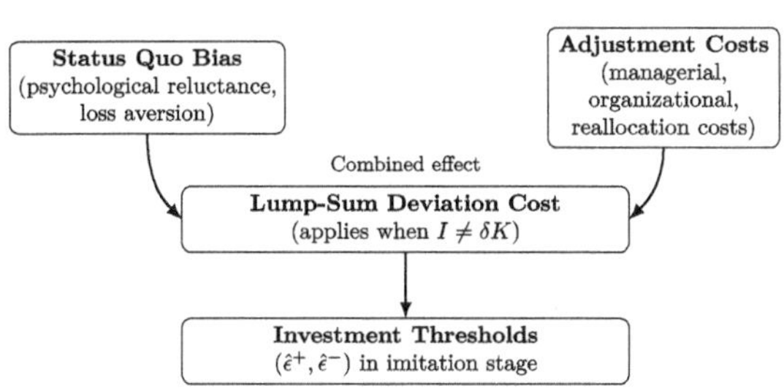


**Fig. 5.3** Behavioral frictions and investment inertia.
Behavioral frictions arise from the joint effect of status quo bias and adjustment costs. Together they act as a lump-sum deviation cost that determines the investment thresholds governing imitation

asymmetry between expansion and contraction. Managers may therefore avoid proposing new capacity even when expected profits are positive, knowing that failure after expansion attracts more criticism than inaction.

Under a symmetric quasi-steady state (SQSS) with stationary capital stock $K$, all firms invest $I_i = \delta K$ each period to maintain capacity. We refer to this rule as the *capital replacement strategy*, and firms adhering to it as *incumbents*. Any firm choosing $I_i \neq \delta K$ is a mutant. Keynes's observation that long-term expectations rest on "flimsy" foundations helps explain why mutants may arise suddenly: Small shifts in confidence can produce abrupt deviations in investment behavior. As emphasized by Kirman and Kirman (1992), the presence of occasional mutants makes macroeconomic dynamics more realistic by allowing aggregate fluctuations to emerge from interaction rather than only from large exogenous shocks.

Mutations may be triggered by diverse idiosyncratic events—balance-sheet pressures, managerial disruptions, or internal governance conflicts. Through strategic complementarity, a sufficiently large group of optimistic mutants during an expansion can induce incumbents to follow, while pessimistic mutants in a downturn can pull aggregate investment downward.

## 5.3.2 *Investment Adjustment Costs*

Beyond psychological reluctance, altering investment levels entails *real adjustment costs*. As emphasized by Penrose (1959), expanding production requires not only physical capital but also the transfer of know-how and the reorganization of workflows. Installing new equipment, for example, demands training, new maintenance routines, and often a redesign of production layouts. These managerial and organizational costs arise in addition to the physical outlays for capital goods.

Following Uzawa (1968, 1969), we represent adjustment costs as a convex, increasing function of deviations from replacement investment. Under this structure, interior deviations are weakly dominated, so that firms' optimal investment choices collapse to a discrete set. The formal implication is stated in Lemma 5.1; a full derivation of the underlying cost function is provided in Appendix 5.B.

For tractability, we approximate these costs as a lump-sum penalty that is incurred whenever a firm deviates from $I_i = \delta K_i$. This lump-sum captures both tangible reorganization expenses and the intangible burden of overcoming status quo bias. Together, these frictions embed *inertia* into the investment process and generate the critical thresholds that govern imitation dynamics.

Table 5.3 summarizes how status quo bias and organizational adjustment costs jointly restrict feasible investment responses, motivating the discrete strategy set used in the mutation stage.

**Organizational Switch Cost**
We assume a size-adjusted lump-sum cost is paid whenever a firm changes its scale of operation, independent of the magnitude of the change.

**Table 5.3** Neoclassical "jelly" capital vs. fixed capital with Penrose–Uzawa adjustment costs

| Neoclassical "jelly" capital | Fixed capital with $\phi(z)$ *(Penrose–Uzawa)* |
|---|---|
| Capital is perfectly divisible, homogeneous, and instantly reconfigurable | Capital goods are specific in form and location once installed; reallocation requires time and resources |
| Adjustment can be made in arbitrarily small increments at zero marginal cost for infinitesimal changes | Adjustment costs are convex, rising rapidly with the magnitude of change; prohibitive beyond a finite capacity limit |
| No inaction range: Even a tiny gap between marginal product and user cost triggers investment or disinvestment | Finite inaction range: Changes occur only when expected returns clear a threshold above (or below) replacement cost |
| Investment is smooth and continuous; under flexible prices, the economy rapidly converges to full-capacity/full-employment output | Investment is lumpy and threshold driven; moderate shocks may not induce adjustment, allowing underemployment to persist |
| Behavioral frictions are absent | Behavioral frictions (e.g., status quo bias) and irreversibility can be incorporated explicitly into $\phi(z)$ |

**Assumption 5.2 (Size-Adjusted Lump-Sum Cost)** A lump-sum adjustment cost is incurred whenever a firm changes its capital scale, independent of the magnitude of the adjustment but proportional to installed capital:

$$C_B(K_t) = \kappa_c\, K_t \cdot \mathbf{1}\{z_t \neq 0\},$$

where $\kappa_c$ denotes the cost per unit of capital stock. For algebraic convenience, we express this cost in replacement units $I^{\text{rep}} = \delta K_t$ by defining $c_b := \kappa_c/\delta$, so that[8]

$$C_B(K_t) = c_b \delta K_t\, \mathbf{1}\{z_t \neq 0\}.$$

In the *expansion* test, the comparison is between the upper branch and the replacement baseline $1 + \delta\varphi$, so $c_b$ does not enter the upward inequality. In the *contraction* test, skipping replacement and reorganizing shifts the break-even intercept to $1 - (1 + c_b)\delta + \delta\varphi$, and hence, $c_b$ enters the downward inequality. For this reason, we refer to $c_b$ as a *downside bias parameter*, capturing the asymmetry between contraction and expansion. In the mutation step, we also cap the proposed expansion by $(1 - c_b)\delta$, reflecting the same organizational burden.

By combining these behavioral frictions with the confidence-driven incentives developed in the previous section, we are now prepared to construct an evolutionary game-theoretic model in which thresholds, imitation, and aggregate coordination jointly determine investment dynamics and the stability of the macroeconomic equilibrium.

[8] This is a normalization, not a dependence on the realized investment $\Delta I$. Larger firms face higher managerial and coordination burdens (training, retooling, redeployment), hence the proportionality to $K_t$. A useful benchmark is to think of retained earnings as sufficient to cover replacement investment, i.e., of the same order as $\delta K$.

## 5.4 Mutation–Imitation Model of Investment Decision

This section develops the evolutionary game-theoretic model that integrates the forces introduced earlier. The analysis unfolds in two stages. First, in the *mutation stage*, firms form tentative investment plans in response to confidence shocks, generating a distribution of optimistic and pessimistic mutants. Second, in the *imitation stage*, firms revise these plans after observing the aggregate share of mutants and the current utilization rate, leading to threshold-driven shifts in the capital stock. By formalizing this two-stage process, we derive conditions under which small shocks dissipate and when they amplify into regime shifts, thereby clarifying how confidence shocks are either absorbed or amplified into regime shifts through coordination and frictions.

We analyze a two-stage investment game in which each firm chooses net investment $z = I - \delta K \in \{-\delta K, 0, \delta K\}$, corresponding, respectively, to (*cut, replace, double*). Because capital chosen at $t$ becomes productive only at $t+1$ and the period-$t$ wage bill is fixed by a common forward contract, actions do not affect period-$t$ profits. Payoffs are therefore realized in the subsequent SQSS and depend on aggregate utilization and the mutant share. Under convex, kinked adjustment costs of the Penrose–Uzawa type, interior investment choices are weakly dominated. As shown in Lemma 5.C, firms' tentative decisions therefore reduce to the three corners $\{-\delta K, 0, +\delta K\}$. We take this discrete strategy set as primitive in the mutation stage.

Figure 5.4 summarizes the logical structure of the two-stage model, highlighting the interplay between confidence, behavioral frictions, coordination thresholds, and aggregate investment.

The following lemma formalizes this mutation step by showing that, under behavioral inertia and adjustment frictions, each firm's tentative investment choice collapses to a discrete threshold rule.

**Lemma 5.1 (Discrete Investment Choices in the Mutation Stage)** *In the presence of one-period gestation, forward labor contracts, and lump-sum adjustment frictions around replacement investment, each firm's optimal net investment choice at the mutation stage lies in*

$$z_i \in \{-\delta K,\ 0,\ +\delta K\}.$$

*Moreover, there exist thresholds $\mu_L < \mu_H$ such that firms with confidence index $\mu_i$ follow a cutoff rule:*

$$z_i = \begin{cases} -\delta K, & \mu_i < \mu_L, \\ 0, & \mu_i \in [\mu_L, \mu_H], \\ +\delta K, & \mu_i > \mu_H. \end{cases} \tag{5.4.1}$$

***Proof (Deferred)*** See Appendix 5.C. □

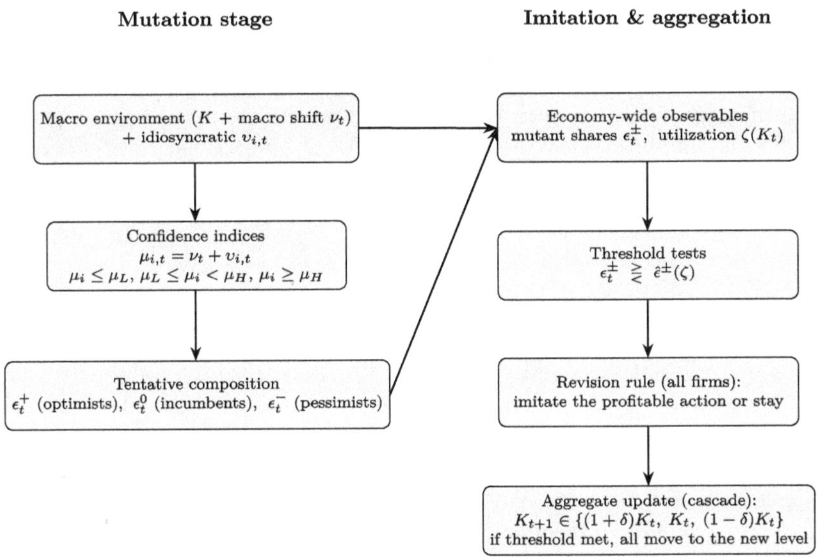


**Fig. 5.4** From confidence to contagion.
Confidence shocks create initial mutants, and imitation spreads their behavior, amplifying local deviations into macro- fluctuations

Empirically, plant-level investment is largely characterized by small replacement expenditures punctuated by occasional large adjustments when the desired-actual capital gap becomes sufficiently wide, motivating the three-way discretization $\{-\delta K, 0, +\delta K\}$ adopted here.[9]

## Mutation Stage

At the start of period $t$, each firm forms a preliminary investment plan based on its confidence index

$$\mu_i = \nu_t + \epsilon_i,$$

where the macroshift $\nu_t$ is common across the economy and $\epsilon_i$ captures idiosyncratic heterogeneity.

As established in Lemma 5.1, investment choices in the mutation stage are discrete and governed by confidence thresholds. Each firm chooses among cutting investment ($-\delta K$), simple replacement (0), or expansion ($+\delta K$), depending on where its $\mu_i$ lies relative to the cutoffs $(\mu_L, \mu_H)$.

These preliminary decisions generate a cross-sectional distribution of *mutants*: firms that tentatively plan to deviate from the replacement strategy, by either expand-

[9] Micro-evidence suggests that capital adjustment is largely lumpy, with small replacement investment most of the time and occasional large jumps when the capital gap becomes large; see Doms and Dunne (1998), Caballero and Engel (1999).

ing or contracting capacity. The resulting mutant shares provide the aggregate signal that feeds into the imitation stage, where firms reassess their plans in light of realized utilization and the observed behavior of others.

The mutation stage captures how dispersed private confidence generates tentative, idiosyncratic investment plans, whereas the imitation stage governs how those plans are revised once firms observe aggregate behavior and capacity utilization, transforming individual deviations into coordinated macroeconomic outcomes.

**Imitation Stage**

After observing the utilization rate and the aggregate share of optimistic or pessimistic mutants, all firms have an opportunity to revise. If the observed share exceeds the relevant threshold $\hat{\epsilon}^{\pm}(\zeta)$, then firms coordinate by aligning with the prevailing direction of deviation: They switch en masse to $+\delta K$ in the expansion case, or to $-\delta K$ in the contraction case. The parameter $\hat{\epsilon}^{\pm}(\zeta)$ governs how large a critical mass must be reached before revision occurs.

A crucial question arises: Why is the second stage of particular significance? Although in principle firms could compute ex ante the aggregate distribution of $\mu_i$ once $\nu_t$ is realized, the two-stage setup captures realistic frictions:

- Information limits
  Firms may not observe $\nu_t$ perfectly, or may be uncertain about adjustment-cost parameters. Observing the realized mutant share aggregates dispersed information and provides a coordination signal.
- Limited attention
  Rather than solving a full fixed point, firms follow a simple rule of thumb: Act on individual confidence in the mutation stage, but revise only if a salient coordination cue appears in the imitation stage.
- Partial commitment
  With one-period gestation, some investment is preplanned but still revisable before resources are irreversibly committed. The imitation stage models this "last look" adjustment window.

The two-stage decision process—where individuals form a tentative plan based on private information and subsequently revise it after observing others—has been studied extensively in the literature on social learning and herding behavior. Banerjee (1992) and Bikhchandani et al. (1992) provide foundational models in which sequential decision-makers, despite possessing independent private signals, may abandon their initial plans once prior choices reveal sufficiently strong public information, giving rise to informational cascades. Chamley and Gale (1994) extend this framework to investment timing, showing how agents may strategically delay commitment to benefit from others' revealed information, a theme further developed in Chamley (2004) synthesis of rational herding.[10]

---

[10] In industrial organization, Milgrom and Roberts (1982) demonstrate how early-stage actions, such as pricing, act as public signals that shape later entrants' choices, effectively linking initial

The model developed below, however, does not capture the full transitional dynamics of macrovariables, but it provides an intuitive and analytically clean microfoundation for threshold-based regime shifts in investment. It will serve as a stepping stone toward more realistic—but analytically demanding—models of transition behavior.

### *5.4.1 Tentative Investment Decision in Mutation Stage*

At the start of each period, an economy-wide sentiment shock $\nu_t \sim U[-V, V]$ is drawn. $V$ is the *sentiment shock amplitude* (half-range) of the aggregate $\nu_t$ (using the organizational switch cost $C_B$ defined above, with $c_b := \kappa_c/\delta$). V controls the volatility (dispersion) of $\nu_t$. Each firm then receives an idiosyncratic disturbance $\epsilon_i \sim U[1 - c_b, 1 + c_b]$, independent across firms. The confidence index is $\mu_i = \nu_t + \epsilon_i$, so $\nu_t$ shifts the entire distribution, while $\epsilon_i$ spreads firms around that shift. Figure 5.5 provides how $\mu_i$ is constructed.

For simplicity, we omit the time subscript $t$ when it is clear from context. The set of players (firms) is [0, 1]. We model investment as a deviation from replacement,

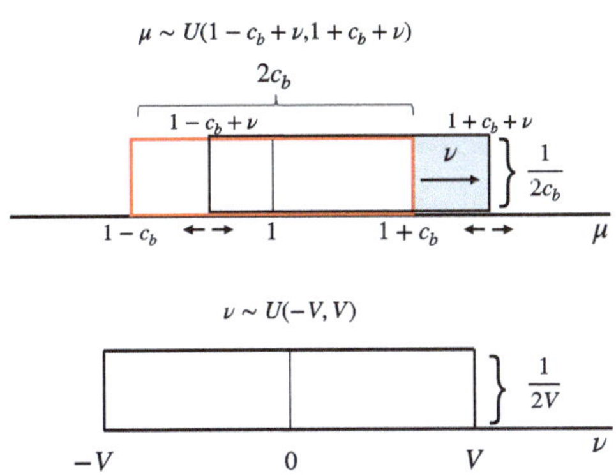


**Fig. 5.5** Proportion of optimistic investors.
Each firm's confidence index is $\mu_i = \nu_t + \epsilon_i$, with $\nu_t \sim U[-V, V]$ and $\epsilon_i \sim U[1 - c_b, 1 + c_b]$ independent across firms, so $\mu_i \sim U[1 - c_b + \nu_t, 1 + c_b + \nu_t]$. Vertical lines mark the status quo threshold 1 and the lemma's decision thresholds $\mu_L = 1 - c_b$ and $\mu_H = 1 + c_b$. The intervals correspond to the three actions: $\mu_i < \mu_L$ (cut, $I_i = 0$), $\mu_i \in [\mu_L, \mu_H]$ (hold, $I_i = \delta K$), and $\mu_i > \mu_H$ (expand, $I_i = 2\delta K$). The shaded area to the right of $1 + c_b$ indicates the share of optimistic mutants, which is equal to $\nu_t/(2c_b)$

---

moves to subsequent revisions. Laboratory evidence by Anderson and Holt (1997) confirms that cascade-like behavior emerges even when revision could enhance decision accuracy, while Ellison and Fudenberg (1993) highlight the role of simple heuristics in guiding updates. Together, this body of work provides a robust theoretical and empirical foundation for understanding how tentative decisions evolve through observation and revision in uncertain environments.

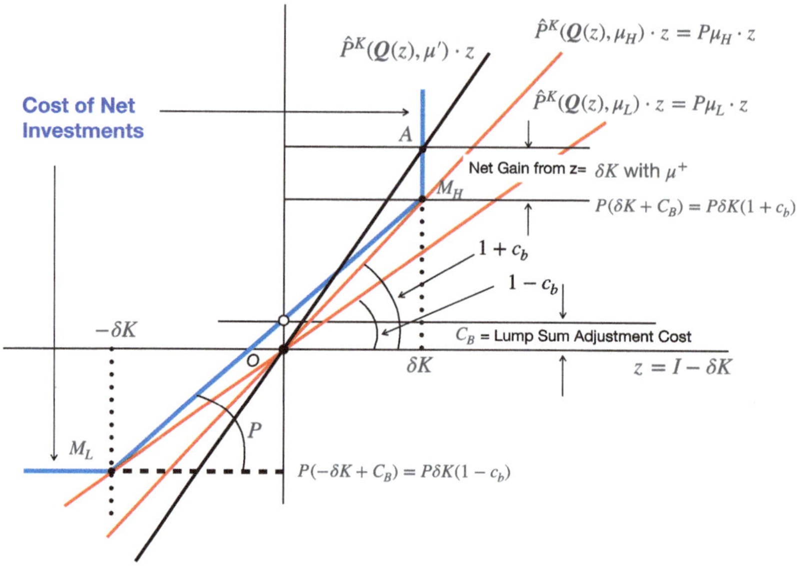


**Fig. 5.6** Adjustment costs, confidence, and tipping thresholds.
Horizontal axis: real net investment $z$. Vertical axis: present value of costs and expected returns. As $z$ increases, adjustment costs rise and become prohibitive at $z = \delta K$ (point $M_H$). For $z < -\delta K$, the cost schedule is flat (left of $M_L$) because resale is impossible; further disinvestment does not reduce costs. At $\mu = \mu_H$, the line $P\mu z$ passes through both $M_H$ and the origin, making the firm indifferent between $z = \delta K$ and $z = 0$, analogously for $\mu = \mu_L$ and $z = -\delta K$

i.e., $z_i = I_i - \delta K$. As noted above, period-$t$ profits are identical across firms. Moreover, the positive or negative *spillovers of aggregate net investment* are shared equally across firms in period $t$, leaving per-firm period-$t$ profits unchanged.

In Fig. 5.6, the red straight line labeled $\widehat{P}^K(Q(z), \mu_H)$ passes through the origin and the point $M_H$. This indicates that, when confidence is at the high threshold $\mu_H$, a firm is indifferent between choosing the replacement strategy ($z = 0$) and expanding capital scale by $z = \delta K$: Both yield zero additional profit. For values $\mu > \mu_H$, the slope of the profit line is higher, but increasing $z$ beyond $\delta K$ is dominated because the adjustment-cost function rises more steeply; thus $\delta K$ remains the optimal choice.

The lower red line $\widehat{P}^K(Q(z), \mu_L)$ passes through the origin and the point $M_L$. At the low threshold $\mu_L$, the firm is indifferent between maintaining capital at $z = 0$ and contracting by $z = -\delta K$. When $\mu < \mu_L$, any additional reduction in investment beyond $I_i = 0$ no longer reduces costs due to the irreversibility of investment while decreasing $\widehat{P}^K(Q(z), \mu)$.

For any $\mu \in [\mu_L, \mu_H]$, the adjustment-cost curve lies everywhere above the net present value line, so any nonzero $z$ produces strictly negative net payoffs. In this middle range, the inaction option ("inaction" here means replacement, i.e., $z = 0$) strictly dominates both contraction and expansion. Thus, the optimal responses lie within the set of strategies: $\{-\delta K, 0, +\delta K\}$, with players restricted to choosing only pure strategies (Lemma 5.1). Figure 5.6 shows how the behavioral and investment adjustment costs lead to the discontinuity at $z = 0$ and kinks in the adjustment-cost function, resulting in the above strategy space. This geometric argument shows how the thresholds $(\mu_L, \mu_H)$ partition the confidence space into three regimes—contraction, replacement, and expansion—corresponding exactly to the case distinctions in Lemma 5.1. (See Appendix 5.C for a formal proof.)

**Lemma 5.2 (Signal-Determined Mutant Shares)** *The proportions of optimistic and pessimistic mutants are determined by the aggregate signal* $\nu$ *as*

$$\epsilon^+(\nu) = \begin{cases} \nu/(2c_b), & \nu \geq 0, \\ 0, & \nu < 0, \end{cases}$$

$$\epsilon^-(\nu) = \begin{cases} -\nu/(2c_b), & \nu \leq 0, \\ 0, & \nu > 0. \end{cases}$$

***Proof (deferred)*** See Appendix 5.D for proof. □

Figure 5.5 illustrates the proportion of optimistic mutants ($\mu_i > 1 + c_b$) is $\nu_t/(2c_b)$ when $\nu_t > 0$ (and zero when $\nu_t \leq 0$). Conversely, the share of pessimistic mutants ($\mu_i < 1 - c_b$) is $-\nu_t/(2c_b)$ when $\nu_t < 0$ (and zero when $\nu_t \geq 0$).

**Remark 5.2 (Convexity of Cost Function)** The adjustment-cost function $\phi$ is convex on each side of $z = 0$ but exhibits an activation jump at $z = 0$. Consequently, it is not globally convex on any interval containing 0, but the local convexity on $(0, \delta K]$ and $[\delta K, 0)$ together with the discontinuity at 0 suffices for the threshold result in Lemma 5.1.

As a result of Lemma 5.C, all firms share the same discrete strategy space for net investment in the imitation stage,

$$Z = \{-\delta K,\ 0,\ \delta K\}.$$

For convenience, we normalize this as

$$\Psi = \{-1,\ 0,\ +1\},$$

where $-1$, 0, and $+1$ correspond respectively to halting investment ($I = 0$), simple replacement ($I = \delta K$), and double replacement ($I = 2\delta K$). In the imitation stage, firm $i$ selects an action $\psi_i \in \Psi$ according to its confidence index:

$$\psi_i(\mu_i) = \begin{cases} -1, & \mu_i < 1 - c_b, \\ 0, & \mu_i \in [1 - c_b,\ 1 + c_b], \\ +1, & \mu_i \geq 1 + c_b. \end{cases}$$

We refer to firms choosing $+1$ as *optimistic mutants*, those choosing $-1$ as *pessimistic mutants*, and those choosing 0 as *incumbents*. The shaded area in Fig. 5.5 represents the proportion of optimistic mutants in the population.

## 5.4.2 Imitation Stage

This subsection examines how firms—mutants and incumbents alike—revise their tentative investment decisions after observing the capital utilization rate and the aggregate pattern of choices in the economy. Given the constraints of bounded rationality (Simon, 1947), only two alternatives come to their mind: Either $\psi_i = 0$, or 1 when $\nu > 0$, and symmetrically $\psi_i = 0$, or $-1$ when $\nu < 0$. They are motivated to select the option with the highest expected profit in the new equilibrium, aligning with Simon (1947)'s concept of "satisfactory choice" under bounded rationality, which he later expanded upon with the *satisficing principle*. Specifically, their focus is on narrowing the gap between expected sales and supply capacity by monitoring two key indices: the current utilization rate and the share of mutants.

The proportion of optimistic or pessimistic mutants, denoted $\epsilon^{\pm}$, serves as a proxy for aggregate demand expectations. Firms also take into account the current excess replacement ratio, $\zeta(K)$, which measures whether existing productive capacity is over- or underutilized. If utilization is low, expanding facilities yield little benefit; if capacity is tight, expansion becomes attractive. Hence, the imitation stage combines two signals: the realized mutant share $\epsilon^{\pm}$ and $\zeta(K)$.

### 5.4.2.1 Decision Environment

**Capital Utilization Rate**
We define

$$\zeta(K) := \frac{K^d(K) - K}{\delta K}, \tag{5.4.2}$$

which measures the deviation of desired capital from the current stock, expressed in units of replacement investment. For clarity, we refer to $\zeta$ as the *excess replacement ratio*. A positive value indicates excess demand for capital beyond replacement, while a negative value reflects underutilization relative to replacement. Here $K^d(K)$ denotes the capital stock required to meet aggregate demand. In the stationary equilibrium, aggregate demand is $X^d = (c + \delta)K$, so cost minimization (derived in Appendix 4.D to Chap. 4) implies

$$K^d(K) = \kappa^{1-\alpha}\,(c + \delta)\,K, \tag{5.4.3}$$

which, by using PM equation, can be expressed as

$$K^d(K) = \frac{\alpha(\eta - 1)}{\eta} \cdot \frac{c + \delta}{r + \delta} \cdot K.$$

Substituting this into (5.4.2) yields the utilization rate:

$$\zeta(K) = \frac{1}{\delta}\Big((c + \delta)\kappa^{1-\alpha} - 1\Big) \tag{5.4.4}$$

$$= \frac{1}{\delta}\left(\frac{\alpha(\eta - 1)}{\eta} \cdot \frac{c + \delta}{r + \delta} - 1\right), \tag{5.4.5}$$

where both the capital–labor ratio $\kappa$ and the user cost of capital $(r + \delta)$ are determined endogenously in equilibrium.

**Interpretation** The replacement–investment gap $\zeta(K)$ measures the excess of desired capital over installed capital, expressed in units of replacement investment. It summarizes the profitability of expansion relative to the existing capital stock and is therefore closely related to the capital–labor ratio. When capital is scarce, the marginal productivity of capital is high, desired capital exceeds installed capital, and $\zeta(K)$ is positive. As capital accumulates, diminishing returns reduce marginal productivity and lower $\zeta(K)$.

Positive values of $\zeta(K)$ indicate that additional capital is profitable, whereas negative values indicate excess installed capital. However, profitability alone does not imply positive aggregate capital drift because firms face inertia arising from the mutation–imitation process and the status-quo bias in investment decisions. Expansionary profitability must therefore be sufficiently strong before aggregate expansion becomes dominant. In the imitation stage, $\zeta(K)$ serves as a macroeconomic profitability signal governing the relative attractiveness of expansionary and contractionary investment strategies.

#### 5.4.2.2 Effect of Mutants on Operational Profits in Period $t$

"Mutants" enter at the outset of period $t$, but they are *incumbent* firms that switch their investment plan; they do not add a new variety. Their entry has *no within-period effect on* $Q_t$. This subsection shows why incumbents and mutants have the same operating payoffs in period $t$.

**Assumption 5.3 (Gestation and Quasi-Fixity)** Investment chosen in period $t$ becomes productive only in $t + 1$, while employment $N_t$ is predetermined for production in $t$[11] Consequently, output and operating profits in period $t$ are independent of mutant entry and investment choices.

Thus, effective capacity and labor input at $t$ are the same for mutants and incumbents. Moreover, because the demand system for varieties is unchanged in period $t$ and firms set the same price on the PM locus, their period-$t$ operating payoffs coincide.

#### 5.4.2.3 Evaluation of Payoffs from Period-$(t + 1)$ Onward

Under Assumption 5.3, payoffs can be written as the period-$(t + 1)$ SQSS profit $\Pi^{t+1}(I;\text{ utilization}, \mu)$, where $\mu$ is the mutant share that maps into next-period utilization and demand.

A positive aggregate signal $\nu > 0$ induces firms with high idiosyncratic confidence to become optimistic mutants in the mutation stage. A fraction $\epsilon^+$ of firms therefore play $\psi = +1$ (double replacement), while the remaining $1 - \epsilon^+$ incumbents play $\psi = 0$. Consequently, firms expect the aggregate capital stock from period $t + 1$ onward to increase to

$$K(\epsilon^+) = (1 + \epsilon^+\delta)\, K.$$

Symmetrically, a negative signal $\nu < 0$ generates a share $\epsilon^-$ of pessimistic mutants playing $\psi = -1$, which reduces the expected capital stock to

$$K(\epsilon^-) = (1 - \epsilon^-\delta)\, K.$$

The corresponding aggregate demand (see Appendix 4.D of Chap. 4) is then

$$X^d(\epsilon^\pm) = (c + \delta)\,(1 \pm \epsilon^\pm\delta)\, K \quad \text{as } \nu \gtrless 0.$$

From (5.4.4), recall that

$$(c + \delta)\kappa^{1-\alpha} = 1 + \delta\zeta. \tag{5.4.6}$$

It follows that the capital stock required to meet demand is

$$K^d(\epsilon^\pm) = \kappa^{1-\alpha}(c + \delta)(1 \pm \epsilon^\pm\delta)\, K.$$

---

[11] *Why $N_t$ is predetermined.* (i) *Search/matching lag.* Vacancies posted at $t$ convert into filled positions only with a delay; matches form stochastically and become productive next period. Production at $t$ therefore uses the beginning-of-period stock $N_t$. (ii) *Contracts/rosters.* Intraperiod labor contracts, shift rosters, and notice requirements make mid-period separations costly or infeasible. Staff levels are set before operations begin; any reduction can be implemented only at the next scheduling/contract renewal date. In both cases, $N_t$ is a predetermined state for production in period $t$.

Substituting (5.4.6) yields

$$K^d(\epsilon^{\pm}) = (1 + \delta\zeta)(1 \pm \epsilon^{\pm}\delta)\,K, \\ \approx (1 + \delta\zeta)\,K \;\pm\; \delta\epsilon^{\pm}K. \qquad (5.4.7)$$

Equation (5.4.7) shows that as the fraction of mutants $\epsilon^{\pm}$ rises, the required capital stock deviates from its baseline level $(1 + \delta\zeta)K$. When the replacement investment gap equals zero, the capital stock is exactly aligned with demand under the fixity-of-capital assumption; in this case, mutants directly generate a demand–supply gap whose size grows with $\epsilon^{\pm}$. Once the gap crosses certain thresholds, the current equilibrium becomes unsustainable and the economy shifts to a new equilibrium. More generally, the excess replacement ratio $\zeta$ mediates these dynamics, as highlighted in (5.4.7).

**Demand Price vs. Risk-Adjusted Demand Price**

We distinguish between the *demand price of capital*, $P^K$, and the *risk-adjusted demand price of capital*, $\widehat{P}^K$. The object $P^K$ is the Keynesian demand price: The present value of expected *gross* returns that supports the *normal* level of investment under the prevailing technology and market structure. With monopolistic competition and finite elasticity of substitution $\eta < \infty$, firms set a markup over marginal cost, so the normal operating point embeds positive profits and satisfies $P^K > P^S = P$. In contrast, $\widehat{P}^K$ captures the *subjective, risk-adjusted* valuation that governs discrete choices between halting investment, replacement, and double replacement; it incorporates confidence and risk perception (our $\mu$), and therefore, rationalizes heterogeneous switches at the thresholds derived in Lemma 5.1.

**Perfect Competition as a Knife-Edge Case**

When $\eta \to \infty$, markups vanish and normal profits are driven to zero, so that the Keynesian demand price of capital coincides with its supply price, $P^K = P^S$. In this limit, the replacement equilibrium becomes knife edge: There is no profit buffer to absorb adverse sentiment, and an arbitrarily small share of pessimistic mutants suffices to overturn replacement investment. Thus, contrary to the conventional view that perfect competition delivers a natural and robust long-run steady state, it is precisely the perfectly competitive benchmark that is most fragile in the presence of confidence shocks and investment irreversibility.

The normal-profit condition governed by $P^K$ pins down the existence of a viable steady state with replacement investment, whereas $\widehat{P}^K$—through the confidence index $\mu$—determines *which* discrete action a firm takes in the neighborhood of that steady state. Thus $P^K$ explains why the economy can sit at replacement with positive profits (for $\eta < \infty$), whereas $\widehat{P}^K$ explains why, given sentiment dispersion, some firms double, while others halt, with the mutant thresholds depending on primitives such as $\eta$, adjustment costs, and $\zeta$.

*Imitation-Stage Decision Rule (Values at $t+1$)* Similar to the mutation stage, firms do not need to focus on current-period profits; instead, they compare the expected profits from period $t + 1$ onward (Assumption 5.3) when revising their tentative plans. Let the final action be $z \in \{-\delta K,\ 0,\ +\delta K\}$ (gross investment $I = \delta K + z$). After the gestation lag, all incumbents retain the installed capital stock $K$, while optimistic mutants hold $(1 + \delta)K$ and pessimistic mutants $(1 - \delta)K$.

*Organizational Switch Cost* In the *expansion* test, we compare the upper branch to the replacement baseline $1 + \delta\varphi$, so $c_b$ does not appear in that inequality; in the *contraction* test, skipping replacement and reorganizing shift the break-even intercept to $1 - (1 + c_b)\delta + \delta\varphi$; hence, $c_b$ enters the downward inequality. In the mutation step, we also cap the proposed expansion by $(1 - c_b)\delta$, reflecting the same organizational burden.

*Economic Meaning of the Threshold Condition* Before defining the tipping thresholds formally, it is useful to clarify the economic meaning of the key comparison. The expression

$$K^{\text{op}}(+\delta K, \epsilon) - K^{\text{op}}(0, \epsilon) \ \geq\ \varphi\,\delta K$$

compares the *effective gain in operating capacity* generated by an expansionary move with its *opportunity cost*, measured in capital units.

The left-hand side captures how much additional capacity a firm can actually put to work when it expands by $\delta K$, given both demand constraints and the presence of other mutants. Because production follows the short-side rule, this gain depends not only on installed capacity but also on whether aggregate demand—shaped by the mutant share $\epsilon$—is sufficient to absorb it. The right-hand side, $\varphi\,\delta K$, represents the real cost of expansion: the replacement-scaled investment outlay required to create the additional capacity.

The inequality therefore states that expansion is viable only if the *realized increase in usable capacity* is large enough to justify its cost. When the condition holds, expanding firms can profitably employ their extra capital; when it fails, additional investment merely creates idle capacity. The tipping threshold $\hat{\epsilon}^{+}$ is precisely the smallest mutant share at which this inequality becomes satisfied. An analogous comparison governs the contraction threshold, where the relevant question is whether savings from avoided replacement outweigh the loss of operating capacity.

**Definition 5.1 (Upward and Downward Tipping Thresholds)** We define $\hat{\epsilon}^{+}$ ($\hat{\epsilon}^{-}$) as the smallest mutant share that makes expansion (contraction) weakly dominate inaction (replacement).

$$\hat{\epsilon}^{+}(\zeta) := \inf\Big\{\epsilon \in [0, 1] :\ K^{\text{op}}(+\delta K, \epsilon) - K^{\text{op}}(0, \epsilon) \ \geq\ \varphi\,\delta K\Big\},$$

and

$$\hat{\epsilon}^{-}(\zeta) := \inf\Big\{\epsilon \in [0, 1] :\ K^{\text{op}}(-\delta K, \epsilon) - K^{\text{op}}(0, \epsilon) \ \geq\ -\varphi\,\delta K\Big\}.$$

**Lemma 5.3 (Upward and Downward Tipping Thresholds)** *Under Assumption 5.2, the upward and downward tipping thresholds are given by*

$$\hat{\epsilon}^{+}(\zeta_t) = \begin{cases} \max\{0,\ \varphi - \zeta_t\}, & \text{if } \varphi \leq 1 - c_b, \\ \text{no solution (expansion never optimal)}, & \text{if } \varphi > 1 - c_b, \end{cases} \tag{5.4.8}$$

*and*

$$\hat{\epsilon}^{-}(\zeta_t) = \max\{\,0,\ 1 + c_b + \zeta_t - \varphi\,\}, \qquad \text{feasible only if } \hat{\epsilon}^{-}(\zeta_t) \leq 1. \tag{5.4.9}$$

***Proof*** See Appendix 5.D. □

The upward threshold exists only if the maximum capacity gain net of organizational costs exceeds the investment cost, whereas the downward threshold always exists but may lie outside the feasible range.

Lemma 5.3 characterizes the critical mass of optimistic and pessimistic mutants required to overturn replacement investment, given the state of capacity utilization and the presence of adjustment frictions. We note that, throughout this section, installed (existing) capacity for incumbents is taken as given, while their actual operation is determined endogenously by the short-side rule as the minimum of capacity and demand.

Figure 5.6 establishes the microfoundation of discrete investment behavior. Because changing scale entails a lump-sum organizational cost, the investment cost schedule is discontinuous at $z = 0$, generating a region of strict inaction. As the confidence index $\mu$ varies, the confidence-adjusted return line pivots. The thresholds $\mu_H$ and $\mu_L$ are the confidence levels at which this return line just meets the investment cost schedule at the discrete corners $z = \pm\delta K$, yielding zero net gain from expansion or contraction. These thresholds determine which firms become optimistic or pessimistic mutants in the mutation stage.

Figure 5.7 compares the effects of optimistic and pessimistic mutants on aggregate demand and capacity at different population shares. Panels (a) and (b)

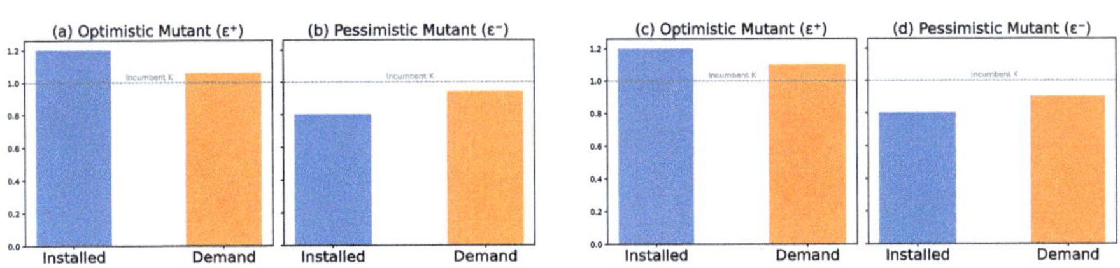


**Fig. 5.7** Short-side effects of optimistic and pessimistic mutants.
The left pair of panels **(a)**–**(b)** shows the effects of a mutant share of $\epsilon = 0.3$, while the right pair **(c)**–**(d)** shows $\epsilon = 0.5$. Within each pair, the left panel depicts optimistic mutants (expansion) and the right panel depicts pessimistic mutants (contraction). In all panels, the incumbents' *installed (existing) capacity* remains fixed at $K$ (horizontal dotted line), while actual operation is determined by the short-side rule as the minimum of capacity and demand-required capital

show the effects of optimistic and pessimistic mutants when the mutant share is 0.3, while panels (c) and (d) report the corresponding outcomes for a share of 0.5. As the mutant share increases, the demand faced by mutants moves closer to their adjusted installed capacity, reducing their own demand–capacity gap, while the corresponding gap for incumbents widens. Consequently, the mutant strategy becomes increasingly attractive as its population share grows, with high utilization favoring expansionary mutants and low utilization favoring contractionary mutants.

The aggregate consequences of these tentative choices are illustrated in Fig. 5.7. As the mutant share $\epsilon$ increases, demand is progressively reallocated toward the mutant group. This reduces the supply–demand mismatch faced by mutants, while widening the mismatch for incumbents, whose installed capacity remains fixed. The figure therefore visualizes the endogenous reinforcement mechanism: The relative payoff of the mutant strategy improves with its population share, creating increasing returns to imitation.

Which type of mutation is reinforced depends on the state of utilization. When the excess replacement ratio $\zeta$ is low, avoiding replacement investment is advantageous, so pessimistic mutants are rewarded. When $\zeta$ is high, additional capacity can be absorbed by demand, making expansion by optimistic mutants profitable. Utilization thus selects the direction in which deviations are amplified.

Figure 5.8 translates this intuition into the closed-form upward threshold. The operating capital of optimistic mutants, $K^{\text{op}}(+, \epsilon)$, reflects both demand constraints and adjustment costs, and is capped at $(1 - c_b)\delta K$. Subtracting the investment cost $\varphi\delta K$ yields the net gain from expansion, which crosses the incumbent benchmark at $\hat{\epsilon}^{+} = \varphi - \zeta$ whenever $\varphi + c_b < 1$. If this condition fails, even large waves of optimism cannot overturn replacement investment, and the equilibrium

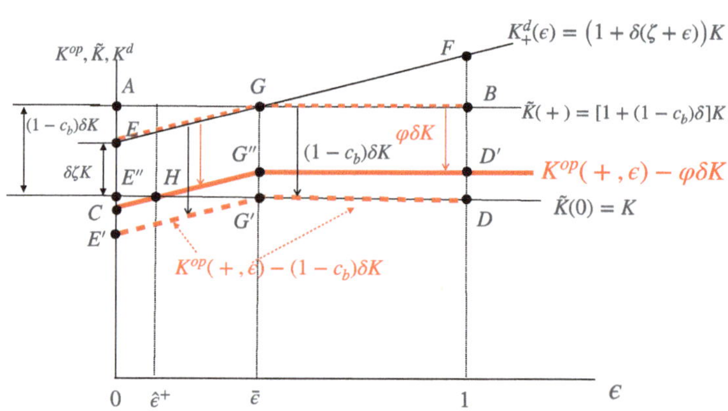


**Fig. 5.8** Closed-form upward thresholds.
The net gain from expansion is given by $K^{\text{op}}(+, \epsilon) - \varphi\delta K$, shown as the dotted red line. The upward kink occurs at $\bar{\epsilon} = 1 - c_b - \zeta$. When the maximum capacity gain $(1 - c_b)\delta K$ covers the investment cost $\varphi\delta K$, this line crosses the incumbent benchmark $K$ at $\hat{\epsilon}^{+} = \varphi - \zeta$. Otherwise, as in the case $\varphi = 1 - c_b$, expansion never yields an advantage, and the equilibrium remains fully robust

is fully robust to upward deviations. The downward threshold follows symmetrically, with organizational frictions entering asymmetrically through the contraction margin.

**Remark 5.3 (Why $c_b$ Matters Only for Contraction?)** In an expansionary episode, the demand side is the binding constraint. Optimistic mutants must generate enough new demand to fully utilize the extra capacity they create. Because their challenge lies in finding buyers rather than overcoming internal frictions, the downward bias parameter $c_b$ has no effect on the profitability test for expansion. By contrast, in a contractionary episode, the supply side becomes the binding constraint. Pessimistic mutants reduce their investment, directly lowering aggregate supply. Here, the presence of status quo and organizational frictions captured by $c_b$ amplifies the difficulty of maintaining the incumbent equilibrium. Thus, $c_b$ explicitly enters the threshold condition for downward tipping, making it easier for pessimistic mutants to disrupt the status quo.

#### 5.4.2.4 Benchmark Case: Upward Robustness and Downward Fragility

We now analyze a benchmark economy that isolates the core asymmetry of the model. Figure 5.9 visualizes this case. We set $\zeta = 0$, $\varphi = 1$, and $c_b = 0$, corresponding to a perfectly competitive jelly-capital equilibrium: The capital market is exactly balanced, firms earn only normal profits, and no organizational or behavioral frictions are present. This stripped-down environment serves as a clean reference point against which the general case can be interpreted.

Panels (a)–(b) plot incumbent and mutant capacity levels, $K$, $(1+\delta)K$, $(1-\delta)K$, together with the demand-required capital curves $K^d_\pm(\epsilon) = (1 \pm \delta\epsilon)K$. The horizontal reference lines $K \pm \varphi\delta K$ mark the profitability margin, highlighted by vertical arrows of length $\varphi\delta K$.

**Upward Robustness**
Panel (a) illustrates the resilience of the equilibrium to optimistic deviations. When optimistic mutants enter, required capital rises to $K^d(\epsilon^+) = (1+\delta\epsilon^+)K$. Even if the entire population becomes optimistic, the resulting operating capacity, net of expansion costs, does not exceed that of incumbents. Because $\zeta = 0$ and $P^S = P^K$, additional demand created by optimistic mutants is shared across firms without generating extra profits. The benchmark equilibrium is therefore fully robust upward: No finite mass of optimists can overturn replacement investment.

**Downward Fragility**
Panel (b) shows a stark contrast. If a fraction $\epsilon^-$ of pessimistic mutants cuts investment, aggregate demand falls to $K^d(\epsilon^-) = (1-\delta\epsilon^-)K$. Even at $\epsilon^- = 0$, halting replacement yields cost savings that place pessimists on par with incumbents. As $\epsilon^-$ increases, this advantage strictly grows. Thus, an arbitrarily

small share of pessimists suffices to destabilize the equilibrium. Cost savings are privately internalized, while the demand shortfall is socialized.

This asymmetry—robustness to optimism but fragility to pessimism—reveals that the perfectly competitive steady state lacks evolutionary stability. In Swinkels' sense, the aggregate environment generated by mutants does not induce best responses that restore incumbency. Instead, once pessimists appear, their advantage persists and the equilibrium collapses.

Figure 5.10 illustrates how this knife-edge property disappears once either markups ($\varphi < 1$), underutilization ($\zeta > 0$), or adjustment frictions ($c_b > 0$) are introduced. The double arrows mark the profitability margin $\varphi\delta K$: Only when the demand-required capital curve crosses this margin does the system tip into a new SQSS. In the benchmark case, this crossing never occurs upward but occurs immediately downward, making the asymmetry transparent at a glance.

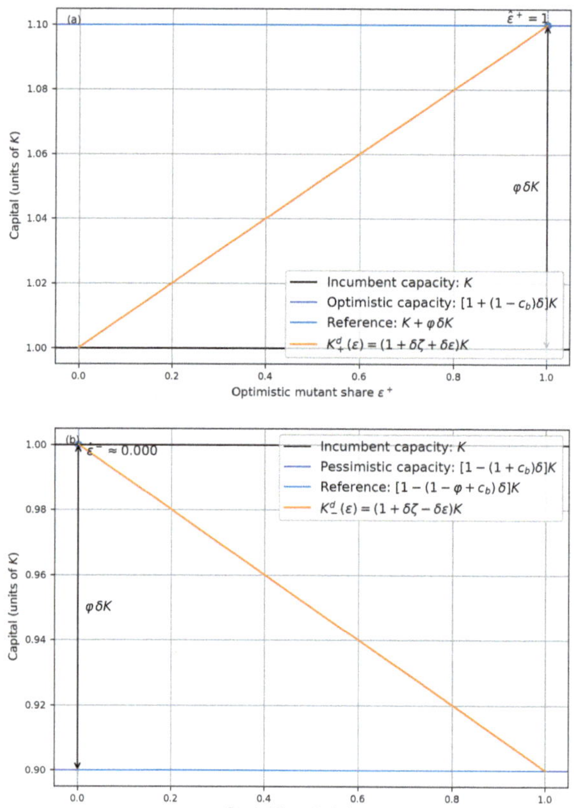


**Fig. 5.9** Benchmark case: upward robustness and downward fragility. In the benchmark economy ($\zeta = 0,\ \varphi = 1,\ c_b = 0,\ \delta = 0.1$), panel (**a**) shows that the incumbent equilibrium is fully robust to upward deviations: Even when the share of optimistic mutants rises to one, expansion never becomes profitable ($\hat{\epsilon}^+ = 1$). Panel (**b**) contrasts this with complete downward fragility: An arbitrarily small mass of pessimistic mutants suffices to overturn replacement investment ($\hat{\epsilon}^- = 0$)

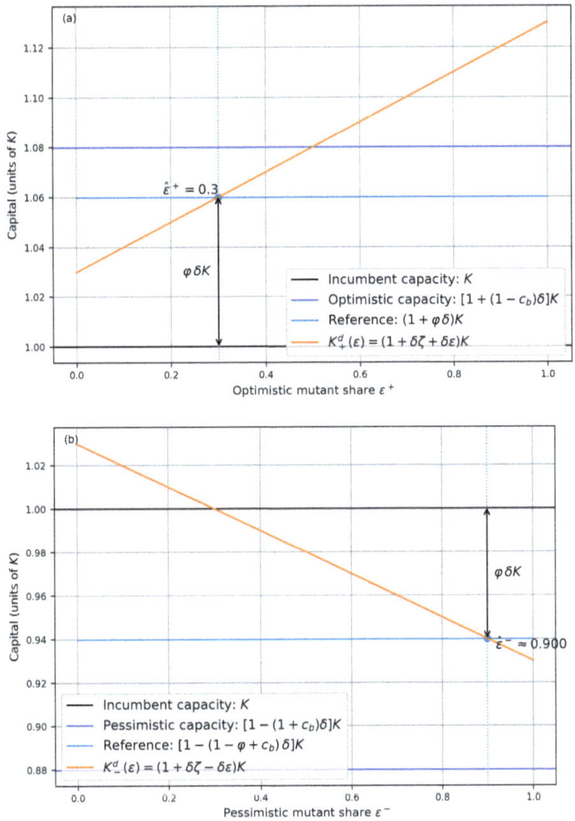


**Fig. 5.10** General case: When small shocks tip the system.
In this general economy ($\zeta = 0.3,\ \varphi = 0.6,\ c_b = 0.2,\ \delta = 0.1$), panel (**a**) shows the expansion case: With positive utilization pressure ($\zeta > 0$) and a profitability margin ($\varphi < 1$), the demand-required capital $K_+^d(\epsilon^+)$ intersects the profitability threshold at a finite mutant share $\hat{\epsilon}^+ = 0.3 < 1$, so that a sufficiently large group of optimistic mutants overturns the incumbent equilibrium. Panel (**b**) shows the contraction case: The demand-required capital $K_-^d(\epsilon^-)$ crosses the contraction profitability margin only when the share of pessimistic mutants exceeds $\hat{\epsilon}^- = 0.9 > 0$, implying that downward deviations require a critical mass and are no longer triggered by arbitrarily small shocks

## 5.4.3 *Equilibrium of the Two-Stage Game*

This section synthesizes the results of the two-stage analysis and establishes the evolutionary game-theoretic equilibrium of the model.

To simplify the exposition, we define the *utilization profitability gap* as

$$x_t := \zeta_t - \varphi.$$

Then, the revision thresholds are given by the following equations:

$$T^{+}(x) := 2c_b(\varphi - \zeta)_+ = 2c_b\, x^-,$$
$$T^{-}(x) := 2c_b(1 + c_b + \zeta - \varphi)_+ = 2c_b(1 + c_b + x)_+,$$

where $x^- := \max\{-x, 0\}$.

**Assumption 5.4 (Investment Viability Condition)** Investment is viable when demand is sufficient: The profitability margin covers the behavioral and organizational frictions, i.e.,

$$1 - \varphi \geq c_b.$$

Combining Lemma 5.2, the threshold conditions (5.4.8)–(5.4.9), and Assumption 5.4, the law of motion for aggregate capital becomes

$$K_{t+1} = \begin{cases} (1 + \varepsilon_t^+(K_t))\, K_t, & \nu_t \geq T^+(x_t), \\ (1 - \varepsilon_t^-(K_t))\, K_t, & \nu_t \leq -T^-(x_t), \\ K_t, & \text{otherwise}, \end{cases} \tag{5.4.10}$$

where the realized proportional jump sizes satisfy

$$0 \leq \varepsilon_t^{\pm} \leq \frac{\delta}{2b}.$$

Equation 5.4.10 describes the aggregate law of motion under the two-stage game. The first line, $K_{t+1} = (1 + \delta)K_t$, obtains when the optimistic mass implied by the aggregate signal is large enough to clear the expansion threshold ($\nu_t/(2c_b) > \varphi - \zeta_t \geq 0$). In that case, imitation makes double replacement viable and the capital stock steps up by $\delta K_t$. The second line, $K_{t+1} = (1 - \delta)K_t$, arises when the pessimistic mass exceeds the contraction threshold ($-\nu_t/(2c_b) > 1 + c_b + \zeta_t - \varphi > 0$): Halting investment becomes the best response and aggregate capital steps down by $\delta K_t$. In all remaining states, neither mutant group reaches its tipping threshold, so firms stick to simple replacement and aggregate capital remains at $K_t$. The thresholds shift with $\zeta_t$ and $\varphi$: A larger utilization gap $\zeta$ lowers the bar for optimists, while raising it for pessimists, making upward movements easier and downward movements harder.

**Remark 5.4 (Swinkels-Style Entrants-Only Revision)** An alternative formulation, closer to Swinkels' notion of "stable entrants," restricts the second-stage revision to the entrants themselves (Swinkels, 1992). Incumbents continue with simple replacement, while only entrants decide whether to remain mutants or renege and revert to replacement. If the realized mutant share $\epsilon^{\pm}$ is below its threshold $\hat{\epsilon}^{\pm}(\zeta)$, entrants fade out and the aggregate stock remains $K_t$. If the share exceeds the threshold, all entrants of that sign survive, and the new aggregate state becomes $K_{t+1} = (1 \pm \delta\,\epsilon^{\pm})K_t$, followed next period by re-homogenization as all firms respond to the new aggregate state. The threshold stability logic is preserved—mutants cannot survive below threshold—while the magnitude of adjustment is proportional to the entrant mass rather than an all-or-nothing jump. This is analytically convenient yet stylized; persistent coexistence and asynchronous adjustment would require simulation or agent-based approaches.

We are now prepared to formally present the revision rule as a unique cascade equilibrium that is driven by the macroshift parameter $\nu_t$. This equilibrium is characterized by the specific dynamics of the macroshift, highlighting its crucial role in determining the system's behavior.

**Proposition 5.4 (Unique Cascade Equilibrium Under the Revision Rule)** *Fix parameters $\delta \in (0, 1)$, $c_b \in (0, 1)$, and let $x_t = \zeta_t - \varphi$. Given the macroshift parameter $\nu_t \in \mathbb{R}$, define the revision rule (imitation/aggregation stage)*

$$s_t(\nu_t, x_t) = \begin{cases} +1, & \nu_t \geq T^+(x_t), \\ -1, & \nu_t \leq -T^-(x_t), \\ 0, & \textit{otherwise.} \end{cases}$$

*Conditional on $s_t = +1$ or $s_t = -1$, aggregate capital evolves as*

$$K_{t+1} = \begin{cases} (1 + \varepsilon_t^+(K_t))K_t, & s_t = +1, \\ (1 - \varepsilon_t^-(K_t))K_t, & s_t = -1, \\ K_t, & s_t = 0, \end{cases}$$

*where the realized proportional jump sizes satisfy the uniform bound*

$$0 \leq \varepsilon_t^+, \varepsilon_t^- \leq \frac{\delta}{2b}.$$

*Then for every $(\nu_t, x_t)$, there exists a* unique cascade equilibrium *in the two-stage game (mutation then imitation). In this equilibrium, all firms choose the same action $s_t \in \{-1, 0, +1\}$ dictated by the rule above (upward cascade, downward cascade, or stay). The selection is monotone in $\nu_t$ and has at most two discontinuities at $\nu_t = T^+(x_t)$ and $\nu_t = -T^-(x_t)$.*

***Proof (Sketch)*** In the imitation stage, the payoff *differences* between the "up," "down," and "stay" actions are common across firms and strictly monotone in the macroshift $\nu_t$ once the state $(K_t, x_t)$ is fixed. By construction, $\nu_t \geq T^+(x_t)$ makes "up" strictly dominant; $\nu_t \leq -T^-(x_t)$ makes "down" strictly dominant; otherwise "stay" weakly dominates both. Because $T^+(x_t)$, $T^-(x_t) \geq 0$, the two dominance regions are disjoint, so at most one cascade can be triggered for any given $\nu_t$. At the knife-edge equalities (thresholds), a standard status quo tiebreak (stay) or an infinitesimal payoff perturbation selects a unique action. Since the imitation map is unanimous and the aggregate law $K_{t+1} = (1 + \delta s_t)K_t$ depends only on $s_t$, the resulting equilibrium is a *cascade* and is unique. Monotonicity in $\nu_t$ follows from the one-sided threshold structure. □

The proposition presented above naturally gives rise to the following corollary, which plays a crucial role in understanding the stability of capital stock dynamics that will be explored in the subsequent chapter.

**Corollary 5.1 (Comparative Statics in the Gap)** *For fixed $\nu_t$, the upward threshold $T^+(x_t)$ is weakly decreasing in $x_t$, while the downward threshold magnitude $T^-(x_t)$ is weakly decreasing in $x_t$. Hence, a larger $x_t$ makes upward cascades easier to trigger and downward cascades harder to trigger.*

*Link to Valuation and Inducement* From Eq. (5.4.5), the utilization gap $\zeta$ is inversely related to the stationary user cost of capital $r + \delta$. Since, in the SQSS, the expected depreciation-adjusted marginal efficiency of investment satisfies $\mathbb{E}[\text{D-MEI}] = \mathbb{E}[\mu](r + \delta)$, higher average D-MEI corresponds to lower $\zeta$ when $\mathbb{E}[\mu]$ is fixed. Via Corollary 5.1, this has a clear implication: Higher valuation pressure makes upward cascades harder to trigger and downward cascades easier to trigger, shifting probability mass away from upward cascades toward inaction or contraction.

Appendix 5.E shows that, under the normalization $\mathbb{E}[\mu] = 1$, the utilization index $\zeta$ is a monotone decreasing function of the benchmark D-MEI, linking valuation tightness to the stationary return requirement.

The two-stage evolutionary game clarifies how confidence shocks, imitation, and adjustment frictions interact to shape macroeconomic dynamics. In the next section, we connect this framework to the broader theory of sunspot equilibria and discuss how agent-based simulations can enrich the analysis of transition paths.

## 5.5 Beyond Rational Coordination—Sunspots, ABS, and Mutants

The central aim of this book is to provide the *microfoundations of Keynesian economics*. Hartley (1997, p. 167) cautions that "(e)xplaining macroeconomics in terms of microeconomic agents is insufficient; it must be explained in terms of microeconomic agents who follow acceptable behavior." In standard usage, "acceptable" behavior is synonymous with "equilibrium behavior." Yet much of modern macroeconomics is built upon *representative-agent* models. A fundamental contradiction lies at the core of this approach: The representative agent is simultaneously the decision-maker and the aggregate environment. Any deviation she undertakes shifts the aggregate in lockstep, leaving no independent margin for "out-of-equilibrium" dynamics. Endogenous instability is thus ruled out by construction. Hartley furthermore observed, for the representative-agent method to be valid, "we would need a micro theory in which interactions among heterogeneous agents played no role in explaining behavior." This limitation motivates the search for alternative perspectives.

### 5.5.1 Sunspot Equilibria

One alternative is the theory of *sunspot equilibria* (Azariadis, 1981; Cass & Shell, 1983), in which purely extrinsic variables—random signals unrelated to fundamentals—can nonetheless coordinate beliefs and move the economy between equilibria. This framework demonstrates the fragility of rational expectations equilibria: Confidence alone can alter outcomes.

**Definition 5.2 (Sunspot Equilibrium)** Given state $(K_t, x_t)$ and a public extrinsic signal $\nu_t$ observed by all firms, a sunspot equilibrium is a measurable selection rule $\sigma : \mathbb{R} \to \Psi$ such that, for each realization of $\nu_t$, all firms choosing $\sigma(\nu_t)$ is a mutual best response in the imitation stage (yielding an economy-wide cascade action $s_t = \sigma(\nu_t)$).

**Corollary 5.5 (Unique Sunspot Equilibrium)** *Fix parameters $\delta \in (0, 1)$, $c_b \in (0, 1)$ and state $(K_t, x_t)$. Let the aggregate revision rule be as in Proposition 5.4 with thresholds $T^+(x_t) \geq 0$ and $T^-(x_t) \geq 0$. Define the selection rule $\sigma^* : \mathbb{R} \to \Psi$ by*

$$\sigma^*(\nu) = \begin{cases} +1, & \nu \geq T^+(x_t), \\ -1, & \nu \leq -T^-(x_t), \\ 0, & \textit{otherwise}. \end{cases}$$

***Proof (Sketch)*** *By Proposition 5.4, for any fixed $(K_t,\ x_t)$ and $\nu_t$, exactly one of the three actions strictly dominates (except possibly at the two knife-edge thresholds). The rule $\sigma^*$ picks that dominant action, yielding unanimous best responses (a cascade). Disjoint dominance regions imply uniqueness; at thresholds, adopt the standard status quo tiebreak (or note uniqueness a.s. when $\nu$ is atomless).* □

In our setup, the macroshift parameter $\nu_t$ serves as a public, extrinsic signal (a sunspot) that acts as a focal device for coordination: It selects a unique cascade action at each state. A deeper puzzle remains, however: Why should otherwise rational agents coordinate on a causally irrelevant signal? As emphasized by Kandori (1997), rationality alone does not justify equilibrium selection; additional elements—such as perturbations, dynamics, or institutions—are needed to explain which equilibrium prevails.

In the framework developed here, those elements are provided by the entry of mutants—firms that deviate from the incumbent replacement strategy. Mutants act as evolutionary perturbations that may, under certain thresholds, shift the economy toward a new focal point. In this way, the model supplies a microfoundation for coordination failures and self-fulfilling transitions that the pure sunspot approach leaves unexplained.

### *5.5.2 Agent-Based Simulation*

A second alternative is *agent-based simulation* (ABS). Unlike representative-agent models, ABS separates the decision-maker from her environment. Each firm interacts with an environment shaped by others, yet retains the capacity to deviate idiosyncratically. If those deviations prove profitable, others imitate them. Ripples of behavior overlap and amplify, producing aggregate booms or busts.

This perspective powerfully captures how endogenous fluctuations can emerge from local interactions without reliance on arbitrary extrinsic signals. However, ABS typically comes at the cost of analytical transparency: The dynamics are often too complex to admit sharp, closed-form explanations. For instance, an ABS model might illuminate how inflation or deflation emerges from firm-level interactions, but isolating which actions trigger a recovery from depression becomes elusive once all agents are densely interconnected.

### *5.5.3 Mutants as a Middle Ground*

The framework developed in this book occupies a middle ground between sunspot theory and ABS. Takahashi's "Artificial Wicksell–Keynes Macroeconomy" constructs a fully closed economy of monopolistically competitive firms, where aggregate demand is distributed with idiosyncratic variation across firms. Relative prices influence allocations, but randomness ensures that no knife-edge exactness holds. Some firms receive unexpectedly high orders, interpret these as signals of rising demand, and invest. Their suppliers then expand in turn. Small ripples interact and reinforce, culminating in aggregate booms; conversely, pessimistic shocks can propagate into collective downturns.

This mutant–incumbent framework distills the essential ABS insight—that local deviations and imitation can trigger macrolevel shifts—into a tractable and analytically rigorous model. With only two agent types, incumbents and mutants, it isolates the thresholds at which minor perturbations tip the system as a whole. It thereby avoids the implausibility of sunspot coordination and the intractability of full-scale simulation, while preserving the core intuition: Fluctuations are endogenous, arising from the fragile interplay of confidence, imitation, and the fixity of capital.

### 5.5.4 Sunspot Interpretation of $v_t$

Within this framework, the macrosignal $\nu_t$ may still be read as a *sunspot*: an extrinsic random variable with no effect on fundamentals, yet capable of coordinating expectations once firms treat it as a salient cue.

Then $\sigma^*$ is a sunspot equilibrium and is *unique* for each realization of $\nu$ (unique almost surely if $\nu$ has a continuous distribution). Consequently, the two-stage game (mutation then imitation) admits a unique cascade outcome $s_t = \sigma^*(\nu_t) \in \{-1, 0, +1\}$ for every $\nu_t$.

**Remark 5.5 (Sunspot Interpretation)** The macro signal $\nu_t$ can be interpreted as a "sunspot": An extrinsic variable that leaves fundamentals unchanged yet coordinates expectations. In our framework, a positive $\nu_t$ makes optimistic mutants focal; a negative $\nu_t$ makes pessimistic mutants focal. The evolutionary thresholds provide the microfoundations for how such signals influence aggregate outcomes.

### 5.5.5 Transition Dynamics and ABS

The theory here emphasizes conditions for regime shifts between quasi-steady states (QSSEs), not the fine-grained transition paths. Modeling the detailed dynamics of prices, capital utilization, or heterogeneous adjustment would require a richer ABS framework. Such an approach was pursued in Takahashi (2021), where stochastic demand shocks at the firm level generated ripple effects that could scale into macrobooms or busts. Those simulations, while not cast in mutation–imitation terms, captured a similar amplification mechanism.

In this book, by contrast, the focus remains on analytical clarity. The transition phase is acknowledged as conceptually important, but abstracted to preserve tractability.

### 5.5.6 Concluding Perspective

To conclude, this chapter shows how the entry of mutants beyond a threshold can shift the SQSS equilibrium to a new level. Repeated over time, such shifts generate evolutionary dynamics that recall the work of Young (1993) and Kandori et al. (1993) on equilibrium selection. They also connect to Gabaix (2014) theory of sparse rationality, in which decision-makers allocate limited attention across many possible actions. In our setting, firms' bounded rationality in coordinating investment magnifies the role of such evolutionary shifts. The next chapter builds on this foundation to develop a macroeconomic model of evolutionary dynamics.

# Appendix

## 5.A Notation

**Table 5.4** Notation (investment theory)

| Symbol | Meaning | Units/notes |
|---|---|---|
| *Stocks, flows, and rates* | | |
| $K$ | Installed capital stock (incumbent baseline) | Capital units |
| $\delta$ | Replacement ratio (depreciation rate per period) | Fraction per period |
| $I_i$ | Gross investment of firm $i$ | Capital units |
| $z := I_i - \delta K$ | Net investment (relative to replacement) | Capital units |
| $K_{i,t+1} = K + z_i$ | One-period gestation law of motion | Capital units |
| $c$ | Consumption–capital intensity in SQSS (from IS) | Fraction of $K$ |
| $r$ | Real interest rate | Per period |
| *Shocks and beliefs* | | |
| $\nu$ | Aggregate "macro shift" (sentiment/fundamental) | Centered, uniform on $[-V, V]$ |
| $\epsilon_i$ | Idiosyncratic confidence shock for firm $i$ | Uniform on $[\,1 - c_b,\ 1 + c_b\,]$ |
| $\mu_i = \nu + \epsilon_i$ | Firm-level confidence index | Dimensionless |
| $\zeta$ | Excess-demand utilization index | $\zeta = (K^d - K)/(\delta K)$ |
| *Choices and strategy sets* | | |
| $Z = \{-\delta K,\ 0,\ \delta K\}$ | Discrete net investment actions (halt, replace, double) | Capital units |
| $\Phi^z = \{-1, 0, +1\}$ | Normalized action set $(-1, 0, +1)$ | Direction only |
| $\phi_i^z \in \Phi^z$ | Firm $i$'s normalized action in imitation stage | |
| $\tilde{K}(+) = \left[1{+}(1{-}c_b)\delta\right]K$ | Installed capacity after expansion | Capital units |
| $\tilde{K}(-) = \left[1{-}(1{+}c_b)\delta\right]K$ | Installed capacity after contraction | Capital units |
| *Prices and profitability* | | |
| $P^K$ | Demand price of capital (present value of gross profits) | Price per capital |
| $\widehat{P}^K(r, \mu)$ | Risk/attitude-adjusted capital price | Equals $P\,\mu$ in reduced form |
| $P^S$ | Supply price (commitment) of new capital | Price per capital |
| $\varphi := P^S/P^K$ | Investment cost in demand-price units | Dimensionless |
| $(1 - \varphi)\delta K$ | Profitability margin (in capital units) for deviating from replacement | Capital units |

(continued)

**Table 5.4** (continued)

| Symbol | Meaning | Units/notes |
|---|---|---|
| *Technology and costs* | | |
| $\phi(z)$ | Penrose–Uzawa adjustment cost (piecewise, with kinks at $z = \pm\delta K$ and a discontinuity at $z = 0$) | Capital units |
| $c_b$ | Behavioral/organizational wedge in adjustment | Dimensionless |
| $\alpha \in (0, 1)$ | Capital share in production | |
| $\kappa = N/K$ | Capital–labor ratio (from cost minimization) | |
| *Demand, operation, and utilization* | | |
| $K^d(\epsilon^{\pm})$ | Demand-required capital (SQSS requirement) | $K^d = (1 + \delta\zeta \pm \delta\epsilon^{\pm})K$ |
| $K^{\text{op}}(\pm, \epsilon^{\pm})$ | Operating capital (short side) | $\min\{\tilde{K}(a), K^d(\epsilon^{\pm})\}$ |
| $X^d$ | Aggregate demand (goods) | $X^d = (c + \delta) K$ in SQSS |
| *Mutant shares and thresholds* | | |
| $\epsilon^+, \epsilon^-$ | Shares of optimistic/pessimistic mutants | in $[0, 1]$ |
| $\hat{\epsilon}^+(\zeta)$ | Expansion threshold | $\max\{0, \varphi - \zeta\}$ (if $\varphi \leq 1 - c_b$) |
| $\hat{\epsilon}^-(\zeta)$ | Contraction threshold | $\max\{0, 1 + c_b + \zeta - \varphi\}$ |
| $\epsilon^{\pm} \geq \hat{\epsilon}^{\pm}(\zeta)$ | Tipping condition in imitation stage | Triggers switch to new SQSS |

## 5.B Derivation of a Penrose–Uzawa Adjustment Cost

This appendix constructs a Penrose–Uzawa style adjustment-cost function that rationalizes the discrete investment choices used in the main text. The cost captures both behavioral frictions—stemming from status quo bias—and standard organizational adjustment costs. Together, these features generate a jump at replacement investment and kinks at $\pm\delta K$, implying corner optimality.

If a firm intends to alter the current operational scale, it must incur additional costs due to behavioral frictions—the sum of costs to overcome status quo bias and standard investment adjustment costs. As stated in Assumption 5.2, we model the real cost induced by status quo bias as a lump-sum independent of $z$ but proportional to the current capital scale, $C_B = c_b\,\delta K$ (with $c_b \in (0, 1)$).[12]

Adjustment costs are assumed to become prohibitive once the change in scale exceeds a threshold. For expositional simplicity, we place the threshold at $z = \delta K$; hence, firms cannot practically choose $z > \delta K$. Because capital is (short-run) fixed

[12] This is plausible: Larger firms incur greater managerial/coordination costs when changing scale.

and illiquid, the resale value of installed equipment is zero, so negative investment does not lower costs beyond forgoing replacement.

Collecting these assumptions, the present value cost function $\phi(z)$ (in the same units as $P$) can be represented by three line segments:

$$\phi(z) = \begin{cases} P(-\delta K + C_B), & z < -\delta K, \\ P(z + C_B), & -\delta K \le z < 0, \\ 0, & z = 0, \\ P(z + C_B), & 0 < z \le \delta K, \\ \infty, & z > \delta K. \end{cases} \tag{5.B.1}$$

(The special case $\phi(0) = 0$ reflects that no adjustment is costless; otherwise, the status quo bias component $C_B$ applies whenever $z \neq 0$.)

Our adjustment-cost specification $\phi(z)$ can be viewed as a Penrose-type function in the sense of Uzawa (1968), adapted to the investment-threshold environment studied here. In line with Penrose (1959) managerial-capacity argument, costs are low for small scale changes but rise rapidly as the magnitude of adjustment increases, and become effectively prohibitive at a finite capacity limit (here, $z = \delta K$). Relative to the smooth, strictly convex forms often used in the Uzawa tradition, we adopt a parsimonious piecewise-linear representation that preserves convexity and delivers transparent threshold behavior. Two departures are deliberate: (i) a lump-sum term capturing status quo bias, which makes any nonzero adjustment costly even for small $|z|$; and (ii) asymmetry on the disinvestment side, reflecting irreversibility and the absence of resale value, which flattens $\phi(z)$ for $z < -\delta K$. These features retain the Penrose intuition-organizational limits that steepen marginal adjustment costs—while tailoring the function to behavioral frictions and capital fixity that are central to our setting.

## 5.C Proof of Lemma 5.1 (Discrete Investment Choices in the Mutation Stage)

The reduced-form threshold rule used in the mutation stage is motivated by three auxiliary assumptions.

**Adjustment-cost structure.** Investment is subject to a Penrose–Uzawa style adjustment-cost schedule, with a discrete jump at replacement and kinks at $z_i = \pm\delta K$.

**Contemporaneous payoff neutrality.** Although aggregate net investment $\sum_{i\in[0,1]} z_i$ shifts goods demand at time $t$, fixed productive capacity and a common labor input imply that a firm's period-$t$ profit is independent of its own $z_i$.

**Gestation and forward-looking payoff evaluation.** Investment chosen at $t$ affects productive capacity only at $t + 1$. Under a common forward labor contract and limited-attention/finite-planning considerations (cf. Woodford 2019), firms therefore evaluate payoffs from $t + 1$ onward, implying that the relevant comparison is across next-period SQSS outcomes.

***Proof*** Let $z := I_i - \delta K \geq -\delta K$. From Eq. (5.2.5), the risk-adjusted present value of gross profits from $z$ is

$$\Delta\Pi^g(z, \mu) = \widehat{P}^K(r, \mu)\, z = P\, \mu\, z.$$

The net present value is

$$\Delta\Pi(z, \mu) = P\, \mu\, z - \phi(z),$$

with $\phi$ satisfying Assumption A: (i) convexity on each side of 0, (ii) an activation jump $F > 0$ at $z \neq 0$, and (iii) higher marginal cost beyond $|z| = \delta K$.

Define $\mu_L$ and $\mu_H$ by the corner indifference conditions

$$\Delta\Pi(-\delta K, \mu_L) = 0, \qquad \Delta\Pi(\delta K, \mu_H) = 0.$$

Using (5.B.1), one obtains

$$\Delta\Pi(-\delta K, \mu) = P\, \delta K\, (1 - \mu - c_b) \;\Rightarrow\; \mu_L = 1 - c_b,$$

$$\Delta\Pi(\delta K, \mu) = P\, \delta K\, (\mu - 1 - c_b) \;\Rightarrow\; \mu_H = 1 + c_b.$$

Hence for $\mu < \mu_L$, the best corner is $z = -\delta K$, and for $\mu > \mu_H$, the best corner is $z = +\delta K$.

For $\mu \in [\mu_L, \mu_H]$, we have $\Delta\Pi(\delta K, \mu) \leq 0$ and $\Delta\Pi(-\delta K, \mu) \leq 0$. On $(0, \delta K]$ (resp. $[-\delta K, 0)$) the function $\Delta\Pi(\cdot, \mu)$ is concave, and $\lim_{z \to 0^\pm} \Delta\Pi(z, \mu) = -F \leq 0$, while $\Delta\Pi(0, \mu) = 0$. Hence on each side, the maximum over the closed interval is attained at $z = 0$, and any $z \neq 0$ yields strictly lower payoff. Therefore $z = 0$ is optimal for all $\mu \in [\mu_L, \mu_H]$.

Combining the three regions delivers the threshold rule Eq. (5.4.1). □

## 5.D Proof of Lemma 5.3 (Upward and Downward Tipping Thresholds)

***Proof of Lemma 5.3*** By Assumption 5.2, the organizational wedge reduces next-period installed capacity (in capital units) to

$$\tilde{K}(z) = K + z - c_b|z|, \qquad z \in \{-\delta K,\ 0,\ +\delta K\}.$$

Given public signals $(\epsilon^{\pm}, \zeta)$, demand-required capital is

$$K^d(\epsilon^{\pm}) = (1+\delta\zeta)K \pm \delta\epsilon^{\pm}K.$$

Operating capacity obeys the short-side rule,

$$K^{\text{op}}(z, \epsilon^{\pm}) = \min\{\tilde{K}(z),\ K^d(\epsilon^{\pm})\}. \tag{5.D.1}$$

Both the demand price of capital $P^K(1)$, evaluated at neutral confidence $\mu = 1$, and the supply price $P^S$ are measured in period-$(t+1)$ value units; see Fig. 5.2. An increase in investment, $z > 0$, entails an additional outlay, whereas a decrease, $z < 0$, generates a saving. The net investment outlay is therefore represented by $P^S z$. Hence, the firm chooses $z$ to maximize period-$(t{+}1)$ value net of investment outlays:

$$z^*(\epsilon^{\pm}, \zeta) \in \arg \max_{z \in \{-\delta K, 0, +\delta K\}} \Big\{ P^K(1)\, K^{\text{op}}(z, \epsilon^{\pm}) - P^S z \Big\}, \tag{5.D.2}$$

with the convention that $z = 0$ is selected when the argmax is not unique. Define the price wedge

$$\varphi := \frac{P^S}{P^K(1)} > 0,$$

so dividing (5.D.2) by $P^K(1)$ yields the equivalent criterion

$$z^*(\epsilon^{\pm}, \zeta) \in \arg \max_{z \in \{-\delta K, 0, +\delta K\}} \Big\{ K^{\text{op}}(z, \epsilon^{\pm}) - \varphi z \Big\}.$$

Using (5.2.1) at $\mu = 1$ gives

$$\varphi = \frac{\alpha(\eta - 1)}{\alpha(\eta - 1) + 1}.$$

**Upward Threshold** Let $\epsilon \geq 0$ denote the share of optimistic mutants and write

$$k_+^d(\epsilon) := \frac{K_+^d(\epsilon)}{K} = 1 + \delta(\zeta_t + \epsilon),$$

where $\zeta_t$ is the utilization gap at time $t$. For brevity, write $K^{\text{op}}(+, \epsilon) := K^{\text{op}}(+\delta K, \epsilon)$ and $K^{\text{op}}(0, \epsilon) := K^{\text{op}}(0, \epsilon)$. Then expansion weakly dominates inaction iff

$$K^{\text{op}}(+, \epsilon) - \varphi\delta K \geq K^{\text{op}}(0, \epsilon),$$

i.e.,

$$\Delta_+(\epsilon) := \frac{K^{\text{op}}(+, \epsilon) - K^{\text{op}}(0, \epsilon)}{K} \geq \varphi\delta.$$

Applying the short-side rule yields

$$\Delta_+(\epsilon) = \begin{cases} 0, & k_+^d(\epsilon) \leq 1, \\ k_+^d(\epsilon) - 1, & 1 < k_+^d(\epsilon) \leq 1 + (1 - c_b)\delta, \\ (1 - c_b)\delta, & k_+^d(\epsilon) > 1 + (1 - c_b)\delta, \end{cases}$$

equivalently,

$$\Delta_+(\epsilon) = \begin{cases} 0, & \zeta_t + \epsilon \leq 0, \\ \delta(\zeta_t + \epsilon), & 0 < \zeta_t + \epsilon \leq 1 - c_b, \\ (1 - c_b)\delta, & \zeta_t + \epsilon > 1 - c_b. \end{cases}$$

Hence, $\Delta_+(\epsilon) \geq \varphi\delta$ has a solution iff $\varphi \leq 1 - c_b$, and in that case the smallest solution is $\hat{\epsilon}^+(\zeta_t) = \max\{0, \varphi - \zeta_t\}$.

**Downward Threshold** Let $\epsilon \geq 0$ denote the share of pessimistic mutants and write

$$k_-^d(\epsilon) := \frac{K_-^d(\epsilon)}{K} = 1 + \delta(\zeta_t - \epsilon).$$

Contraction weakly dominates inaction iff

$$K^{\text{op}}(-, \epsilon) + \varphi\delta K \geq K^{\text{op}}(0, \epsilon),$$

i.e.,

$$\Delta_-(\epsilon) := \frac{K^{\text{op}}(-, \epsilon) - K^{\text{op}}(0, \epsilon)}{K} \geq -\varphi\delta.$$

Evaluating $K^{\text{op}}(-, \epsilon)$ under the short-side rule yields the following piecewise expression for $\Delta_-(\epsilon)$, from which solving $\Delta_-(\epsilon) \geq -\varphi\delta$ gives

$$\hat{\epsilon}^-(\zeta_t) = \max\{0,\ 1 + c_b + \zeta_t - \varphi\},$$

which is feasible only if $\hat{\epsilon}^-(\zeta_t) \leq 1$, so that the required mass of pessimistic mutants does not exceed the population.

□

## 5.E Link Between Utilization and the Benchmark D-MEI

**Lemma 5.1 (Monotone $\zeta$–D-MEI Link)** *In the stationary benchmark, suppose firms evaluate marginal investment projects using a perceived internal discount rate $\hat{\rho}$ and define*

$$\textit{D-MEI} \;\equiv\; \hat{\rho} + \delta.$$

*Under the normalization $\mathbb{E}[\mu] = 1$, there exists a constant $\bar{\kappa} > 0$ such that*

$$1 + \delta\zeta \;=\; \frac{\bar{\kappa}}{\mathbb{E}[\,\textit{D-MEI}]},$$

*and hence*

$$\frac{\partial \zeta}{\partial\, \mathbb{E}[\,\textit{D-MEI}]} \;<\; 0.$$

***Proof*** From Eq. (5.4.5), there exists a scale constant $\kappa_0 > 0$ such that

$$1 + \delta\zeta \;=\; \frac{\kappa_0}{r + \delta}.$$

In the stationary benchmark, the perceived norm of discounting coincides on average with the economy-wide financing and profit benchmark, so that

$$\mathbb{E}[\,\text{D-MEI}] = \mathbb{E}[\hat{\rho} + \delta] = \mathbb{E}[\mu]\,(r + \delta).$$

Under the maintained normalization $\mathbb{E}[\mu] = 1$, this reduces to

$$\mathbb{E}[\,\text{D-MEI}] = r + \delta.$$

Substituting into the first expression yields

$$1 + \delta\zeta = \frac{\kappa_0}{\mathbb{E}[\,\text{D-MEI}]} \equiv \frac{\bar{\kappa}}{\mathbb{E}[\,\text{D-MEI}]},$$

where $\bar{\kappa} := \kappa_0 > 0$. Since the right-hand side is strictly decreasing in $\mathbb{E}[\,\text{D-MEI}]$, it follows that $\partial\zeta/\partial\,\mathbb{E}[\,\text{D-MEI}] < 0$. □

## References

Anderson, L. R., & Holt, C. A. (1997). Information cascades in the laboratory. *American Economic Review, 87*(5), 847–862.

Azariadis, C. (1981). Self-fulfilling prophecies. *Journal of Economic Theory 25*(3), 380–396. https://doi.org/10.1016/0022-0531(81)90020-5

Banerjee, A. V. (1992). A simple model of herd behavior. *Quarterly Journal of Economics, 107*(3), 797–817.
Bikhchandani, S., Hirshleifer, D., & Welch, I. (1992). A theory of fads, fashion, custom, and cultural change as informational cascades. *Journal of Political Economy, 100*(5), 992–1026.
Blanchard, O. J., & Summers, L. H. (1986). Hysteresis and the European unemployment problem. In S. Fischer (Ed.), *NBER macroeconomics annual* (pp. 15–78). MIT Press. https://doi.org/10.3386/w1950
Caballero, R. J., & Engel, E. M. (1999). Explaining investment dynamics in us manufacturing: A generalized (S, s) approach. *Econometrica, 67*(4), 783–826. https://doi.org/10.1111/1468-0262.00053
Cass, D., & Shell, K. (1983) Do sunspots matter? *Journal of Political Economy, 91*(2), 193–227. https://doi.org/10.1086/261139
Chamley, C. (2004). *Rational herds: Economic models of social learning*. Cambridge University Press. https://doi.org/10.1017/CBO9780511616372
Chamley, C., & Gale, D. (1994). Information revelation and strategic delay in a model of investment. *Econometrica, 62*(5), 1065–1085.
Cooper, R., & John, A. (1988). Coordinating coordination failures in Keynesian models. *The Quarterly Journal of Economics, 103*(3), 441–463. https://doi.org/10.2307/1885539
Crawford, V. P., & Sobel J (1982), Strategic information transmission. *Econometrica, 50*(6), 1431–1451. https://doi.org/10.2307/1913390
Doms, M., & Dunne, T. (1998). Capital adjustment patterns in manufacturing plants. *Review of Economic Dynamics, 1*(2), 409–429. https://doi.org/10.1006/redy.1998.0011
Ellison, G.,& Fudenberg, D. (1993). Rules of thumb for social learning. *Journal of Political Economy, 101*(4), 612–643. https://doi.org/10.1086/261890
Fehr, E., Kirchsteiger, G., & Riedl, A. (1993). Does fairness prevent market clearing? An experimental investigation. *The Quarterly Journal of Economics, 108*(2), 437–459. https://doi.org/10.2307/2118338
Fisher, I. (1930). *The theory of interest: As determined by impatience to spend income and opportunity to invest it*. Macmillan
Gabaix, X. (2014). A sparsity-based model of bounded rationality. *The Quarterly Journal of Economics, 129*(4), 1661–1710. https://doi.org/10.1093/qje/qju024
Harsanyi, J. C. (1975). Can the maximin principle serve as a basis for morality? A critique of John Rawls's theory. *American Political Science Review, 69*(2), 594–606. https://doi.org/10.2307/1959090
Hartley, J. E. (1997). *The representative agent in macroeconomics*. Routledge.
Kahneman, D., & Tversky, A. (1982). The psychology of preferences. *Scientific American, 246*(1), 160–173. https://doi.org/10.1038/scientificamerican0182-160
Kahneman, D., Knetsch, J. L., & Thaler, R. (1986). Fairness as a constraint on profit seeking: Entitlements in the market. *The American Economic Review, 76*(4), 728–741. https://doi.org/10.2307/j.ctvcm4j8j.13
Kandori, M. (1997). *Evolutionary game theory in economics* (pp. 243–277). Cambridge University Press. https://doi.org/10.1017/CCOL521580110.008
Kandori, M., Mailath, G. J., & Rob, R. (1993). Learning, mutation, and long run equilibria in games. *Econometrica: Journal of the Econometric Society* 29–56. https://doi.org/10.2307/2951777
Keynes, J. M. (1936). *The general theory of employment, interest and money*. Macmillan.
Keynes, J. M. (1937). The general theory of employment. *The Quarterly Journal of Economics, 51*(2), 209–223. https://doi.org/10.2307/1882087
Keynes, J. M. (2013). *A treatise on probability*. Courier Corporation.
Kirman, P., & Kirman, A. P. (1992). Whom or what does the representative individual represent. *The Journal of Economic Perspectives, 6*(2), 117–136. https://doi.org/10.1257/jep.6.2.117
Lerner, A. P. (1944). The relation of wage policies and price policies. *The Review of Economic Studies, 12*(1), 54–65. https://doi.org/10.2307/2296100

Milgrom, P., & Roberts, J. (1982). Limit pricing and entry under incomplete information: An equilibrium analysis. *Econometrica, 50*(2), 443–459.
Minsky, H. P. (1975). *John Maynard Keynes*. Columbia University Press.
Penrose, E. T. (1959). *The theory of the growth of the firm*. Basil Blackwell.
Samuelson, W., & Zeckhauser, R. (1988). Status quo bias in decision making. *Journal of Risk and Uncertainty, 1*(1), 7–59. https://doi.org/10.1007/bf00055564
Selten, R. (1975). Reexamination of the perfectness concept for equilibrium points in extensive games. *International Journal of Game Theory, 4*, 25–55. https://api.semanticscholar.org/CorpusID:119759202
Simon, H. A. (1947). *Administrative behavior: A study of decision-making processes in administrative organization*. Macmillan.
Swinkels, J. M. (1992). Evolutionary stability with equilibrium entrants. *Journal of Economic Theory, 57*(2), 306–332. https://doi.org/10.1016/0022-0531(92)90038-j
Takahashi, I. (2021). *An Artificial Wicksell–Keynes Macroeconomy: Integrating business cycle and cumulative process*. Springer. https://doi.org/10.1007/978-981-16-6839-5
Uzawa, H. (1968). The Penrose effect and optimum growth. *The Economic Studies Quarterly (Tokyo 1950), 19*(1), 1–14. https://doi.org/10.1017/cbo9780511664496.017
Uzawa, H. (1969). Time preference and the Penrose effect in a two-class model of economic growth. *The Journal of Political Economy, 77*(4), 628–652. https://doi.org/10.1086/259554
Woodford, M. (2019). Monetary policy analysis when planning horizons are finite. *NBER Macroeconomics Annual 33*(1), 1–50. https://doi.org/10.1086/700892
Young, H. P. (1993). The evolution of conventions. *Econometrica: Journal of the Econometric Society* 57–84. https://doi.org/10.2307/2951778

# Chapter 6
# Stochastic Dynamics and Ergodic Long-Run Equilibrium

*It is an outstanding characteristic of the economic system in which we live that while it is subject to severe fluctuations in respect of output and employment, it is not violently unstable. Fluctuations tend to wear themselves out before proceeding to extremes and eventually to reverse themselves.*

— John Maynard Keynes, *The General Theory*, Ch. 22

## 6.1 Introduction

This chapter formalizes the stochastic dynamics generated by the evolutionary investment mechanism developed in the previous chapter, treating the capital stock as a Markov process whose long-run behavior is characterized by its invariant distribution. A key stabilizing force in the stochastic dynamics of capital adjustment arises from the fact that the desired capital stock grows less than proportionally with the installed stock. Economically, as firms accumulate capital, existing productive capacity becomes more fully utilized, so desired replacement and expansion grow less than proportionally with $K$. This behavior is captured by the excess replacement ratio $\zeta(K)$, which measures the extent of over- or underutilization of productive capacity and is strictly decreasing in $K$. As $\zeta(K)$ declines, upward investment adjustments become less likely, while downward adjustments become more frequent. Large capital stocks therefore exhibit negative expected drift, whereas sufficiently small stocks experience positive drift. The chapter formalizes this stability mechanism through the elasticity condition $\varepsilon_{K^d,K} < 1$ and the monotonicity of $\zeta(K)$, a key structural property established in Appendix 6.B.1. It then shows how these properties, combined with a Lyapunov analysis, imply global ergodicity and a unique invariant distribution for the capital-stock process.

The results established in this chapter provide the probabilistic foundation for the analysis of stochastic long-run equilibrium developed in the following chapters, where the invariant distribution of the capital-stock process becomes the central object of study for macroeconomic stability and comparative statics.

 
I. Takahashi, *Microfoundations of Keynesian Economics*, Evolutionary Economics and Social Complexity Science 37, https://doi.org/10.1007/978-981-95-9390-3_6

The remainder of the chapter proceeds as follows. Section 6.2 introduces the branch probabilities governing upward and downward adjustments of the capital stock. Section 6.3 develops the evolutionary adjustment mechanism, describing how mutation events, profit comparisons, and industry-level heterogeneity generate stochastic movements in aggregate capital. Section 6.4 then establishes the probabilistic foundations of the model by defining the Markov transition kernel and verifying the Feller property, which ensures the existence of an invariant distribution. Section 6.5 strengthens these results by proving irreducibility and regeneration properties that imply the uniqueness of the invariant distribution. Section 6.6 applies the Foster–Lyapunov criterion to characterize the drift structure of the process and to identify the neutral point governing long-run dynamics. Section 6.7 concludes by establishing ergodicity and describing the resulting stochastic long-run equilibrium.

**Standing Assumption**
Throughout this chapter, we impose the exponent-matching restriction

$$\beta = \alpha/(1 - \alpha).$$

This assumption ensures that the induced functions $r(K)$ and $\zeta(K)$ are globally monotone, which is essential for the stochastic stability analysis below, including the drift and irreducibility arguments.

### *6.1.1 Dynamics of Capital Adjustment*

This chapter provides the formal stochastic dynamics underlying the adjustment mechanism introduced in Chap. 5. To avoid repetition, we do not restate the economic narrative here; instead, we define the Markov transition kernel for $K$, establish its regularity properties, and characterize the invariant distribution and ergodic long-run behavior.

### *6.1.2 Microindustry–Macrolayers*

To formalize these dynamics, we construct a model with three interconnected layers linking microeconomic behavior to aggregate outcomes. At the microlevel, individual firms evaluate investment returns under idiosyncratic sentiment shocks, reflecting bounded rationality and limited attention. At the industry level, these heterogeneous signals are aggregated into a uniform random disturbance that determines whether investment expands or contracts within each sector. At the macrolevel, the distribution of sectoral outcomes produces a common confidence index governing the evolution of aggregate capital.

This layered structure provides an analytically tractable way to show how dispersed microshocks generate coordinated macrofluctuations. It also highlights an important distinction: With a single industry, aggregate investment adjusts discontinuously once macrosentiment crosses a threshold; with many industries, sectoral heterogeneity creates a continuum of threshold crossings that smooth the adjustment probabilities $p_{\pm}(K)$. The industry layer thus transforms a binary, threshold-driven response into a smooth and probabilistic evolution of aggregate capital.

### 6.1.3 Evolutionary Adjustment and Investment Mutants

The remainder of the chapter embeds this micro-industry– macrostructure within an evolutionary framework in which investment strategies diffuse only when they outperform incumbents. Following Young (1993) and Kandori et al. (1993), mutants enter sporadically and trigger transitions between SQSS positions once their profitability crosses the relevant selection threshold. The resulting capital-stock process is a piecewise-deterministic Markov process (PDMP): It evolves deterministically between investment opportunities and experiences discrete jumps when selection events occur. This perspective clarifies the economic interpretation of the stochastic structure: Macrofluctuations arise from continual rebalancing among heterogeneous investment strategies rather than from undifferentiated aggregate shocks.

The following sections develop these dynamics formally, specify the branch probabilities governing upward and downward investment transitions, and derive the stochastic evolution of the capital stock that forms the basis for the ergodic analysis to come.

### 6.1.4 Real-Term Accounting and Price-Setting Convention

Before formalizing the stochastic transition mechanism, we briefly describe the price-setting and accounting conventions that allow the analysis to be conducted entirely in real terms.

Within each period, nominal prices are preset, so the intra-period inflation rate is zero. Between periods, however, the price level $P_t$ may adjust in response to aggregate conditions. To avoid holding-gain complications arising from such inter-period adjustments, we adopt a one-period rollover convention and record all flows in real terms. Under 100% reserves and zero bank equity, funding equals loans each period, which implies that household net worth coincides with the capital stock: $NW_{H,t} = K_t$. These conventions allow the stochastic evolution of $K_t$ to be analyzed entirely in real terms, simplifying the construction of the transition kernel used in the remainder of the chapter.

## 6.2 Branch Probabilities

Investment adjustments occur through discrete upward or downward innovations triggered by the entry and selection of mutants. For each capital level $K$, we denote by $p_+(K)$ and $p_-(K)$ the probabilities that the next innovation is expansionary or contractionary, respectively. These probabilities summarize the aggregation of idiosyncratic firm-level sentiment, sectoral disturbances, and the macrolevel confidence index derived from the SQSS environment.

**Definition 6.1 (Branch Probabilities)** For each state $K \in \mathcal{K} := [0, \bar{K}]$, the probabilities of upward and downward innovations satisfy

$$p_+(K) \geq 0, \qquad p_-(K) \geq 0, \qquad p_+(K) + p_-(K) \leq 1.$$

The residual probability

$$p_0(K) \;:=\; 1 - p_+(K) - p_-(K)$$

corresponds to a period with no aggregate jump in the capital stock.

Later, when applying the Foster–Lyapunov criterion, we will restrict attention to the subset $S := [1, \bar{K}] \subset \mathcal{K}$.

The presence of the no-jump mass $p_0(K)$ reflects industry-level inertia: Although individual sectors experience idiosyncratic shocks each period, their effects need not aggregate into a sufficiently strong signal to overturn the incumbent SQSS position. Only when the confidence-weighted profitability differential crosses the relevant threshold does an economy-wide upward or downward innovation occur.

In the multi-industry economy, heterogeneity across sectors smooths the aggregate adjustment probabilities: Even if the underlying profitability signal changes continuously, only a fraction of industries cross their thresholds in any given period, producing continuous variations in $p_\pm(K)$. In contrast, in a single-industry economy, threshold crossings generate discontinuous, binary transitions, and $p_\pm(K)$ collapse to step functions. The multi-industry formulation thus provides both greater realism and analytical tractability.

Later we specify the functional form of $p_\pm(K)$ as clipped-affine functions of the excess replacement ratio $\zeta(K)$. This structure captures the stabilizing mechanism described above: As $K$ rises and $\zeta(K)$ falls, upward innovations become less likely and downward innovations more likely, ensuring the negative drift required for the ergodicity results established in Sect. 6.7.

## 6.3 Evolutionary Adjustment of Capital

Throughout the analysis, the feasible set of symmetric quasi-steady states is compact. Recall $\mathcal{K} = [0, \bar{K}]$. This upper bound reflects feasibility constraints arising from labor endowment and the PM–LM tangency ceiling (see Lemma 6.2

in Appendix 6.B). Capital accumulation thus evolves within a bounded region, but the path taken within this region is shaped by an evolutionary selection process rather than by continuous optimization.

### 6.3.1 Selection-Based Capital Dynamics

Firms do not choose investment levels by solving an intertemporal optimization problem. Instead, capital evolves through an evolutionary mechanism: Investment strategies that yield higher profitability than the incumbent level diffuse through imitation, while inferior strategies vanish. If a capital scale proposed by a mutant type earns a higher net return than the incumbent scale, the mutant attracts followers and the aggregate capital stock increases in the next period. Conversely, if a pessimistic mutant proposing a lower capital scale achieves a higher return, the aggregate stock contracts. When neither mutant beats the incumbent, the capital stock remains unchanged.

This evolutionary adjustment mechanism aligns with the spirit of Young (1993) and Kandori et al. (1993), where transition events occur only when a sufficiently profitable deviation arises. Here, the deviations correspond to discrete up- or down-scaling of the capital stock.

The detailed economic foundations of this mechanism are developed in Chap. 5; here we retain only the induced transition logic needed to study the stochastic evolution of capital.

### 6.3.2 Mutation Events and Profit Comparisons

In each period, at most one mutant type enters: either an *optimistic* mutant proposing a higher capital scale or a *pessimistic* mutant proposing a lower scale. Let $\pi^{\text{inc}}(K)$ denote the incumbent profit rate at the SQSS associated with $K$, and let $\pi^{\text{opt}}(K)$ and $\pi^{\text{pes}}(K)$ denote the mutant profit rates when evaluated at the same price system $(r^*(K), \omega^*(K))$.

The aggregate capital change in period $t$ is therefore

$$\Delta K = \begin{cases} +\Delta K_{\text{opt}} > 0, & \text{if an optimistic mutant enters and } \pi^{\text{opt}}(K) > \pi^{\text{inc}}(K), \\ -\Delta K_{\text{pes}} < 0, & \text{if a pessimistic mutant enters and } \pi^{\text{pes}}(K) > \pi^{\text{inc}}(K), \\ 0, & \text{otherwise.} \end{cases} \tag{6.3.1}$$

Equivalently,

$$\Delta K_t = \mathbf{1}_{\{\text{opt enters}\}} \mathbf{1}_{\{\pi^{\text{opt}} > \pi^{\text{inc}}\}} \Delta K_{\text{opt}} - \mathbf{1}_{\{\text{pes enters}\}} \mathbf{1}_{\{\pi^{\text{pes}} > \pi^{\text{inc}}\}} \Delta K_{\text{pes}}.$$

Thus $K$ rises only when an expansionary mutant outperforms the incumbent and falls only when a contractionary mutant yields a higher return.

### 6.3.3 Sectoral Heterogeneity and Smoothing of Transitions

Confidence shocks and idiosyncratic disturbances influence profitability through the inducement margin of the MEI. With many industries, local differences in profitability produce a continuum of threshold crossings each period, which smooths the adjustment probabilities $p_{\pm}(K)$ and the magnitudes of jumps. This multi-industry environment transforms the binary "switch-or-not" mechanism of a single-industry economy into a smooth probabilistic evolution of aggregate capital.

**Remark 6.1 (Single-Industry Case)** When the economy consists of a single industry, investment adjusts discontinuously once macrosentiment crosses its threshold. The model nevertheless satisfies the conditions required for the drift and minorization arguments developed below, since its transition kernel retains an absolutely continuous component on an interior block.

### 6.3.4 Continuous and Jump Dynamics

Two additional forces shape the adjustment of capital. Depreciation and replacement requirements create a stabilizing pull toward previous levels of $K$: Even without sentiment changes, firms must reinvest merely to maintain capacity. Installation frictions further dampen the response of investment to short-lived optimism or pessimism.

Together, these stabilizing forces and the discrete mutation mechanism generate a piecewise-deterministic Markov process (PDMP): Capital evolves deterministically between investment opportunities and experiences discrete jumps when profitability-based selection triggers a regime change.[1]

This evolutionary PDMP forms the foundation of the stochastic model developed in the remainder of the chapter. The next section introduces the branch probabilities that govern the direction of jumps and link the selection mechanism to the aggregate Markov process for the capital stock. For the purposes of drift analysis, an equivalent continuous–jump representation is developed in Appendix 6.G.

A detailed summary of the underlying stochastic transition mechanism, including its continuous and jump components, is provided in Appendix 6.A.

[1] A PDMP follows a deterministic flow between random jump times and undergoes discontinuous transitions according to a specified kernel; see Davis (1984).

### 6.3.5 *Macrotipping and Industry-Level Adjustment*

The transition rule for aggregate investment combines two distinct layers of randomness. The first operates at the macrolevel through the aggregate sentiment variable $\nu_t$, while the second operates across industries through idiosyncratic draws $U \sim \text{Unif}[0, 1]$. It is useful to separate these two mechanisms, since they play fundamentally different economic roles.

**(1) Macrotipping via Aggregate Sentiment**
In the Investment chapter (see Eq. (5.4.10) in Chap. 5), we assume that the sentiment shock $\nu_t \sim \text{Unif}[-V, V]$.

When $\nu_t > 0$, a share $\nu_t/(2c_b)$ of firms becomes optimistic mutants. Incumbents adopt the optimistic investment rule only if this share exceeds the tipping threshold $\varphi - \zeta(K_t)$. Hence, upward activation occurs when

$$\nu_t > \nu^{\mathrm{U}}(K_t), \qquad \nu^{\mathrm{U}}(K) := 2c_b(\varphi - \zeta(K)).$$

Similarly, downward activation occurs when

$$\nu_t < \nu^{\mathrm{D}}(K_t), \qquad \nu^{\mathrm{D}}(K) := -2c_b(1 + c_b + \zeta(K) - \varphi).$$

Define the normalized sentiment parameter

$$b := \frac{c_b}{V}.$$

These conditions determine whether a given sentiment shock makes an investment regime shift feasible.

**(2) Industry-Level Activation via Idiosyncratic Draws**
Conditional on the sign of $\nu_t$, industries respond heterogeneously. Each industry draws

$$U \sim \text{Unif}[0, 1],$$

representing the realized salience of sentiment.

Define cutoff functions

$$\hat{p}(K) := 1 - 2b(\varphi - \zeta(K)), \qquad \hat{q}(K) := 1 - 2b(\zeta(K) + 1 + c_b - \varphi).$$

Activation occurs when

$$U < \hat{p}(K_t) \quad \text{(optimistic)}, \qquad U < \hat{q}(K_t) \quad \text{(pessimistic)}.$$

Since $U$ is uniform, the conditional activation probabilities are

$$\hat{p}_+(K) := \Pr(U < \hat{p}(K) \mid v_t > 0), \qquad \hat{p}_-(K) := \Pr(U < \hat{q}(K) \mid v_t < 0).$$

Because $\Pr(v_t > 0) = \Pr(v_t < 0) = 1/2$, the unconditional branch probabilities used in the transition kernel are

$$p_+(K) = {}^1\!/\!{}_2\,\hat{p}_+(K), \qquad p_-(K) = {}^1\!/\!{}_2\,\hat{p}_-(K).$$

**Equivalent Probabilistic Representation of Tipping Conditions**
Let $\epsilon^{\pm}(v_t)$ denote mutant shares. Upward activation occurs when

$$v_t > 2c_b\hat{\epsilon}^+(\zeta_t),$$

and downward activation when

$$-v_t > 2c_b\hat{\epsilon}^-(\zeta_t).$$

Writing $v_t = V(2u_t - 1)$ with $u_t \sim \text{Unif}[0, 1]$ yields equivalent conditions

$$u_t > \frac{1}{2}\left(1 + \frac{2c_b\hat{\epsilon}^+(\zeta_t)}{V}\right), \qquad u_t < \frac{1}{2}\left(1 - \frac{2c_b\hat{\epsilon}^-(\zeta_t)}{V}\right).$$

These conditions imply feasibility restrictions

$$2c_b\hat{\epsilon}^{\pm}(\zeta_t) \leq V.$$

**(3) Combining the Layers**
Adjustment therefore proceeds in two steps. First, aggregate sentiment selects the feasible regime through the tipping conditions above. Second, conditional on that regime, industries activate probabilistically through idiosyncratic draws $U$.

Aggregation of these industry responses produces the transition probabilities and jump sizes governing the stochastic evolution of $K_t$.

**Remark 6.2 (Connection to the Reduced-Form Transition Kernel)** The two-layer structure described above provides the microfoundation for the state-dependent probabilities $p_+(K)$ and $p_-(K)$ that appear in the Markov transition kernel of $K_t$. The macrosentiment $v_t$ determines whether an optimistic or pessimistic shift is feasible in principle, while the industry-level draws $U$ determine the fraction of industries that actually adjust. The effective up- and down-probabilities $p_{\pm}(K)$ therefore represent the *realized* activation rates of these branches once both macro- and industry-level randomness are accounted for. This interpretation underlies the clipped-affine functional forms introduced below.

For later use, we introduce the *unclipped* or *raw* transition probabilities associated with upward and downward activation:

$$p_+^{\mathrm{raw}}(\zeta) \;=\; 0.5 + b\,(\zeta - \varphi), \qquad p_-^{\mathrm{raw}}(\zeta) \;=\; 0.5 - b\,\big(\zeta + 1 + c_b - \varphi\big). \qquad (6.3.2)$$

These represent the linear response of mutation intensities to changes in the investment sentiment $\zeta(K)$ prior to the bound constraints that ensure probabilities remain in $[0, 0.5]$.

To obtain the actual transition probabilities, we apply a clipping operator to restrict the raw expressions to the admissible interval:

$$p_+(K) \;=\; \mathrm{clip}\big[p_+^{\mathrm{raw}}(\zeta(K));\; 0, 0.5\big], \qquad p_-(K) \;=\; \mathrm{clip}\big[p_-^{\mathrm{raw}}(\zeta(K));\; 0, 0.5\big]. \qquad (6.3.3)$$

**Lemma 6.1 (Clipping Thresholds Implied by Monotone Investment Sentiment)** *Let $\{K_t\}$ be the Markov chain with transition kernel* (6.5.1), *where $\Gamma_+(K, j) = K\gamma_+^j$ and $\Gamma_-(K, j) = K\gamma_-^j$ with $\gamma_+ > 1 > \gamma_- > 0$ and $j \in \{0, 1, \ldots, J_{\max}\}$. Let $p_\pm(K)$ be defined by the clipped forms* (6.3.3) *based on the raw probabilities* (6.3.2). *Assume that $\zeta(K)$ is continuous and strictly decreasing in $K$ (as established in Lemma 6.23).*

*Let $K_{p0}^{(0.5)}$, $K_{q0}^{(0.5)}$, $K_{p0}$, $K_{q0}$ be the thresholds in Table 6.1. Then the thresholds are ordered as*

$$K_{p0}^{(0.5)} < K_{q0} < K_{p0} < K_{q0}^{(0.5)},$$

*and moreover*

$$p_+(K) = 0 \text{ for all } K \geq K_{p0}, \qquad p_-(K) = 0 \text{ for all } K \leq K_{q0}.$$

*Define*

$$\underline{K} := K_{q0}\gamma_-^{J_{\max}}, \qquad \overline{K} := K_{p0}\gamma_+^{J_{\max}}, \qquad S^* := [\underline{K}, \overline{K}].$$

*Then $S^*$ is forward-invariant: If $K_t \in S^*$, then $K_{t+1} \in S^*$ almost surely. Consequently, any invariant probability measure $\pi$ of $\{K_t\}$ satisfies*

$$\mathrm{supp}(\pi) \subseteq S^*.$$

**Table 6.1** Thresholds defined by maximal and minimal shift probabilities

| Quantity | Condition | Interpretation |
|---|---|---|
| $K_{p0}^{(0.5)}$ | $\zeta(K_{p0}^{(0.5)}) = \varphi$ | $p_+ = 0.5$ |
| $K_{q0}^{(0.5)}$ | $\zeta(K_{q0}^{(0.5)}) = \varphi - (1 + c_b)$ | $p_- = 0.5$ |
| $K_{p0}$ | $\zeta(K_{p0}) = \varphi - \frac{1}{2b}$ | $p_+ = 0$ |
| $K_{q0}$ | $\zeta(K_{q0}) = \frac{1}{2b} + \varphi - (1 + c_b)$ | $p_- = 0$ |

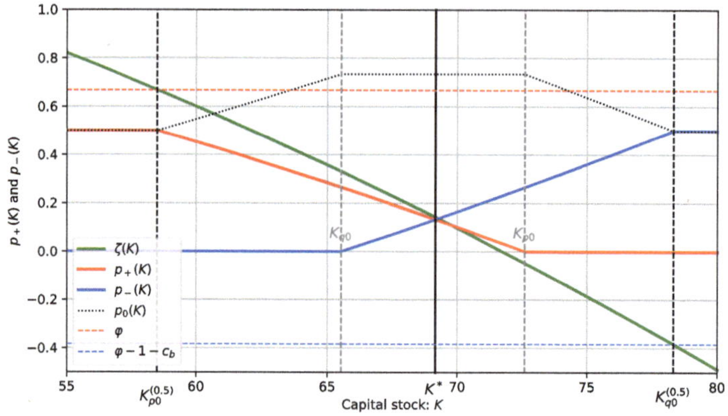


**Fig. 6.1** State-dependent branch probabilities $p_{\pm}(K)$ and residual probability $p_0(K)$. The reduced-form transition probabilities are clipped-affine functions of the profitability index $\zeta(K)$. These probabilities remain continuous even when aggregate jump sizes are discretized (Parameters for illustration: $\delta = 0.06$, $b = 0.7$, $c_b = 0.05$, $\varphi = 2/3$, $\alpha = 1/3$, $\eta = 7$, $c = 0.175$, $W = 25$, $M = 1000$, and $\lambda = 0.1$)

***Proof*** See Appendix 6.B.2. □

Four threshold values of $K$ play a central role in describing the transition between interior and clipped regimes. They are determined by the points at which $p_+$ or $p_-$ attains its upper or lower bound. Because $\zeta(K)$ is decreasing in $K$, these thresholds are strictly ordered (see Lemma 6.1). These thresholds, and their defining conditions, are summarized in Table 6.1.

The subscripts $p$ and $q$ correspond to the up- and down-branch probabilities $p_+$ and $p_-$, respectively. To avoid coexistence of optimistic and pessimistic mutants, the clipping rule is imposed at $p_+ = 0.5$; we therefore denote by $K_{p0}^{(0.5)}$ the interior activation threshold at which the down branch becomes operative. The same symmetric notation applies to the downward-branch threshold $K_{q0}^{(0.5)}$.[2]

Figure 6.1 illustrates the shape of the transition probabilities over a domain that encompasses the interval $D := [K_{q0}, K_{p0}]$. The down-branch probability $p_-(K)$ vanishes at $K_{q0}$ and the up-branch probability $p_+(K)$ vanishes at $K_{p0}$, so these points determine the admissible boundaries of movement. The continuous branch probabilities shown here reflect the underlying microlevel foundations; later analytic sections discretize the jump sizes, but the probabilities $p_{\pm}(K)$ remain as shown. The interior probability $p_0(K) = 1 - p_+(K) - p_-(K)$ fills the remainder of the domain.

In this figure, the upper and lower capital thresholds are defined by the conditions $K_{p0}^{(0.5)}$: $\zeta(K) = \varphi$ and $K_{q0}^{(0.5)}$: $\zeta(K) = \varphi - (1 + c_b)$, as summarized in Table 6.1. Since $\varphi := P/P^K$, a decline in $\varphi$ corresponds to an improvement in profitability. A fall in $\varphi$ therefore shifts both the upward and downward transition probability

[2] The subscript "0" is an index labeling the reference boundary of the admissible window; it does not indicate that the capital stock is zero.

schedules, $p_+(K)$ and $p_-(K)$, rightward, implying that the admissible $K$-window moves to higher levels of capital.

Throughout the remainder of the book, we refer to the interval $[K_{q0}, K_{p0}]$ as the admissible $K$-window. This object coincides with what we elsewhere describe more informally as the economy's stability corridor: a bounded region of the state space within which the stochastic dynamics remain well behaved. The terminology "window" emphasizes the admissible range of capital stocks, while "corridor" emphasizes the resulting evolution of trajectories over time.

By contrast, an increase in the downside bias parameter $c_b$ shifts the lower threshold line $\varphi - (1 + c_b)$ downward, which raises the $p_-(K)$ schedule for given $K$. This strengthens the downward adjustment tendency without affecting the upper threshold, thereby introducing a systematic asymmetry in capital dynamics.

Having specified the economic structure of investment adjustment, we now formalize the induced stochastic dynamics in terms of a Markov transition kernel and establish its basic regularity properties.

## 6.4 Foundations: Transition Kernel and Feller Property (Continuous Version)

A distinctive feature of the framework is that, despite its rich institutional and behavioral structure, the stochastic evolution of the economy can be represented in terms of a single-state variable—the aggregate capital stock $K$.

### *6.4.1 The Markov Transition Kernel for Capital*

We begin by identifying the effective domain of the capital process along the locally stable branch.

**Lemma 6.2 (Compact Feasibility Set for $K$ on the Stable Branch)** *Along the stable PM–LM branch $r(K) \in (0, \sqrt{\delta})$ with $r'(K) > 0$, the induced labor demand*

$$N^d(K) = \frac{1-\alpha}{\alpha} \frac{K}{\theta_{\mathrm{PM}}} (r(K) + \delta)^{\frac{1}{1-\alpha}}$$

*is continuous and strictly increasing in $K$, with $N^d(0) = 0$. Let $K_{\bar{N}}$ denote the unique solution to $N^d(K_{\bar{N}}) = \bar{N}$. Then feasible SQSS capital stocks must satisfy $K \leq K_{\bar{N}}$. Let $K_{\mathrm{crit}}$ denote the PM–LM tangency ceiling. The effective state space is therefore*

$$\mathcal{K} = [0, \bar{K}], \qquad \bar{K} = \min\{K_{\bar{N}}, K_{\mathrm{crit}}\},$$

*which is compact.*

***Proof (Sketch)*** On the stable branch $r'(K) > 0$. Since the map $x \mapsto x^{1/(1-\alpha)}$ is increasing and $r \mapsto r + \delta$ is increasing, the composite $(r(K) + \delta)^{1/(1-\alpha)}$ is increasing in $K$. Multiplying by $K$ yields a strictly increasing, continuous $N^d(K)$. The boundary $N^d(0) = 0$ follows from $r(K) \downarrow 0$ as $K \downarrow 0$. By the intermediate value theorem, there is a unique $K_{\bar{N}}$ with $N^d(K_{\bar{N}}) = \bar{N}$. Feasibility also requires the PM–LM intersection to exist (stable branch), which holds only for $K < K_{\text{crit}}$. The intersection of these two constraints is the closed interval $[0, \overline{K}]$. □

To apply the Foster–Lyapunov criterion, we work with the Lyapunov function $V(K) = K$. The drift inequality is required only outside a fixed compact "small" set. Accordingly, we take $C := [0, 1]$ and verify the Foster–Lyapunov bound on $S := [1, \bar{K}] \subseteq \mathcal{K}$. No restriction $K \geq 1$ is imposed on the state space; rather, the region $C$ is absorbed into the small set on which the drift condition is not required. The choice of the numerical threshold 1 is a normalization and can be replaced by any strictly positive constant.

The capital-stock process is modeled as a time-homogeneous Markov chain on the compact state space $S$,

$$\mathcal{B}(S) = \text{Borel } \sigma\text{-algebra.}$$

Given $K \in S$, the next state $K'$ takes one of three values, corresponding to upward, downward, or no adjustment:

$$K' \in \{K(1 + \varepsilon^+),\ K(1 - \varepsilon^-),\ K\},$$

with branch probabilities $p_+(K)$, $p_-(K)$, and $p_0(K) = 1 - p_+(K) - p_-(K)$. Conditional on selecting a branch, the jump magnitudes $\varepsilon^\pm$ have distributions $\theta_\pm(\cdot; K)$. In the reduced two-branch law used for the stability analysis, $\theta_\pm(\cdot; K)$ are degenerate at a single point; in the continuous mutation model of Chap. 5, they are smooth densities. Both specifications fit the kernel representation below.

The transition kernel is

$$P(K, A) = p_+(K)\Gamma_+(K, A) + p_-(K)\Gamma_-(K, A) + p_0(K)\mathbf{1}_{\{K \in A\}}, \tag{6.4.1}$$

where the pushforward measures are defined by

$$\begin{aligned} \Gamma_+(K, A) &= \theta_+(\{e > 0 : K(1 + e) \in A\}; K), \\ \Gamma_-(K, A) &= \theta_-(\{e > 0 : K(1 - e) \in A\}; K). \end{aligned} \tag{6.4.2}$$

**Lemma 6.3 (Operator–Kernel Equivalence)** *Let the transition operator be defined for bounded measurable f by*

$$
\begin{aligned}
(Pf)(K) &= (1 - p_+(K) - p_-(K))f(K) \\
&\quad + p_+(K)\int_{\vartheta_+(K)}^{1} f(\Gamma_+(K, j))\,\frac{dj}{1-\vartheta_+(K)} \\
&\quad + p_-(K)\int_{\vartheta_-(K)}^{1} f(\Gamma_-(K, j))\,\frac{dj}{1-\vartheta_-(K)}.
\end{aligned}
\tag{6.4.3}
$$

*Then $(Pf)(K) = \int f(K')P(K, dK')$, and $P$ is a Markov transition kernel.*

***Proof*** Let $J_+ \sim \mathrm{Unif}[\vartheta_+(K), 1]$ denote the conditional distribution of the up-branch jump. Then

$$
\int_{\vartheta_+(K)}^{1} f(\Gamma_+(K, j))\,\frac{dj}{1-\vartheta_+(K)} = \int f(K(1+e))\,\theta_+(de; K),
$$

and analogously for the down branch. Substituting these identities into (6.4.3) yields $(Pf)(K) = \int f(K')P(K, dK')$, which is a convex combination of probability measures and hence a probability measure. □

This explicit form of $Pf$ shows that continuity of the primitives $p_\pm$, $\vartheta_\pm$, and $\Gamma_\pm$ will imply the Feller property.

### 6.4.2 Feller Property of the Capital-Stock Process (Continuous Mutation Law)

To verify the Feller property, we establish continuity of the kernel's building blocks. The key structural input is the continuity and strict monotonicity of the equilibrium mapping $K \mapsto \zeta(K)$ along the locally stable PM–LM branch, which follows from the PM–LM system (see Lemma 6.23 in Appendix 6.B).

To prepare for the stochastic stability analysis below, we establish continuity of the clipped branch probabilities and the associated update maps. Proofs are provided in Appendix 6.C.

**Lemma 6.4 (Clipped-Affine Continuity)** *If $g$ is continuous, then $h(K) = \mathrm{clip}[a+b\,g(K); \underline{c}, \bar{c}]$ is continuous. If $g$ is $L$-Lipschitz, then $h$ is $|b|L$-Lipschitz.*

From this, we obtain continuity of the branch probabilities and update maps.

**Lemma 6.5 (Continuity of Primitives)** *Under continuity of $\zeta(K)$, the branch probabilities*

$$
p_+(K) = \mathrm{clip}[0.5 + b(\zeta(K) - \varphi); 0, 0.5],
$$

$$
p_-(K) = \mathrm{clip}[0.5 - b(\zeta(K) - \varphi + 1 + c_b); 0, 0.5]
$$

*are continuous. Consequently, $\vartheta_\pm(K) = 1 - p_\pm(K)$ and the update maps $\Gamma_\pm(K)$ are continuous on $\mathcal{K}$.*

**Remark 6.3 (Economic Intuition for the Update Maps)** The update maps $\Gamma_{\pm}(K)$ reflect aggregation from an industry layer to the aggregate capital stock. Conditional on a given branch being selected, only a fraction of industries revise their investment plans, while the remaining industries follow the replacement rule.

When the probability of an up-branch $p_{+}(K)$ is high, a larger share of industries is already inclined to expand, so the aggregate capital adjustment associated with a realized up-jump can be achieved through smaller per-industry revisions. Conversely, when $p_{+}(K)$ is low, only a small subset of industries adjusts, and each adjusting industry must undertake a larger revision for the aggregate effect to be comparable.

This mechanism is encoded in the factors $\delta\big(1 - p_{\pm}(K)/2\big)/(2b)$: The realized jump remains proportional to the current scale $K$, but its magnitude is inversely related to the prevalence of the corresponding branch. The maps $\Gamma_{\pm}$ therefore embody the idea that aggregate investment dynamics arise from the interaction between economy-wide scale and the extensive margin of industry-level adjustment, rather than from fixed, state-independent jumps.

With the continuity of the primitives $p_{\pm}$, $\vartheta_{\pm}$, and $\Gamma_{\pm}$ in hand, we can now verify the Feller property of the transition operator. Intuitively, continuity of these building blocks ensures that small changes in the current state $K$ lead to small changes in the entire one-step law of motion, and hence in the expectation $(Pf)(K)$ for any bounded continuous test function $f$.

We present two formulations of the Feller property. The first is derived from the continuous mutation specification introduced in the Investment chapter, where the distribution of adjustment magnitudes depends smoothly on $K$. In this setting, the transition kernel admits a density with respect to Lebesgue measure whose parameters vary continuously in $K$, and the Feller property follows directly.

The second formulation applies to the reduced two-branch mutation law used throughout the Foster–Lyapunov analysis in this chapter. In this specialized Markov kernel, $K$ transitions to exactly two possible multiplicative levels, with probabilities $p_{+}(K)$ and $p_{-}(K)$ that are continuous in $K$. Since the transition kernel has only finitely many points of support and the branch probabilities are continuous, the discrete kernel is also Feller.

The Foster–Lyapunov argument and the uniqueness of the invariant distribution rely solely on the discrete version of the kernel. The continuous version is presented for completeness and for interpretive clarity, as it provides the microfoundation for the activation probabilities and the investment propensity $\zeta(K)$.

The following proposition is now a direct consequence of the continuity lemma and a standard dominated convergence argument.

**Proposition 6.1 (Feller Property (Continuous-Jump Kernel))** *Assume Lemma 6.5. Then the transition kernel $P$ in* (6.4.1) *is Feller on $\mathcal{K}$; that is, for every $f \in C_b(\mathcal{K})$, the map*

$$K \;\mapsto\; Pf(K) \;:=\; \int_{\mathcal{K}} f(K')\, P(K, dK')$$

*is continuous on $\mathcal{K}$.*

***Proof*** Fix $f \in C_b(\mathcal{K})$ and let $K_n \to K$ in $\mathcal{K}$. By Lemma 6.5, the maps $K \mapsto p_\pm(K)$, $\vartheta_\pm(K)$, and $\Gamma_\pm(K, j)$ are continuous. Hence for each $j \in \{0, 1, \ldots, J_{\max}\}$, we have $f(\Gamma_\pm(K_n, j)) \to f(\Gamma_\pm(K, j))$, while $|f(\Gamma_\pm(K_n, j))| \le \|f\|_\infty$. Since the kernel is a finite mixture (finite sum) over $j$ with continuous weights, we may pass to the limit term by term to obtain $Pf(K_n) \to Pf(K)$. See Ethier and Kurtz (1986, Section 4.2) for the general argument. □

Continuity of the transition operator is the only requirement needed to invoke the Krylov–Bogoliubov theorem, which provides existence of an invariant distribution; we turn to this next.

### *6.4.3 Existence of an Invariant Distribution: Krylov–Bogoliubov*

Because $P$ is Feller and the state space $\mathcal{K} = [0, \bar{K}]$ is compact, the hypotheses of the Krylov–Bogoliubov theorem are satisfied. This subsection formalizes this conclusion and records the resulting existence of an invariant distribution.

**Definition 6.1 (Empirical Measures)** For a Markov process $(K_t)_{t\ge 0}$ on $\mathcal{K}$, the $n$-th empirical measure starting from $K_0 = K$ is the random probability measure:

$$\mu_n^K(A) := \frac{1}{n}\sum_{t=0}^{n-1} \mathbf{1}_{\{K_t \in A\}}, \qquad A \in \mathcal{B}(S).$$

Equivalently, $\mu_n^K = \frac{1}{n}\sum_{t=0}^{n-1} \delta_{K_t}$, where $\delta_x$ denotes the Dirac measure at $x$.

**Remark 6.4 (Tightness on Compact State Space)** Since $\mathcal{K} = [0, \bar{K}]$ is compact, the sequence $\{\mu_n^K\}_{n\ge 1}$ is automatically tight for every initial state $K \in \mathcal{K}$.

The Krylov–Bogoliubov theorem then provides the key existence result:

**Theorem 6.1 (Krylov–Bogoliubov)** *Let $(K_t)_{t\ge 0}$ be a Feller Markov process on a compact metric space $\mathcal{K}$. Then for each initial state $K \in \mathcal{K}$, the sequence of empirical measures $\{\mu_n^K\}_{n\ge 1}$ admits weakly convergent subsequences, and every limit point is an invariant probability measure.*

***Proof (Sketch)*** Compactness of $\mathcal{K}$ implies that the sequence $\{\mu_n^K\}$ is tight and hence has a weakly convergent subsequence $\mu_{n_k}^K \Rightarrow \pi$. The Feller property ensures that for every $f \in C_b(\mathcal{K})$ the map $K \mapsto Pf(K)$ is continuous, so

$$\int f\, d\mu_{n_k}^K \longrightarrow \int f\, d\pi, \qquad \int Pf\, d\mu_{n_k}^K \longrightarrow \int Pf\, d\pi.$$

A standard telescoping argument shows that

$$\left| \int f \, d\mu_n^K - \int Pf \, d\mu_n^K \right| \leq \frac{2\|f\|_\infty}{n} \longrightarrow 0,$$

and taking limits along the subsequence $\{n_k\}$ yields

$$\int f \, d\pi = \int Pf \, d\pi, \qquad \forall f \in C_b(\mathcal{K}),$$

so $\pi$ is invariant. The existence of invariant measures follows by the Krylov–Bogolyubov theorem; see Einsiedler and Ward (2011, Corollary 4.2) for a classical formulation of the Krylov–Bogolyubov theorem. □

**Corollary 6.6 (Existence of an Invariant Distribution)** *The capital-stock process* $(K_t)$ *admits at least one invariant probability measure on the compact state space* $[0, \bar{K}]$. *This follows from the Feller property of the transition kernel and compactness, via the Krylov–Bogoliubov theorem.*

***Proof*** Since $P$ is Feller by Proposition 6.1 and the state space $\mathcal{K}$ is compact, the claim follows immediately from Theorem 6.1. □

**Remark 6.5 (Interpretation)** Economically, the theorem ensures that the stochastic dynamics of investment and aggregate adjustment settle into a well-defined long-run distribution, even though each period's adjustment depends on the interaction of aggregate sentiment and industry-level randomness. The invariant distribution characterizes the ergodic behavior of the model and forms the basis for the convergence and stability results developed in the next sections.

These results provide the probabilistic foundation for the stability and long-run analysis that follows, while preserving the parsimonious, single-state representation of the economy.

Having established the existence of invariant probability measures for the capital-stock process, and having interpreted this result in economic terms, we now turn to the analytical structure required for studying uniqueness and the long-run behavior of the system. Although the underlying mutation mechanism can be formulated with a continuous distribution of jump sizes, the key forces shaping aggregate investment dynamics are captured by the two effective directions of adjustment—upward and downward—together with their state-dependent activation probabilities. To prepare for the irreducibility and Foster–Lyapunov arguments developed later, it is therefore useful to adopt a discrete representation of aggregate jumps that retains the essential economic content of the model while providing a more tractable transition structure. The next subsection explains why such a discretization is natural and how it links back to the continuous formulation introduced above.

## 6.5 Existence and Uniqueness of the Invariant Distribution

The preceding section established that the continuous mutation model yields a Feller transition kernel on the compact state space, guaranteeing the existence of at least one invariant distribution. However, proving uniqueness and global convergence requires additional structural ingredients—most notably a small set, $\phi$-irreducibility, aperiodicity, and a Foster–Lyapunov drift—that do not arise as transparently in the continuous formulation. For this reason, the remainder of the chapter adopts a discretized version of the aggregate jump mechanism. This discrete kernel preserves the economic content of the investment-selection process while providing a tractable structure in which the required minorization and drift conditions can be verified cleanly. We therefore begin by explaining why such a discretization is natural and how it connects back to the continuous model analyzed above.

### *6.5.1 Rationale for Discretizing Aggregate Jumps*

At the microlevel, each industry faces a small menu of adjustment options: a modest upward revision of its capital stock, a modest downward revision, or no adjustment at all. When the number of industries is large, these idiosyncratic adjustments aggregate into a smooth response at the macrolevel, with the overall change in capital determined by the state-dependent activation probabilities governing upward and downward moves. From the aggregate perspective, however, the fine details of these microscopic adjustment magnitudes matter far less than the directional imbalance between upward and downward investment pressures. This observation motivates the use of a parsimonious two-branch representation of aggregate jumps, which distills the continuous mutation mechanism into the essential forces governing the evolution of $K_t$.

Under the law of large numbers (LLN), the fraction of industries that activate in each direction converges, in each period, to a deterministic function of macrosentiment and the underlying idiosyncratic activation rates. As a result, the aggregate response of the economy within each branch (up, down, or no change) behaves as a *continuous* function of the state $K$. The smoothness comes not from the adjustment of individual firms—each of which moves in small, discrete steps—but from the averaging effect of a large population of industries whose behavior is governed by state-dependent activation probabilities. In this sense, the continuous transition mechanism of the macrolevel process emerges naturally from a microscopically discrete environment.

Yet for the purpose of analyzing long-run dynamics—irreducibility, drift behavior, and uniqueness of invariant distributions—the continuous kernel can be replaced by an equivalent two-branch discrete representation that captures the essential upward and downward forces acting on the capital stock.

For example, in the up-branch (denoted by +), the LLN-implied aggregate multiplier takes the form

$$m_+(K) = 1 + a_+(K), \qquad a_+(K) \in [0, \bar{a}],$$

where $a_+(K)$ represents the fraction of industries that activate and the magnitude of their individual adjustments. The down-branch multiplier $m_-(K)$ has an analogous expression.

Although this continuous representation is natural from the microfoundation, it creates two difficulties for the analysis of the Markov process. First, the transition kernel would involve densities whose values can become arbitrarily small on subsets of the state space, making it impossible to obtain the uniform minorization properties required for irreducibility and the Foster–Lyapunov criterion. Second, allowing a continuum of jump sizes obscures the geometric structure of multiplicative transitions and leads to a technically burdensome kernel.

To retain both tractability and fidelity to the microlevel economics, the continuous aggregate responses $m_\pm(K)$ are approximated by a finite family of multiplicative jump factors:

$$m_\pm(K) \approx \gamma_\pm^j, \qquad j \in \{0, 1, \dots, J_{\max}\}.$$

Each index $j$ represents a discretized intensity level of aggregate investment adjustment, not the number of industries adjusting. The integer $J_{\max}$ determines the resolution of this discretization and is chosen so that the values $\gamma_\pm^j$ provide a sufficiently fine approximation of the continuous LLN multipliers $m_\pm(K)$ on the bounded domain $D = [K_{q0}, K_{p0}]$, and hence on any compact interior subset $D_\eta := [K_{q0} + \eta, K_{p0} - \eta]$. We let $\delta_\pm(K, j)$ denote the state-dependent conditional weights assigned to the $j$-th discretized jump level, conditional on selecting the $\pm$ branch at state $K$.

This discretization has two important consequences. First, it yields a transition kernel with a finite number of atoms in each branch, allowing uniform lower bounds on transition probabilities over compact interior subset $D_\eta = [K_{q0} + \eta,\ K_{p0} - \eta]$. Second, it preserves the multiplicative structure essential for the dynamics of $K_t$ while enabling rigorous proofs of irreducibility and geometric ergodicity. In this way, the discretized jump-size specification offers a tractable reduced-form representation that remains consistent with the underlying LLN microfoundation.

These observations justify working with a reduced representation of the aggregate adjustment mechanism: Although the microlevel jump sizes may be drawn from a continuous distribution, their macrolevel impact is summarized by the two directional forces generated by upward and downward activations. The precise magnitudes of microscopic adjustments are therefore immaterial for understanding the evolution of the aggregate capital stock; what drives the dynamics is the imbalance between the two activation probabilities, $p_+(K)$ and $p_-(K)$. This motivates the discrete jump formulation used in the remainder of this chapter, where the capital-stock process is modeled as a Markov chain whose transitions reflect only the essential upward and downward movements implied by the underlying microstructure.

We now formalize this reduced representation by specifying the Markov transition kernel associated with the discrete up–down update mechanism.

### 6.5.2 *Discrete Transition Kernel and Its Feller Property*

To analyze irreducibility and long-run stability, it is useful to work with a discretized representation of investment mutations. This discrete kernel preserves the economic content of the continuous model while allowing for a transparent verification of the small-set property and the Foster–Lyapunov drift.

Recall that the state space is represented as a time-homogeneous Markov chain on the compact state space

$$\mathcal{K} = [0, \bar{K}].$$

For the purposes of the drift and irreducibility analysis below, we will work on the restricted set

$$S := [1, \bar{K}] \subset \mathcal{K},$$

treating $[0, 1]$ as a small set that can be absorbed into the Foster–Lyapunov constant term. Given the state $K$, the economy moves to one of two possible updated levels, corresponding to the aggregate consequences of upward activation or downward activation across industries. These transitions occur with state-dependent probabilities $p_+(K)$ and $p_-(K)$, which encode the underlying microlevel activation behavior described earlier. The resulting transition mechanism captures all essential features of the continuous mutation process while providing a tractable structure for the analysis of irreducibility and long-run stability.

**Discrete Update Mechanism**

Given $K \in S$, the next-period capital stock $K'$ is generated in three steps:

1. **Branch selection:** With probabilities

$$p_+(K), \qquad p_-(K), \qquad p_0(K) := 1 - p_+(K) - p_-(K),$$

   the process selects the up-, down-, or no-change branch.
2. **Discrete adjustment intensity:** Conditional on selecting the up- or down-branch, an integer

$$J_\pm \in \{0, 1, \ldots, J_{\max}\}$$

is drawn from a clipped-affine distribution $\delta_\pm(K, j)$. These probabilities are continuous in $K$ and uniformly bounded away from zero on the interior interval $D = [K_{q0}, K_{p0}]$.

3. **Multiplicative update:**

$$K' = \begin{cases} \Gamma_+(K, J_+) := K\,\gamma_+^{J_+} & \text{with prob. } p_+(K)\delta_+(K, J_+), \\ K, & \text{with prob. } p_0(K), \\ \Gamma_-(K, J_-) := K\,\gamma_-^{J_-} & \text{with prob. } p_-(K)\delta_-(K, J_-), \end{cases}$$

where $\gamma_+ > 1 > \gamma_- > 0$. Accordingly, the up-branch multipliers $\gamma_+^j$ increase with $j$, whereas the down-branch multipliers $\gamma_-^j$ decrease with $j$.

**Transition Kernel**

For any Borel set $A \subseteq S$, the kernel takes the form

$$P(K, A) = p_0(K)\mathbf{1}_A(K) + p_+(K)\sum_{j=0}^{J_{\max}} \delta_+(K, j)\,\mathbf{1}_A(\Gamma_+(K, j))$$
$$+\, p_-(K)\sum_{j=0}^{J_{\max}} \delta_-(K, j)\,\mathbf{1}_A(\Gamma_-(K, j)), \tag{6.5.1}$$

where $\delta_x$ denotes the Dirac point mass at $x$ (i.e., $\delta_x(A) = \mathbf{1}_A(x)$). Equivalently, the law of $K_{t+1}$ given $K_t = K$ is

$$P(K, \cdot) = p_0(K)\,\delta_K \,+ p_+(K)\sum_j \delta_+(K, j)\,\delta_{\Gamma_+(K,j)}$$
$$+\, p_-(K)\sum_j \delta_-(K, j)\,\delta_{\Gamma_-(K,j)}.$$

Although the following result is formally straightforward, it plays an important bookkeeping role by linking the discrete update rule to the Markov operator used in the Feller and Foster–Lyapunov analysis.

**Lemma 6.7 (Operator–Kernel Equivalence: Discrete Update)** *For any bounded measurable $f: S \to \mathbb{R}$,*

$$(Pf)(K) := p_0(K)\,f(K) + p_+(K)\sum_{j=0}^{J_{\max}} \delta_+(K, j)\,f\big(\Gamma_+(K, j)\big)$$
$$+\, p_-(K)\sum_{j=0}^{J_{\max}} \delta_-(K, j)\,f\big(\Gamma_-(K, j)\big)$$

*satisfies*

$$(Pf)(K) = \int_S f(K')\, P(K, dK').$$

*Hence $P(K, \cdot)$ is a Markov transition kernel.*

***Proof*** Since $p_0 + p_+ + p_- = 1$ and $\sum_j \delta_\pm(K, j) = 1$, the expression is a convex combination of Dirac measures; integrating $f$ against this measure produces the stated identity. □

**Continuity of the Primitives**

The Feller property rests on the continuity of the basic building blocks.

**Lemma 6.8 (Continuity of Transition Primitives)** *If $\varsigma \in C([1, \bar{K}])$, then for every $j$:*

*(i)* $p_\pm(K) = \text{clip}_{[0,1]}(0.5 \pm b(\varsigma(K) - \varphi_\pm))$ *are continuous.*
*(ii)* $\delta_\pm(K, j)$ *are clipped-affine in $K$ and therefore continuous.*
*(iii)* $\Gamma_\pm(K, j) = K\, \gamma_\pm^j$ *are continuous in $K$.*

***Proof*** Clipping preserves continuity. Parts (ii)–(iii) follow directly from the definitions. □

**Feller Property**

**Lemma 6.9 (Feller Property (Discrete-Jump Kernel))** *Under Lemma 6.8, the discrete transition kernel* (6.5.1) *is Feller; that is, for every $f \in C_b(S)$,*

$$K \longmapsto Pf(K) = \int_S f(K'), P(K, dK')$$

*is continuous on $S$.*

***Proof (sketch)*** For $K_n \to K$, continuity of $p_\pm$, $\delta_\pm$, and $\Gamma_\pm$ implies

$$f(\Gamma_\pm(K_n, j)) \to f(\Gamma_\pm(K, j))$$

for each $j$, and all coefficients converge. The number of terms is finite, so $Pf(K_n) \to Pf(K)$. □

**Remark 6.6 (Economic Role of the Discrete Update Maps)** The grid $\gamma_\pm^j$ provides a tractable approximation of the continuous range of feasible jump magnitudes. Higher branch probabilities $p_\pm(K)$ shift mass toward smaller $j$ (milder adjustments), while lower $p_\pm(K)$ place mass on larger $j$ (stronger adjustments). Thus the update maps encode how aggregate adjustment intensity varies with the likelihood of activation, preserving the key economic interpretation of the continuous model.

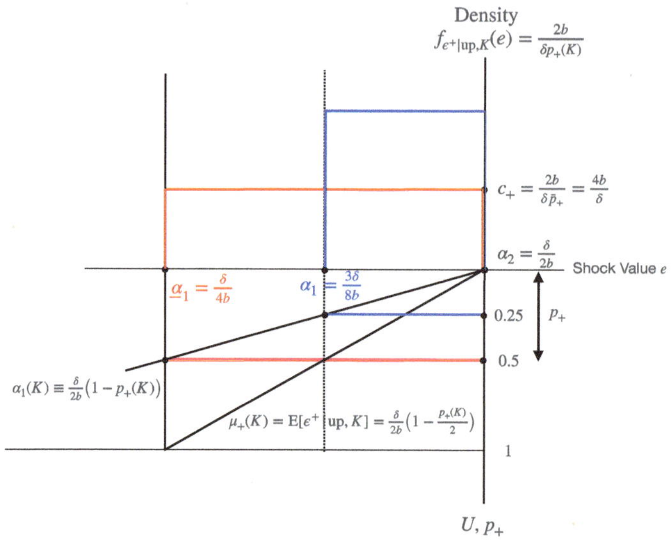


**Fig. 6.2** Distributional structure of up-branch jump magnitudes.
At the microfoundation level, the conditional jump magnitude $\varepsilon^+$ is supported on a state-dependent range determined by profitability signals. The conditional up-shock is uniformly distributed on $[\underline{\alpha}_1(K), \alpha_2]$. Only the lower endpoint depends on $p_+(K)$; a higher $p_+(K)$ reduces $\underline{\alpha}_1(K)$ and widens the support downward. For analytical tractability, the continuous support is replaced by a finite grid of discrete jump levels $\gamma_+^j$ in the stochastic kernel

This explicit representation makes it immediately clear that all up- and down-jumps take place on a finite discrete grid of the form $K\gamma_{\pm}^j$, a feature that will be crucial for the irreducibility and regeneration arguments below.

Figure 6.2 illustrates the continuous benchmark specification in which, conditional on selecting the up-branch, the jump size is drawn from a uniform density over a state-dependent interval. As the macrolevel switching probability $p_+(K)$ increases, the lower endpoint $\underline{\alpha}_1(K)$ moves rightward toward the fixed upper bound $\alpha_2 = \frac{\delta}{2b}$, while the global lower bound $c_+$ remains unchanged. Thus the conditional support $[\underline{\alpha}_1(K), \alpha_2]$ widens within the fixed interval determined by $c_+$ and $\alpha_2$, and the height of the density adjusts as $2b/(\delta\, p_+(K))$. In the stochastic analysis below, the underlying continuous range of feasible jump magnitudes is approximated by a finite grid of discrete jump magnitudes, $\gamma_{\pm}^j$. This grid, along with its state-dependent weights $\delta_{\pm}(K, j)$, can be interpreted as a discretization of the continuous range. This preserves the economic meaning of gradual or strong adjustments while delivering a finite-atom Markov kernel that is convenient for our ergodic analysis.

Although the law of large numbers would collapse independent shocks to a single deterministic average if all industries responded identically, the investment adjustments in our model are mediated by heterogeneous idiosyncratic activations and state-dependent thresholds. These produce a distribution of feasible aggregate jump magnitudes conditional on each branch. The discretized multipliers $\gamma_{\pm}^j$ and their weights $\delta_{\pm}(K, j)$ serve as a tractable representation of this cross-sectional aggregation and are not intended as the limit of an LLN collapse to a singleton jump size.

Having established the discrete transition kernel and verified its Feller continuity, we now possess a well-behaved Markovian framework in which small perturbations of the current capital stock translate smoothly into corresponding changes in the one-step law of motion. Continuity alone, however, does not ensure that the process can reach all economically relevant regions of the state space or that it revisits these regions frequently enough to support long-run stability. The next task is therefore to analyze the connectivity structure of the chain: We must determine whether the multiplicative updates generated by the discrete jump grid allow the process to move, with positive probability, between neighborhoods of any two admissible states. This analysis yields the $\phi$-irreducibility and aperiodicity properties that, together with the Foster–Lyapunov drift, form the structural core of the ergodicity argument developed in the subsequent section.

### 6.5.3 *Irreducibility*

Having established the Feller property and the existence of at least one invariant distribution, we now show that the Markov process satisfies a global irreducibility condition with respect to Lebesgue measure. The argument exploits the block structure developed in Sect. 6.5.4, which provides uniform access to the interior block and mixing within that block. Combined, these two features ensure that the chain can reach any non-null subset of the state space from every starting point in $\mathcal{K}$.

This section proves that the Markov process $\{K_t\}_{t\geq 0}$ generated by the investment dynamics is $\phi$—irreducible with respect to Lebesgue measure. The essential fact is that, although the probability of no adjustment is large (at least 0.5 uniformly in $K$), the probabilities of upward and downward adjustments remain strictly positive throughout the interior of the compact domain $D = [K_{q0}, K_{p0}]$, where $p_0(K) \geq 0.5$ for all $K$ and $p_+(K), p_-(K) > 0$ on $D$. The compact domain $D$ can always be chosen so that both $p_+(K)$ and $p_-(K)$ remain strictly positive on $D$, while $p_0(K)$ stays uniformly bounded away from zero.[3] Together with the interior support condition on adjustment magnitudes established below, these uniformly positive branch probabilities imply that every interior interval of $D$ is reachable from every initial point in $D$.

[3] In the continuous specification, the branch probabilities satisfy $p_+(K_{p0}) = 0$ and $p_-(K_{q0}) = 0$ at the extremal SQSS levels. In the present discretized formulation, we work on a compact subdomain $D_\eta = [K_{q0} + \eta,\ K_{p0} - \eta]$ chosen strictly inside the natural feasibility region. With a suitable choice of clipping threshold (the "notch"), the $\{p_{\pm}(K)\}$ remain strictly positive on $D_\eta$, while $p_0(K) \geq 0.5$ holds uniformly. This retains the economic interpretation of the extremal states while ensuring the uniform positivity needed for irreducibility and Lyapunov arguments below.

### 6.5.4 Block Structure and Regeneration

Throughout this subsection, our analysis employs the discrete transition kernel Eq. (6.5.1). We now summarize the block structure of the state space and introduce three key ingredients for the subsequent regeneration analysis: (i) an $M$-step push-up from the lower block, (ii) a minorization condition on the interior block, and (iii) a bridge-entry condition from the upper block.

Recall the compact state space $S = [1, \bar{K}]$ and the compact bridge strip $D = [K_{q0}, K_{p0}]$ constructed from the branch probabilities. We use the notation

$$L := [1, K_{q0}), \qquad U := (K_{p0}, \bar{K}],$$

so that $S = L \cup D \cup U$. The closed bridge strip, $D = [K_{q0}, K_{p0}]$, includes the clipping thresholds at its endpoints. Since uniform positivity may fail arbitrarily close to these thresholds, all uniform lower bounds are imposed on compact interior subsets $D_\eta \subset D_0 = (K_{q0}, K_{p0})$. The key requirement is the existence of uniform lower bounds for both branch probabilities and admissible adjustment masses on compact interior subsets of the bridge strip.

To facilitate the Foster–Lyapunov and small-set arguments that follow, we impose a set of block-wise regularity conditions on the switching probabilities and the support of investment adjustments.

In our specification, it holds automatically on any compact subinterval strictly contained in $(K_{q0}, K_{p0})$. Indeed, $p_\pm(K)$ are continuous in $K$ and strictly positive away from the clipping thresholds $K_{q0}$ and $K_{p0}$, and the conditional allocation over interior SQSS configurations assigns strictly positive mass to the indices $j^*_\pm$ by construction. Therefore, both $\underline{p}_\pm$ and $\underline{\delta}_\pm$ are strictly positive by the extreme-value theorem applied on compact subsets of the interior.

To apply standard Markov stability arguments, we impose a small number of local regularity conditions on the behavior of the switching probabilities across different regions of the state space. When uniform bounds are required, we work on a shrunken interior set $D_\eta = [K_{q0} + \eta, \; K_{p0} - \eta] \subset D$ to stay away from the clipping thresholds.

**Assumption 6.1 (Switching Probabilities on a One-Sided Region)** Let $\theta$ denote the fixed structural parameter vector $(\delta, c_b, \lambda, c_y, \dots)$, and let $p_+(K)$ and $p_-(K)$ be the switching probabilities defined from $\zeta(K)$ in Sect. 6.3.5. Assume that the SQSS region of interest is contained in $C := [1, K_{p0}]$ for some $K_{p0} > 1$. We impose the following block conditions:

1. **Lower block** There exist $K_{q0} \in (1, K_{p0})$ and a constant $\underline{p}_+ > 0$ such that

$$p_+(K) \ \geq \ \underline{p}_+ \qquad \text{for all } K \in [1, K_{q0}].$$

2. **Interior block** There exist $\eta > 0$ and constants $0 < \underline{p} < \bar{p} < 1$ such that on

$$D_\eta \ := \ [K_{q0} + \eta, \ K_{p0} - \eta],$$

both switching probabilities satisfy

$$\underline{p} \le p_+(K),\ p_-(K) \le \bar{p} \qquad \text{for all } K \in D_\eta.$$

3. **Continuity** The functions $K \mapsto p_+(K)$ and $K \mapsto p_-(K)$ are continuous on $C$. This holds automatically because $\zeta(K)$ is continuous, and the switching rules are continuous in $\zeta$ for each fixed $\theta$.

**Assumption 6.2 (Interior Bounds on Branch Probabilities and Jump Support)** There exists a nonempty compact interval $I \subset D_0$, where $D_0 = (K_{q0}, K_{p0})$, such that the following hold:

(i) (Strict positivity of branch probabilities) The up- and down-branch probabilities satisfy

$$\underline{p}_+ := \inf_{K \in I} p_+(K) > 0, \qquad \underline{p}_- := \inf_{K \in I} p_-(K) > 0.$$

(ii) (Interior support of conditional jump magnitudes) Under either branch $\pm$, the conditional law of the jump magnitude admits strictly positive mass (respectively, density) on a compact interval of "strong" interior up- or down-configurations. In the discrete-grid specification, this assumption is equivalent to requiring that

$$\underline{\delta}_+ := \inf_{K \in I} \mathbb{P}\{ j \ge j_+^* \mid \text{up-branch at } K\} > 0,$$

$$\underline{\delta}_- := \inf_{K \in I} \mathbb{P}\{ j \ge j_-^* \mid \text{down-branch at } K\} > 0,$$

and hence, the one-step probabilities of a "strong" up- or down-jump are uniformly positive on $I$:

$$\inf_{K \in I} \mathbb{P}\{ K_1 \ge \gamma_\uparrow K \} \ge \underline{p}_+ \underline{\delta}_+ > 0, \qquad \inf_{K \in I} \mathbb{P}\{ K_1 \le \gamma_\downarrow K \} \ge \underline{p}_- \underline{\delta}_- > 0.$$

**Assumption 6.3 (Interior Atoms on the Upper Block)** We impose the following conditions on the upper block $U = (K_{p0}, \bar{K}]$:

(i) (Strict positivity of branch probabilities) The down-branch probability satisfies

$$\underline{p}_- := \inf_{K \in U} p_-(K) > 0.$$

(ii) (Strict positivity of conditional strong down-jumps) Under the down-branch, the conditional law of the jump magnitude assigns strictly positive mass (respectively, has density bounded below) to configurations that generate a "strong" down-adjustment. In the discrete-grid specification, this is equivalent to requiring that

$$\underline{\delta}_- := \inf_{K \in U} \mathbb{P}\{ j \geq j_-^* \mid \text{down-branch at } K \} > 0.$$

In the continuous-density version, the same property holds with the conditional probabilities replaced by lower bounds on the conditional densities over a compact interior subinterval of admissible down-jump magnitudes (corresponding to multipliers at most $\gamma_\downarrow < 1$).

**Assumption 6.4 (Common Interior Regeneration Atom)** Let $D_\eta := [K_{q0} + \eta, K_{p0} - \eta]$ be a compact interior sub-interval of the feasible region, with $\eta > 0$ chosen so that $p_\pm(K) > 0$ on $D_\eta$. Assume that the discretized transition kernel contains a common interior regeneration atom: there exist $\xi_0 \in D_\eta$ and a constant $c > 0$ such that

$$P(K, \{\xi_0\}) \geq c \qquad \text{for all } K \in D_\eta.$$

**Notation** For later use, we introduce the following notation. Let $\gamma_\uparrow > 1$ (respectively, $\gamma_\downarrow < 1$) denote the minimal multiplicative factor of proportional increase (respectively, decrease) in the capital stock that constitutes a sufficiently "strong" up-adjustment (respectively, low adjustment).

**Definition 6.2 (Minimal Multiplicative Up-Factor and Threshold Index)** Let $\Gamma_{+,1} \leq \Gamma_{+,2} \leq \cdots \leq \Gamma_{+,J}$ denote the discrete multiplicative factors associated with the interior SQSS configurations under the up-branch. Fix $\gamma_\uparrow > 1$ such that $\gamma_\uparrow^M K_{p0}^{(0.5)} \geq K_{q0} + \eta$ for the value of $M$ in Lemma 6.11. Define the minimal index

$$j_+^* := \min\{ j : \Gamma_{+,j} \geq \gamma_\uparrow \}.$$

Then any selection of an up-configuration with index $j \geq j_+^*$ guarantees a proportional increase of size at least $\gamma_\uparrow$.

**Remark 6.7 (On the State Dependence of Reachability)** Even away from the clipping region, the admissible jump magnitudes generated by the multipliers $\gamma_\pm^j$ remain interior. However, the effective reachability of these interior configurations becomes state dependent, since the branch probabilities $p_\pm(K)$ may become small as $K$ approaches the endogenous turning points $K_{q0}$ and $K_{p0}$. As a result, while the underlying adjustment mechanism continues to point toward interior outcomes, the probability mass allocated to these movements can shrink near the boundaries.

This state-dependent thinning motivates the introduction of a common interior regeneration mechanism in Assumption 6.4, which restores a uniform lower bound on transition probabilities across the interior region.

**Remark 6.8 (Economic Interpretation of the Common interior Atom)** Assumption 6.4 posits that within the compact interior band $D_\eta$ there exists a common "benchmark" capital level $\xi_0$ that is reached with uniformly positive probability in one period, regardless of the current state $K \in D_\eta$. Economically, this captures the idea that, in addition to state-dependent adjustments driven by the multipliers $\gamma_{\pm}^{j}$, there is a small but persistent probability of coordinated reallocation toward a representative interior configuration. Such events may be interpreted as episodes of synchronized adjustment, rebalancing, or institutional realignment that temporarily override the usual state-dependent dynamics.

From a technical perspective, this common interior atom provides a regeneration mechanism: the process has a uniform chance to "reset" to $\xi_0$ from any interior state. This property underpins the one-step minorization condition and ensures that the interior block $D_\eta$ constitutes a small set for the transition kernel, which is crucial for the irreducibility and stability arguments developed below.

**Lemma 6.10 (Interior Minorization: Discrete Kernel Form)** *Let $D_\eta \subset \mathcal{K}$ be a compact interval. Suppose there exist*

- *a finite set of interior points $G = \{\xi_1, \ldots, \xi_N\} \subset D_\eta$,*
- *a constant $c > 0$,*

*such that for every $K \in D_\eta$ and every $i = 1, \ldots, N$,*

$$P(K, \{\xi_i\}) \;\geq\; c.$$

*Define the atomic probability measure $\lambda_M$ on $\mathcal{K}$ by*

$$\nu_G(A) := \frac{1}{N}\sum_{i=1}^{N} \mathbf{1}_A(\xi_i), \qquad A \in \mathcal{B}(\mathcal{K}).$$

*Then there exists $\varepsilon > 0$ (namely $\varepsilon = cN$) such that*

$$P(K, A) \;\geq\; \varepsilon\, \nu_G(A) \quad \textit{for all } K \in D_\eta \textit{ and all Borel } A \subseteq \mathcal{K}.$$

*In particular, $D_\eta$ is a one-step small set for $P$ with reference measure $\nu_G$.*

***Proof (Deferred)*** See Appendix 6.D.1. □

Interior minorization reflects the fact that, within the intermediate block, the discrete transition kernel retains a uniformly random component that does not depend on the precise current level of capital. In this region, neither upward nor downward adjustments are suppressed by clipping, and the discrete jump magnitudes admit strictly positive probability on a fixed set of interior grid points. As a result, from any starting point in the intermediate block, there is a uniformly positive probability of jumping to the same finite collection of capital levels in the next period.

From a probabilistic perspective, this means that the one-step transition law can be decomposed into two parts: a state-dependent component and a state-independent component. The latter assigns a fixed minimum mass to a common reference distribution supported on interior points. This uniform overlap across all states in the block is exactly what the minorization condition captures.

Economically, interior minorization corresponds to a regime in which investment behavior is sufficiently heterogeneous across industries. Even when aggregate conditions are similar, a non-negligible fraction of industries undertakes discrete, sizable revisions—both upward and downward—so that the aggregate capital stock can move to a common set of configurations with positive probability. This creates genuine mixing in the dynamics: Once the process enters the intermediate block, it partially "forgets" where it came from. In short, Lemma 6.10 states that $D_\eta$ is locally small—establishing a lower bound—on the transition probability condition on $D$. But this applies only when the chain is already in $D_\eta$.

The next lemma records the $M$-step push-up from the lower block into the interior block. (The underlying multi-step growth mechanism is developed in Appendix 6.E; see Lemma 6.24 and Corollary 6.25.)

**Lemma 6.11 (M-Step Push-Up from the Lower Block)** *Fix* $1 < K_{p0}^{(0.5)} < K_{q0}$ *and choose* $\eta > 0$ *so that* $[K_{q0}, K_{q0} + \eta] \subset D$. *Suppose that there exist constants* $\gamma > 1$ *and* $c_* > 0$ *such that for every* $K \in [K_{p0}^{(0.5)}, K_{q0} + \eta]$ *(so the condition applies along the entire path up to the target level),*

$$\mathbb{P}_K\left\{ K_1 \geq \gamma K \right\} \geq c_*. \tag{6.5.2}$$

*Define*

$$M := \left\lceil \frac{\log\left((K_{q0} + \eta)/K_{p0}^{(0.5)}\right)}{\log\gamma} \right\rceil. \tag{6.5.3}$$

*Then, for every* $K \in [K_{p0}^{(0.5)}, K_{q0}]$,

$$\mathbb{P}_K\left\{\tau_{[K_{q0}+\eta,\infty)} \leq M\right\} \geq c_*^M,$$

*where*

$$\tau_{[K_{q0}+\eta,\infty)} := \inf\{t \geq 0 : K_t \geq K_{q0} + \eta\}.$$

***Proof (deferred)*** See Appendix 6.D.2. □

The lower block corresponds to a region in which downward adjustments are effectively shut down by clipping, while upward revisions remain feasible with uniformly positive probability. Economically, this reflects a situation in which investment pessimism has been exhausted: Further contraction is no longer operative, but expansionary revisions—however small—still occur. The M-step push-up lemma formalizes the implication of this asymmetry. Starting from any point in the lower block, a sequence of sufficiently many upward revisions occurs with positive probability, and because each revision scales capital multiplicatively, their cumulative effect is to lift the capital stock into the intermediate region within a finite number of steps. In short, the lower block is transient in the upward direction: The dynamics cannot remain there indefinitely.

A symmetric argument applies to the upper block, using repeated down-jumps of the standard level $j^*_-$.

**Lemma 6.12 (Bridge Entry from the Upper Block)** *Under Assumptions 6.2 and 6.3, there exist an integer $M_U \geq 1$, a constant $\delta_U > 0$, and an interval $D_U \subset D_\eta$ such that*

$$P^{M_U}(K, D_U) \geq \delta_U \qquad \textit{for all } K \in U.$$

***Proof (deferred)*** See Appendix 6.D.3. □

The upper block represents the opposite regime. Here, upward revisions are muted by clipping, while downward adjustments remain active. Economically, this corresponds to a state of overexpansion in which further optimism no longer translates into higher investment, but negative revisions are still possible. The bridge-entry lemma shows that, from any point in this region, there is a uniformly positive probability of experiencing a sufficiently strong downward adjustment that carries the capital stock across the boundary into the intermediate block. Thus, the upper block is transient in the downward direction: Although trajectories may linger, they cannot remain isolated from the interior dynamics.

Taken together, these two results show that the lower and upper blocks act as directional funnels rather than absorbing regions. Capital stocks that fall too low are probabilistically pushed upward, while capital stocks that rise too high are eventually pulled back down. The intermediate block therefore serves as a common meeting ground for trajectories originating anywhere in the state space. This structure is what ultimately enables the construction of a small set and a regeneration mechanism: Regardless of where the process starts, it is funneled into a region where both upward and downward adjustments are active and sufficiently random.

Combining the interior minorization with the push-up and bridge-entry lemmas gives a two-sided regeneration mechanism for the process. It upgrades local minorization on an interior set $D$ into a global small-set property on the entire state space $\mathcal{K}$ through a two-stage argument.

**Proposition 6.13 (Two-Stage Regeneration and Small Set)** *Under Assumptions 6.1–6.4, there exist an integer $M \geq 1$, a constant $\delta > 0$, and a probability measure $\nu$ supported on an interval contained in $D_\eta$ such that*

$$P^{M}(K, \cdot) \geq \delta\, \nu(\cdot) \qquad \textit{for all } K \in D_\eta.$$

*Thus $D_\eta$ is an $M$-small set. In particular, the chain admits a two-stage regeneration structure based on returns to a nonempty small set $C \subseteq D_\eta$.*

***Proof (deferred)*** See Appendix 6.D.4. □

The role of Proposition 6.13 is to upgrade the local minorization available on the interior set $D_\eta$ to a global small-set property on the entire state space $\mathcal{K} = [0, K]$. The minorization lemma guarantees that points in $D_\eta \subset D$ admit a uniform one-step regeneration. The push-up and bridge-entry lemmas show that every point in $\mathcal{K}$ reaches $D_\eta$ with probability uniformly bounded away from zero in a uniformly bounded number of steps. By combining these two stages, the chain admits a uniform minorization after $M$ steps from any initial state, which is precisely the small-set condition required by the general ergodic theory of Markov chains.

**Lemma 6.14 (Smallness of the Communicating Class $C$)** *Let $C = [K_{q0}, K_{p0}]$. Suppose there exist an interior interval $J \subset D_\eta$, an integer $M \geq 1$, and a constant $\delta > 0$ such that*

$$P^M(K, J) \geq \delta \qquad \textit{for all } K \in C.$$

*Suppose further that the interior minorization lemma holds on $J$: there exist a probability measure $\nu_G$, an integer $n_0 \geq 1$, and a constant $c > 0$ such that*

$$P^{n_0}(K, A) \geq c\, \nu_G(A) \qquad \textit{for all } K \in J, \quad A \in \mathcal{B}(\mathcal{K}).$$

*Then $C$ is a small set. In particular,*

$$P^{M+n_0}(K, A) \geq \delta c\, \nu_G(A) \qquad \textit{for all } K \in C, \quad A \in \mathcal{B}(\mathcal{K}).$$

***Proof (deferred)*** See Appendix 6.D.5. □

Intuitively, the set $C$ is small because, once the capital stock enters this region, the adjustment mechanism exhibits uniformly positive randomness that is largely independent of the precise value of $K$. In economic terms, $C$ corresponds to a range of capital stocks in which neither upward nor downward investment pressures are muted by clipping, so that both types of adjustments occur with strictly positive probability. As a result, from any starting point in $C$, the process has a uniformly positive chance of producing a sufficiently strong adjustment that effectively "resets" the dynamics. This uniform mixing property is exactly what the small-set condition captures.

### 6.5.5 Lebesgue Minorization and Global $\phi$-Irreducibility

The results of the preceding subsection show that the process cannot become trapped in either boundary region: From any point in the lower or upper block, there is a uniformly positive probability of reaching an interior part of the communicating class within a bounded number of steps. In other words, the interior region $C$ acts as a recurrent "regeneration corridor": Once the capital stock enters this region, the adjustment mechanism exhibits sufficient uniform randomness that the process can probabilistically reset its dependence on past states. This structural property allows us to restrict attention to a strict interior interval, on which the transition kernel exhibits a nondegenerate density uniformly bounded away from zero. By combining this local minorization with thc uniform cntry property established earlier, we can construct a suitable candidate measure and verify that the transition probabilities assign positive mass to every Lebesgue-positive subset of $C$ after finitely many steps. These ingredients lead to the $\phi$-irreducibility of $C$ in the sense of Meyn–Tweedie, completing the second stage of the argument.

To establish $\phi$-irreducibility, we introduce a natural candidate measure based on the local minorization structure.

**Lemma 6.15 (Local Lebesgue–Minorization on an Interior Interval)** *Let $D = [K_{q0}, K_{p0}]$ and $D_0 = (K_{q0}, K_{p0})$, and suppose that Assumption 6.2 holds. Then there exist an open interval $J \subset D_\eta$ with $\overline{J} \subset D_\eta$, an integer $n_0 \geq 1$, and a constant $c_0 > 0$ such that*

$$P^{n_0}(K, A) \geq c_0\, \lambda(A \cap J) \qquad \text{for all } K \in J \text{ and all } A \in \mathcal{B}(D_\eta).$$

***Proof (deferred)*** See Appendix 6.D.6. □

We now fix the candidate irreducibility measure. Let $J \subset D_\eta$ be the interval supplied by Lemma 6.15.

**Definition 6.3 (Candidate irreducibility measure)** Define a finite measure $\phi$ on $D_\eta$ by

$$\phi(A) := \lambda(A \cap J), \qquad A \in \mathcal{B}(D_\eta),$$

where $\lambda$ denotes Lebesgue measure on $\mathbb{R}$.

Thus $\phi$ is simply Lebesgue measure restricted to the interior interval $J \subset D_\eta$.

**Corollary 6.1 (Uniform Entry into $J$)** *Let $C = L \cup D \cup U$ be the communicating class, and let $J \subset D_\eta$ be the interior interval from Lemma 6.15. Suppose that Lemma 6.11, Lemma 6.12, and the interior entry result into $J$ hold. Then there exist an integer $M \geq 1$ and a constant $\delta > 0$ such that*

$$\mathbb{P}_K\{\tau_J \leq M\} \geq \delta \qquad \text{for all } K \in C.$$

***Proof*** See Appendix 6.D.7. □

The interior minorization lemma ensures that, once the chain is inside a strict interior interval $J \subset D_\eta$, its $n_0$-step transition kernel dominates a fixed multiple of Lebesgue measure on $J$. The push-up and bridge-entry lemmas guarantee that, from any starting point in the communicating class $C := D = [K_{q0}, K_{p0}]$, the chain enters $J$ with uniformly positive probability within a bounded number of steps. Combining these two ingredients yields $\phi$-irreducibility with respect to Lebesgue measure restricted to $J$.

**Proposition 6.16 ($\phi$-Irreducibility on $C$)** *Let $C = [K_{q0}, K_{p0}]$ be the communicating class, and let $\phi(A) := \lambda(A \cap J)$, $A \in B(K)$, where $J \subset D_\eta$ is the interval given by Lemma 6.15. Suppose that Lemma 6.15 and Corollary 6.1 hold. Then the Markov chain $\{K_t\}$ is $\phi$-irreducible on $C$.*

***Proof*** Let $A \in \mathcal{B}(C)$ satisfy $\phi(A) > 0$, i.e. $\lambda(A \cap J) > 0$. By Lemma 6.15, there exist $n_0 \geq 1$ and $c_0 > 0$ such that, for all $K' \in J$,

$$P^{n_0}(K', A) \;\geq\; c_0\, \lambda(A \cap J) \;>\; 0.$$

By Corollary 6.1, there exist $M \geq 1$ and $\delta > 0$ such that $P^M(K, J) \geq \delta$ for all $K \in C$.

Fix any $K \in C$. By the Chapman–Kolmogorov equation,

$$P^{M+n_0}(K, A) \;=\; \int_{\mathcal{K}} P^{n_0}(K', A)\, P^M(K, \mathrm{d}K') \;\geq\; \int_J P^{n_0}(K', A)\, P^M(K, \mathrm{d}K').$$

Using the lower bounds on $J$,

$$P^{M+n_0}(K, A) \;\geq\; \inf_{K' \in J} P^{n_0}(K', A)\, P^M(K, J) \;\geq\; c_0\, \lambda(A \cap J)\, \delta \;>\; 0.$$

Thus, for every $K \in C$ and every Borel set $A \subset C$ with $\phi(A) > 0$, there exists $n \geq 1$ such that $P^n(K, A) > 0$. Hence the chain is $\phi$-irreducible on $C$. □

Irreducibility follows from the fact that the capital process cannot remain confined to any one-sided region of the state space. Capital stocks that fall too low are pushed upward with positive probability through repeated expansionary adjustments, while capital stocks that rise too high are eventually pulled back downward by contractionary revisions. Both mechanisms funnel trajectories into a common interior region in which two-sided adjustments are active and sufficiently random. Once inside this region, the process can reach any subset of the state space with positive probability. In this sense, the dynamics are globally connected: No admissible capital level is probabilistically isolated from the rest.

### 6.5.6 Concluding Remark

The $\phi$-irreducibility of $C$ ensures that the process cannot avoid any Lebesgue-positive region of the communicating class in the long run: Every such set is reached with strictly positive probability within a finite number of transitions, regardless of the starting state. This property will be essential in the next stage of the analysis, where it interacts with the Foster–Lyapunov drift condition. In particular, once a suitable Lyapunov function is identified, the combination of $\phi$-irreducibility and an appropriate negative-drift estimate outside a compact set guarantees the existence and uniqueness of an invariant distribution. We turn to this argument in the next section.

## 6.6 Foster–Lyapunov Criterion

The $\phi$-irreducibility established in the previous section guarantees that the Markov chain repeatedly enters a strict interior region of the capital domain. Irreducibility alone, however, does not determine whether the chain is stable or whether it admits a unique long-run distribution. To address these questions, we complement irreducibility with a Lyapunov drift argument.

The central task is to identify a function $V(K)$ that grows with the capital stock and whose one-step expected change,

$$PV(K) - V(K),$$

is uniformly negative outside a compact region. Once such a drift inequality is verified, the Foster–Lyapunov theorem ensures that the Markov chain possesses a unique invariant distribution and is attracted to it from every initial state.

The capital process takes values in the compact state space $\mathcal{K} = [0, \bar{K}]$. To apply the Foster–Lyapunov criterion with $V(K) = K$, we verify the drift inequality on the restricted set

$$S := [1, \bar{K}] \subset \mathcal{K},$$

treating $[0, 1]$ as a small set on which no drift condition is required.

Before constructing the Lyapunov function, we first analyze the geometry of the drift. In particular, the sign of the drift is governed by the difference between upward and downward mutation probabilities, which in turn is driven by the investment-sentiment variable $\zeta(K)$. This leads naturally to a distinguished value—the neutral point—at which the raw upward and downward mutation pressures balance. Identifying this point and showing that it is realized by a unique capital level $K^*$ is a key preparatory step toward the drift analysis.

### *6.6.1 Drift Analysis and the Neutral Point*

Throughout recall that $\zeta(K)$ is strictly decreasing in $K$. Let $b > 0$, $\varphi \in \mathbb{R}$ be the confidence threshold, and $c_b \geq 0$ the status-quo wedge. The (upward, downward) transition probabilities are given by

$$\begin{aligned} p_+(K) &= \text{clip}[\, 0.5 - b\,(\varphi - \zeta(K),\, );\; 0,\; 0.5]\,, \\ p_-(K) &= \text{clip}[\, 0.5 - b\,(\zeta(K) + 1 + c_b - \varphi\,);\; 0,\; 0.5]\,. \end{aligned} \tag{6.6.1}$$

We now impose the following assumption on parameters to ensure that the investment propensity $\zeta(K)$ behaves consistently with the mutation probabilities defined below.

**Assumption 6.5 (Upward Bias at the Left Boundary)** We impose the boundary condition

$$\zeta(0) \;>\; \varphi - \frac{1+c_b}{2}, \qquad \text{where} \quad \varphi := \frac{1}{1+\frac{1}{\alpha(\eta-1)}}.$$

This ensures that, at very low capital stocks, the raw upward mutation probability exceeds the raw downward mutation probability. Equivalently,

$$p_+^{\text{raw}}(0) \;>\; p_-^{\text{raw}}(0).$$

The condition provides the necessary left-end sign for the drift: It guarantees that the mutation pressures favor upward adjustment near $K = 0$, thereby ensuring that the neutral point $K^*$ (where these pressures balance) lies strictly in the interior of the state space.

With these conditions on $\zeta(\cdot)$, there exists a unique notional level of $\zeta$ at which the raw mutation pressures are symmetric and the process exhibits no systematic bias toward upward or downward adjustment. This "neutral point" serves as the intrinsic reference around which the sign of the drift is determined. The following assumption and definition formalize this notion and establish conditions under which the balancing level is realized by a unique capital stock $K^*$ in the interior (no-clipping) domain.

**Definition 6.4 (Neutral Point)** The *neutral point* of the mutation dynamics is the value $\zeta^*$ at which the expected per-unit drift of the capital-stock process is zero.

Under the present symmetric mutation specification, this condition is equivalent to

$$p_+^{\text{raw}} = p_-^{\text{raw}}.$$

Solving yields

$$\zeta^* = \varphi - \frac{1+c_b}{2}. \tag{6.6.2}$$

If $\zeta(K)$ is strictly monotone in $K$, then the corresponding capital stock level $K^*$ satisfying $\zeta(K^*) = \zeta^*$ is called the *neutral capital level*.

The equivalence follows because the upward and downward drift components enter through the same increasing function

$$p \mapsto p\Big(1 - \frac{p}{2}\Big).$$

Under the continuum-industry specification, the quantity

$$p\Big(1 - \frac{p}{2}\Big)$$

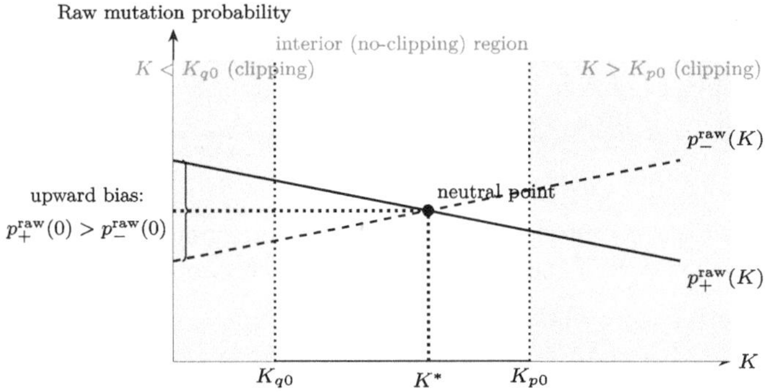


**Fig. 6.3** Raw upward and downward mutation probabilities with clipping regions. The shaded areas indicate the left and right clipping domains $K < K_{q0}$ and $K > K_{p0}$. In the interior region, the raw probabilities are affine in $\zeta(K)$ and intersect once at the neutral point $K^*$

represents the expected aggregate contribution of a given mutation direction. By the law of large numbers, a direction activated with probability $p$ affects a fraction $p$ of industries, while the average normalized mutation magnitude over the activated interval equals $1 - p/2$. Hence the expected per-unit drift contribution is proportional to

$$p\left(1 - \frac{p}{2}\right).$$

Figurc 6.3 providcs a schcmatic illustration of the raw mutation probabilities and highlights the role of the neutral point at which upward and downward mutation pressures balance.

**Remark 6.9 (Economic Interpretation of Neutral Point)** The activation probabilities $p_+(K)$ and $p_+(K)$ respond in opposite directions to changes in the investment propensity $\zeta(K)$: A higher $\zeta$ increases the likelihood of upward mutation and decreases that of downward mutation. For $\zeta(K)$ above this point, upward mutation is more likely; for $\zeta(K)$ below it, downward mutation is more likely. This **balance point** is central in understanding the local behavior of the investment dynamics, particularly in the interior (no-clipping) regime.

The role of the neutral capital level $K^*$, at which the drift vanishes, bears a conceptual resemblance to the normalization properties familiar from general equilibrium theory. In the Walras–Arrow–Debreu framework, excess demand functions are homogeneous of degree zero in prices (see, e.g., (Debreu, 1959)), so that only relative prices matter and equilibrium is determined up to scale. While the present model operates in a different setting, the existence of a neutral point similarly reflects a balance condition that organizes the system's dynamics around an invariant reference level.

**Remark 6.10 (Downward Wedge Fixed from Investment Theory)**
As in the Investment chapter, the downward mutation wedge enters the activation threshold through a constant parameter

$$w_- = 1 + c_b, \qquad c_b \geq 0.$$

No $K$ dependence is introduced here, ensuring that the pre-clipped downward activation probability remains affine in the investment propensity $\zeta(K)$.

**Assumption 6.6 (Small Slope Near the Neutral Point)** Let $b$ satisfy $b < 1/(1 + c_b)$.

Under Assumption 6.6, the balance point $\zeta^* = \varphi - \frac{1+c_b}{2}$ lies in the interior of the domain at which the pre-clipped activation probabilities

$$p_+^{\text{raw}} = 0.5 + b(\zeta - \varphi), \qquad p_-^{\text{raw}} = 0.5 - b(\zeta + 1 + c_b - \varphi)$$

are strictly between 0 and 0.5. Hence, there exists a nonempty neighborhood of $\zeta^*$ on which no clipping occurs.

**Proposition 6.17 (Existence (and Uniqueness) of the Neutral Capital Level $K^*$)** *Let $\zeta(K)$ be strictly decreasing and continuous on $[0, \bar{K}]$, and let $p_\pm(K)$ be defined by*

$$p_+(K) = \text{clip}\big[\, 0.5 + b(\zeta(K) - \varphi)\, ;\ 0, 0.5\big], \tag{6.6.3}$$

$$p_-(K) = \text{clip}\big[\, 0.5 - b(\zeta(K) + 1 + c_b - \varphi)\, ;\ 0, 0.5\big]. \tag{6.6.4}$$

*Define the (interior) raw difference*

$$d(K) := p_+^{\text{raw}}(K) - p_-^{\text{raw}}(K) = b\Big(2(\zeta(K) - \varphi) + 1 + c_b\Big). \tag{6.6.5}$$

*In addition to the lower-endpoint condition in Assumption 6.5, we impose the corresponding upper-endpoint condition*

$$\zeta(\bar{K}) < \varphi - \frac{1 + c_b}{2}.$$

*Together, these endpoint restrictions imply*

$$(i) \quad \zeta(0) > \varphi - \tfrac{1+c_b}{2}, \tag{6.6.6}$$

$$(ii) \quad \varphi > \tfrac{1+c_b}{2}. \tag{6.6.7}$$

*Then $d(0) > 0$ and $d(\bar{K}) < 0$, so by the intermediate value theorem, there exists at least one $K^* \in (0, \bar{K})$ such that*

$$p_+(K^*) = p_-(K^*) \quad \Longleftrightarrow \quad d(K^*) = 0.$$

*Moreover, on any interval on which clipping is inactive (i.e.,* $0 < p_\pm(K) < 0.5$*), we have* $d'(K) = 2b\,\zeta'(K) < 0$*, so* $d(\cdot)$ *is strictly decreasing and the root* $K^*$ *is unique.*

Existence and uniqueness follow immediately from continuity and strict monotonicity of the equilibrium mapping.

In the interior regime, where neither arm is clipped, the drift $f(K)$ inherits its sign from $p_+(K) - p_-(K)$, which is proportional to $\zeta(K) - \zeta^*$. Thus, states with $\zeta(K) > \zeta^*$ exhibit positive expected capital adjustment, while those with $\zeta(K) < \zeta^*$ exhibit negative adjustment. Consequently, $\zeta^*$ represents the intrinsic balancing level of the mutation–imitation process.

### 6.6.2 Lyapunov Function and Foster–Lyapunov Drift Condition

The identification of the neutral point $K^*$ clarifies the qualitative structure of the mutation dynamics: To the left of $K^*$, the upward force dominates, while to the right the downward force prevails. To convert this local sign geometry into a global stability result, we now introduce a Lyapunov function that captures the direction of expected movement in the capital stock. The idea is to measure whether, on average, the process tends to push the state back toward the interior when it wanders too far to the right (and, symmetrically, to control its behavior when it drifts toward the left boundary).

To convert the qualitative drift structure identified above into a formal stability argument, we now introduce a Lyapunov function defined on the entire state space $S = [1, \bar{K}]$. A natural and economically meaningful choice is the linear function $V(K) = K$. Its expected one-step increment,

$$f(K) \;:=\; \mathbb{E}[\,V(K_{t+1}) - V(K_t) \mid K_t = K\,],$$

records the direction of movement of the capital stock: Positive values indicate upward drift, while negative values indicate mean reversion toward the interior of the state space. For notational convenience, we will also write this drift as

$$\Delta V(K) \equiv f(K)$$

and use the two notations interchangeably.

Figure 6.4 provides a geometric overview of the drift argument: The expected drift curve $\Delta V(K)$ crosses zero at $K^*$, indicated by point G, and remains below the Foster–Lyapunov line outside a bounded region, anticipating the formal verification below.

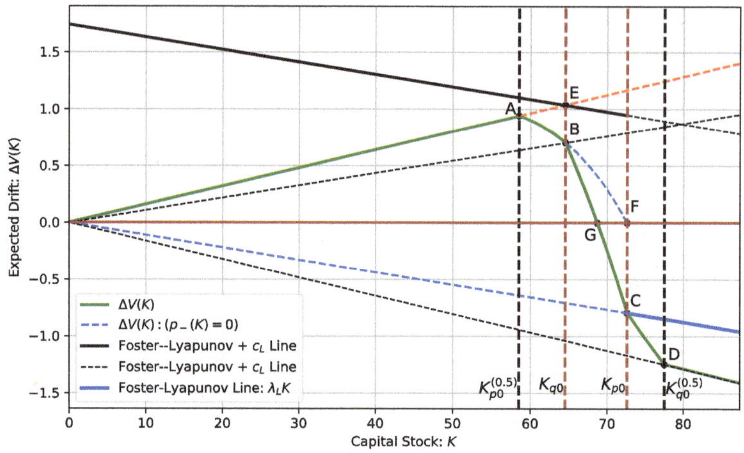


**Fig. 6.4** Geometry of the expected drift and the Foster–Lyapunov bound.
The solid curve plots the expected drift $\Delta V(K)$, and the straight line is the Foster–Lyapunov bound $\lambda_L K + c_L$. Point A corresponds to the maximum of the drift in the positive-drift region, while point G marks the location where the drift crosses zero. For sufficiently large $K$ (points C and D), the drift is strictly negative. Points B and C correspond to the thresholds $K_{q0}$ and $K_{p0}$ that delimit the compact region on which the Foster–Lyapunov inequality is checked. At $K_{p0}$, the point C represents the actual drift $\Delta V(K_{p0})$ (with $p_-(K_{p0}) > 0$), whereas the point F denotes the counterfactual drift obtained under the restriction $p_-(K) = 0$

Our objective is to show that this drift becomes uniformly negative outside a compact region $C_0 \subset S$, which we take to be $C_0 = [1, K_{p0}]$. Establishing this verifies the Foster–Lyapunov drift condition needed for uniqueness of the invariant distribution.

To implement this program, we proceed in two steps. First, we examine the per-unit drift $h(K) = f(K)/K$ and show that it is strictly decreasing on the interior regime $[1, K_{q0}]$. This identifies a maximal feasible growth rate at the lower endpoint and bounds the drift from above by a simple linear function. Second, we construct a Lyapunov line with slope $\lambda_L < 0$,

$$g(K) := f(K) - \lambda_L K,$$

and determine its intercept so that it cuts the upper envelope of feasible drifts at a strategically chosen point. Recall that $K_{p0}^{(0.5)}$ is the point at which $f(K)$ attains its maximum on $[1, K_{q0}]$, and that the interval $[K_{q0}, K_{p0}]$ marks the region in which the branch probabilities begin to clip. It is therefore useful to construct the Lyapunov intercept in advance, before combining these ingredients into the Foster–Lyapunov inequality. To keep the main argument focused, the supporting stability lemmas underlying the Foster–Lyapunov step are deferred to Appendix 6.E. These results provide the multi-step push-up, the uniform entry bound above $K_{q0}$, and the forward-invariance property used below.

The following lemma records the basic geometric facts needed for the Lyapunov-line construction.

**Lemma 6.18 (Construction of the Lyapunov Intercept)** *Let $f(K)$ denote the drift function and let*

$$h(K) := \frac{f(K)}{K}$$

*be the associated per-unit-capital growth rate. Suppose that h is strictly decreasing on the interval $[1, K_{q0}]$. Let*[4]

$$m := h(1) > 0,$$

*and define the constant-growth line*

$$G(K) := mK, \qquad K \geq 0,$$

*which is a ray from the origin with positive slope.*

*Fix $1 < K_{p0}^{(0.5)} < K_{q0}$ and define the points (see Fig. 6.4):*

$$A := \big(K_{p0}^{(0.5)}, G(K_{p0}^{(0.5)})\big), \qquad B := \big(K_{q0}, f(K_{q0})\big), \qquad E := \big(K_{q0}, G(K_{q0})\big).$$

*Define the Lyapunov intercept*

$$c_L := G(K_{q0}) + \mu_L K_{q0},$$

*for some $\mu_L > 0$, and let the corresponding Lyapunov line be*

$$L(K) := c_L - \mu_L K.$$

*Then L passes through E, which lies strictly above both A and B.*

***Proof*** By construction,

$$L(K_{q0}) = c_L - \mu_L K_{q0} = \big(G(K_{q0}) + \mu_L K_{q0}\big) - \mu_L K_{q0} = G(K_{q0}),$$

so $E = (K_{q0}, L(K_{q0}))$ lies on both the Lyapunov line $L$ and the constant-growth line $G$.

Since $G(K) = mK$ with $m > 0$ and $K_{q0} > K_{p0}^{(0.5)} > 0$, we have

$$\text{height}(E) = G(K_{q0}) = mK_{q0} > mK_{p0}^{(0.5)} = G(K_{p0}^{(0.5)}) = \text{height}(A),$$

so $E$ lies strictly above $A$.

---

[4] The slope $m$ corresponds to the maximal drift contribution $0.5(1 - 0.5/2) = 0.375$ attained when $p_+ = 0.5$ (and $p_- = 0$).

Next, for any $K \in [1, K_{q0}]$,

$$f(K) = Kh(K) \; < \; Kh(1) = mK = G(K),$$

where the inequality is strict for $K > 1$ because $h$ is strictly decreasing. In particular,

$$\text{height}(B) \;=\; f(K_{q0}) \;<\; G(K_{q0}) \;=\; \text{height}(E),$$

so $E$ also lies strictly above $B$. Thus $E$ lies strictly above both $A$ and $B$ in the $(K, \Delta V(K))$-plane. □

**Remark 6.11 (Upper Envelope for the Drift)** On $[1, K_{p0}^{(0.5)}]$, the branch probabilities satisfy $p_+ = 0.5$, $p_- = 0$, so the process evolves at the maximal per-unit growth rate $h(1) = m$. The line $G(K) = mK$ therefore represents a hypothetical upper envelope for the drift: Point $E$ is the level that would be attained at $K_{q0}$ if the maximal growth regime were sustained all the way, whereas the actual drift at $K_{q0}$ is strictly lower.

We first ensure that the admissible region $D$ is forward-invariant under the dynamics. This property, established in Appendix 6.F (Lemma 6.26), guarantees that the process remains in the domain on which the drift is analyzed.

**Lemma 6.19 (Monotonicity of the Per-unit Drift)** *Let $p_+^{\text{raw}}(K)$ and $p_-^{\text{raw}}(K)$ be given by*

$$p_+^{\text{raw}}(K) \;=\; \tfrac{1}{2} - b\big(\varphi - \varsigma(K)\big), \qquad p_-^{\text{raw}}(K) \;=\; \tfrac{1}{2} - b\big(\varsigma(K) + 1 + c_b - \varphi\big),$$

*and suppose $\varsigma(K)$ is strictly decreasing in $K$. Then on any interval $D$ on which the branch probabilities are interior and hence unclipped,*

$$p_+(K) \;=\; p_+^{\text{raw}}(K), \qquad p_-(K) \;=\; p_-^{\text{raw}}(K),$$

*with*

$$0 \;<\; p_+(K),\, p_-(K) \;<\; \tfrac{1}{2} \qquad \textit{for all } K \in D,$$

*the per-unit drift*

$$h(K) \;:=\; \frac{\delta}{2b}\Big(p_+(K)\Big(1 - \frac{p_+(K)}{2}\Big) - p_-(K)\Big(1 - \frac{p_-(K)}{2}\Big)\Big)$$

*is strictly decreasing in $K$ on $D$.*

*Proof Idea.* Since $\varsigma(K)$ is strictly decreasing, $p_+(K)$ is strictly decreasing and $p_-(K)$ is strictly increasing on the interior region where the probabilities are unclipped. The map $g(x) = x(1 - x/2)$ is strictly increasing on $[0, 1/2]$, so $g(p_+(K))$ falls in $K$, while $g(p_-(K))$ rises in $K$. Hence their difference—and thus $h(K)$—is strictly decreasing.

***Proof*** Because $\zeta(K)$ is strictly decreasing, we have

$$p_+^{\text{raw}}(K) = \tfrac{1}{2} - b\varphi + b\,\zeta(K) \quad \Longrightarrow \quad \frac{d}{dK}p_+^{\text{raw}}(K) = b\,\zeta'(K) \; < \; 0,$$

and

$$p_-^{\text{raw}}(K) = \tfrac{1}{2} - b\big(1 + c_b - \varphi\big) - b\,\zeta(K) \quad \Longrightarrow \quad \frac{d}{dK}p_-^{\text{raw}}(K) = -b\,\zeta'(K) \; > \; 0.$$

On the interior interval $D$, we have $p_\pm = p_\pm^{\text{raw}}$, so $p_+(K)$ is strictly decreasing and $p_-(K)$ is strictly increasing on $D$, with

$$0 < p_+(K),\; p_-(K) < \tfrac{1}{2}.$$

Define

$$g(x) \;:=\; x\Big(1 - \frac{x}{2}\Big), \qquad x \in [0, 1].$$

Then

$$g'(x) = 1 - x,$$

so $g'(x) > 0$ for $x < 1$ and in particular for all $x \in [0, \frac{1}{2}]$. Hence, $g$ is strictly increasing on $[0, \frac{1}{2}]$. On the interval $D$, we therefore have:

- Since $p_+(K)$ is strictly decreasing and $g$ is strictly increasing, the composite $g\big(p_+(K)\big)$ is strictly decreasing in $K$.
- Since $p_-(K)$ is strictly increasing and $g$ is strictly increasing, the composite $g\big(p_-(K)\big)$ is strictly increasing in $K$, so $-g\big(p_-(K)\big)$ is strictly decreasing.

It follows that the sum

$$g\big(p_+(K)\big) - g\big(p_-(K)\big)$$

is strictly decreasing in $K$ on $D$. Multiplying by the positive constant $\delta/(2b)$ does not change monotonicity, so $h(K)$ is strictly decreasing on $D$. □

The condition $\zeta'(K) < 0$, which drives the monotonicity of the drift, admits a transparent economic interpretation in terms of the elasticity of desired capital with respect to installed capital.

**Remark 6.12 (Elasticity Interpretation)** The fundamental stabilizing mechanism in the model rests on the condition $\zeta'(K) < 0$, which can be expressed in a more economically transparent form. Since $g(K) = K^d(K)/K$ is strictly decreasing, the elasticity of desired capital with respect to installed capital is

$$\varepsilon_{K^d,K} = \frac{d\log K^d}{d\log K} = 1 + \frac{d\log g(K)}{d\log K} < 1.$$

Thus the desired capital stock grows *less than proportionally* with the installed stock. This shortfall in proportionality is equivalent to the condition that guarantees $\zeta(K)$ is strictly decreasing, and hence, that the drift responds negatively to increases in $K$. In economic terms, the system stabilizes because capital demand does not keep pace with capital supply.

Let

$$\lambda_L := h(K_{p0}) < 0, \quad \text{and} \quad c_L := G(K_{q0}) + \lambda_L K_{q0}, \tag{6.6.8}$$

where $K_{q0}$ is the smallest $K$ (to the left of $K_{p0}$) where the interior expressions still apply for both $p_+$ and $p_-$, but the downward arm is about to clip at $p_- = 0.5$. (Any point in $[0, K_{p0}]$ could be used to define $c_L$; picking $K_{q0}$ just tightens the bound on the middle segment.)

Then our goal is to show that

$$f(K) \le \lambda_L K + c_L \qquad \text{for all } K \in [0, K_{p0}],$$

and

$$f(K) \le \lambda_L K \qquad \text{for all } K \in [K_{p0}, \bar{K}].$$

**Proposition 6.20 (Drift Condition with Wedge $c_b$)** *Suppose $\zeta(K)$ is strictly decreasing in $K$. With $p_+$ and $p_-$ defined in Eqs.* (6.6.3)–(6.6.4) *and $V(K) = K$, there exist $\lambda_L < 0$ and $c > 0$ such that*

$$\begin{cases} f(K) \le \lambda_L K + c, & K \in [1, K_{p0}], \\ f(K) \le \lambda_L K, & K \in [K_{p0}, \bar{K}], \end{cases}$$

*where one admissible choice is $(\lambda_L, c) = (\lambda_L, c_L)$ as defined in Lemma 6.18.*

***Proof*** Let $\mu_L > 0$ be the slope parameter from Lemma 6.18 and set $\lambda_L := -\mu_L < 0$. Then the Lyapunov line $L(K) = c_L + \lambda_L K$ goes through $E$ so that it dominates $f$ on $[1, K_{p0}]$ and becomes a strict linear upper bound on $[K_{p0}, \bar{K}]$; monotonicity of $h(K) = f(K)/K$ (Lemma 6.19) ensures $f$ remains below the constant-growth ray, yielding the stated piecewise inequality.

On the interior region, where $p_\pm(K) = p_\pm^{\text{raw}}(K)$, we have

$$p_+(K) + p_-(K) = 1 - b(1 + c_b), \qquad p_+(K) - p_-(K) = b\big[2(\zeta(K) - \varphi) + 1 + c_b\big].$$

Hence,

$$h(K) = \frac{\delta}{2b}\Big[(p_+ - p_-) - \tfrac{1}{2}(p_+^2 - p_-^2)\Big] \tag{6.6.9}$$

$$= \frac{\delta}{2b}(p_+ - p_-)\Big(1 - \tfrac{1}{2}(p_+ + p_-)\Big) \tag{6.6.10}$$

$$= \frac{\delta}{2b} \cdot b[2(\zeta(K) - \varphi) + 1 + c_b] \cdot \tfrac{1}{2}\big[1 + b(1 + c_b)\big] \tag{6.6.11}$$

$$= \frac{\delta}{4}[2(\zeta(K) - \varphi) + 1 + c_b][1 + b(1 + c_b)]. \tag{6.6.12}$$

In view of Table 6.2, $\zeta(K_{p0}) - \varphi = -1/(2b)$, which yields

$$\lambda_L := h(K_{p0}) = -\frac{\delta}{4b}\big[1 - b^2(1 + c_b)^2\big] < 0.$$

For $K \geq K_{p0}$, we are in the upper regime with $p_+(K) = 0$, so

$$h(K) = \frac{f(K)}{K} = -\frac{\delta}{2b}p_-(K)\Big(1 - \frac{p_-(K)}{2}\Big).$$

Because $p_-(K)$ is increasing in $K$ and the function $x \mapsto x(1 - x/2)$ is strictly increasing on $[0, 1/2]$, the composite $p_-(K)\big(1 - p_-(K)/2\big)$ is increasing in $K$, and hence, $h(K)$ is strictly decreasing in $K$ on $[K_{p0}, \bar{K}]$. In particular,

$$h(K) \leq h(K_{p0}) = \lambda_L \qquad \text{for all } K \in [K_{p0}, \bar{K}],$$

and multiplying by $K > 0$ gives

$$f(K) = K\,h(K) \leq \lambda_L K, \qquad K \in [K_{p0}, \bar{K}].$$

On the lower and interior region $[1, K_{p0}]$, the per-unit drift $h(K) = f(K)/K$ is bounded above by the constant-growth line $G(K) = mK$, as shown in Lemma 6.19. By Lemma 6.18, the Lyapunov line $K \mapsto \lambda_L K + c_L$ lies above $G(K)$ on $[1, K_{p0}]$. Combining these bounds yields

$$f(K) = Kh(K) \leq KG(K) \leq \lambda_L K + c_L, \qquad K \in [1, K_{p0}].$$

Taking $c = c_L$ completes the proof.

Collecting the bounds obtained above, we have established a global Foster–Lyapunov drift inequality of the form

$$PV(K) - V(K) \leq \lambda_L K + c\,\mathbf{1}_C(K), \qquad K \in S,$$

with $V(K) = K$, $\lambda_L < 0$, and $C = [1, K_{p0}]$. This verifies the Foster–Lyapunov criterion for stability. In particular, the Markov process possesses a unique invariant distribution on the compact state space $S = [1, \bar{K}]$, and the distribution of $K_t$ converges to this invariant law from every initial condition. □

The Foster–Lyapunov drift condition derived above establishes the key stabilizing force that keeps the capital-stock process from drifting indefinitely toward the upper boundary of the state space. To obtain existence and uniqueness of the invariant distribution, however, this negative-drift property must be complemented by two additional structural ingredients: irreducibility and aperiodicity. These are not extra assumptions but consequences of the model's multiplicative jump mechanism, which ensures that the process can enter any subinterval of $S$ with positive probability and cannot become trapped in deterministic cycles.

By combining the drift inequality with these regularity properties, the general ergodic theorem for Markov chains applies directly. The next section assembles these elements and derives the resulting ergodicity and long-run behavior of the capital-stock process.

## 6.7 Ergodicity and Long-Run Behavior

Having established a Foster–Lyapunov drift condition together with $\phi$-irreducibility, aperiodicity, and the existence of a small set, we now examine the long-run behavior of the capital-stock process. Taken together, these ingredients place the model within the classical framework of ergodic Markov chains on compact state spaces. In such settings, negative drift outside a compact region and recurrent returns to that region imply not only the existence of an invariant distribution, but also its uniqueness and the global convergence of transition probabilities toward this limiting law. Our aim in this section is to make these implications explicit and to clarify the sense in which the economy eventually forgets its initial condition.

To visualize how the structural components interact to produce geometric ergodicity, Fig. 6.5 summarizes the logical dependencies among drift, irreducibility, aperiodicity, and the small-set property.

Existence of an invariant measure follows from the Krylov–Bogoliubov theorem (Hairer 2018, Thm. 4.1). Uniqueness and geometric ergodicity follow from the drift and minorization conditions established above, together with irreducibility and aperiodicity (see Hairer 2018, Thm. 4.4 or Meyn & Tweedie 1993, Thm. 15.0.1).

**Corollary 6.21 (Ergodicity and Uniqueness of the Invariant Distribution)** *Under Assumptions 6.2 and the conditions established in Proposition 6.20, the capital-stock process $\{K_t\}_{t\geq 0}$ admits a unique invariant probability measure $\pi$ on the compact state space $C = [1, \bar{K}]$. Moreover, for every initial state $K_0 \in S$,*

$$\|P^t(K_0, \cdot) - \pi(\cdot)\|_{\mathrm{TV}} \longrightarrow 0 \qquad \text{as } t \to \infty.$$

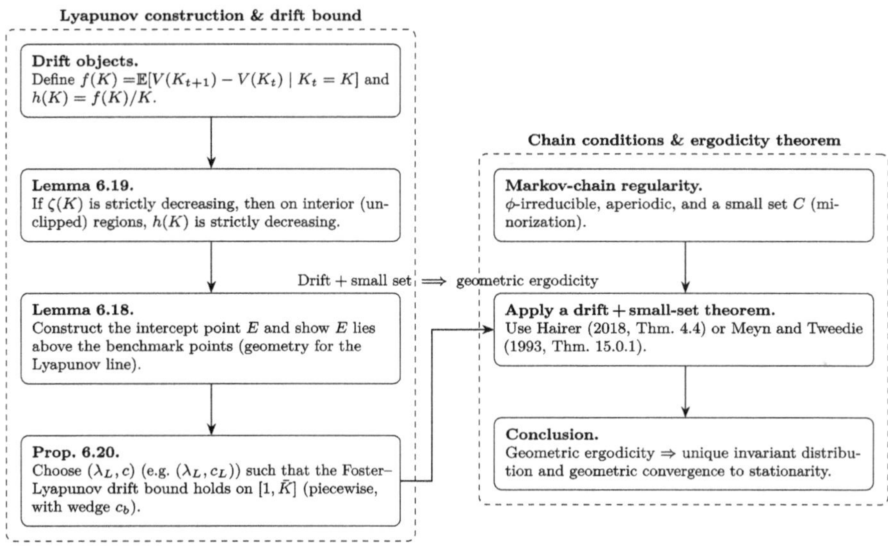


**Fig. 6.5** Logical structure of the ergodicity argument.
Under $\phi$-irreducibility and aperiodicity, the Foster–Lyapunov drift condition and the existence of a small set $C$ (minorization) imply geometric ergodicity and uniqueness of the invariant distribution (Hairer 2018, Thm. 4.4; Meyn & Tweedie 1993, Thm. 15.0.1)

*In particular, the process is positive Harris recurrent and ergodic.*

***Proof (sketch)*** Compactness of $C$ and the Feller property of the transition kernel imply existence of an invariant probability measure via the Krylov–Bogoliubov theorem. Proposition 6.20 establishes a Foster–Lyapunov drift inequality with Lyapunov function $V(K) = K$, negative drift outside the small set $C = [1, K_{p0}]$, and a uniform bound on the expected return time to $C$. Together with the $\phi$-irreducibility and aperiodicity established in Sect. 6.5.3, these conditions imply positive Harris recurrence, uniqueness of the invariant measure, and convergence of transition probabilities in total variation; see, for example, Meyn and Tweedie (1993, Thm. 13.0.1; Thm. 15.0.1) or Hairer (2018, Thm. 4.4). □

The corollary implies that, despite the heterogeneity and randomness of industry-level adjustments, the aggregate capital stock settles into a well-defined long-run statistical equilibrium. The invariant distribution $\pi$ records the frequency with which the economy occupies different regions of the capital space and provides a probabilistic description of the model's stationary macroeconomic environment.

**Remark 6.13 (Stabilizing Mechanism)** This stability result ultimately reflects the elasticity condition $\varepsilon_{K^d,\,K} < 1$ introduced earlier (Remark 6.12), which ensures that the drift turns negative as $K$ increases.

As capacity becomes increasingly slack at high values of $K$, the excess replacement ratio $\zeta(K)$ declines, reducing the probability of upward adjustment and strengthening downward pressure. This elasticity-based mechanism generates the negative drift required for stability and keeps the long-run distribution concentrated within a finite range.

Ergodicity further guarantees that these long-run frequencies do not depend on initial conditions: Regardless of its starting point, the economy eventually displays the same stationary pattern of capital accumulation and adjustment. The invariant distribution therefore serves as a natural summary of the model's long-run macroeconomic behavior.

The ergodic structure developed here provides the foundation for analyzing how macroeconomic parameters affect the shape and location of the invariant distribution. In later chapters, we examine how the mode $K^*$ and the dispersion of $\pi$ respond to changes in monetary conditions, labor-market fundamentals, and nominal wage rigidities, and how these comparative dynamics illuminate the underlying mechanisms of the model.

Having established geometric ergodicity, we next derive bounds on the support of the resulting invariant distribution.

**Lemma 6.22 (Bounds on the Support of the Invariant Distribution)** *Consider the capital-stock Markov chain defined above, and let $K_{q0} < K_{p0}$ denote the clipping points defined earlier.*

*Then for any invariant probability measure $\pi$, its support is contained in the closed interval*

$$\mathrm{supp}(\pi) \subseteq \Big[\big(1 - \tfrac{\delta}{2b}\big)K_{q0},\ \big(1 + \tfrac{\delta}{2b}\big)K_{p0}\Big]. \tag{6.7.1}$$

*Moreover, every open subinterval that intersects* supp($\pi$) *has positive $\pi$-measure.*

*Proof.* See Appendix 6.D.8.

The interval $D = [K_{q0}, K_{p0}]$ serves as the drift-control ("smal") set in the Foster–Lyapunov argument, ensuring tightness and recurrence, but it does not coincide with the support of the invariant distribution. Because the Markov jump mechanism can move the state outside $D$, the actual support expands to $[\big(1 - \delta/(2b)\big)K_{q0}, \big(1 - \delta/(2b)\big)K_{p0}]$.

Figure 6.6 illustrates the transition probabilities and the resulting invariant distribution of the capital-stock process under the baseline assumptions. It revisits the transition probabilities introduced earlier and stacks them with the associated invariant distribution to illustrate how the transition probabilities shape long-run fluctuations.

The distribution is sharply centered around the neutral point $K^*$, reflecting the balance between upward and downward adjustment probabilities in the vicinity of this level. As predicted by Lemma 6.22, the density remains confined to the interval $\big[(1 - \delta/2b)K_{q0},\ (1 + \delta/2b)K_{p0}\big]$, and no mass accumulates outside this range.

Although the thresholds $K_{q0}$ and $K_{p0}$ play a central role in shaping the dynamics, they do not constitute hard bounds on the support of the invariant distribution. Because transition probabilities remain strictly positive even at the edges of the

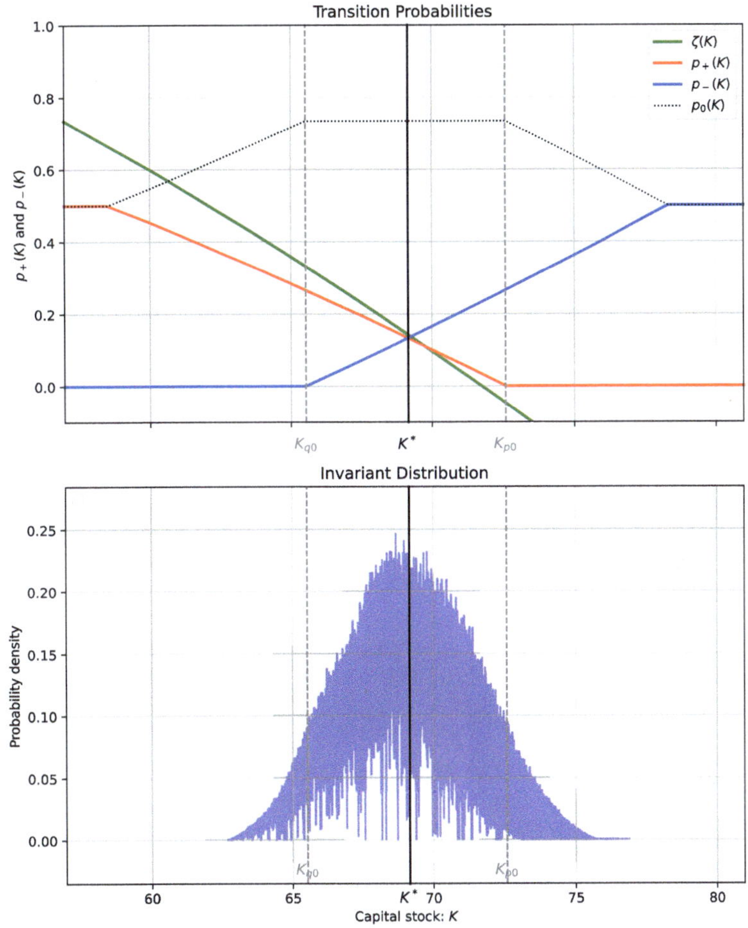


**Fig. 6.6** Transition probabilities and invariant distribution of the capital-stock process. The figure displays a numerically computed invariant distribution for the baseline Markov process defined by the clipped transition probabilities $p_{\pm}(K)$ and the multiplicative jump structure $(1 \pm \delta)K$. The density is unimodal and tightly concentrated near the neutral point $K^*$, where the average drift vanishes. Outside this region, the distribution decays rapidly and remains contained within the theoretical support interval $\left[(1-\delta/2b)K_{q0},\ (1+\delta/2b)K_{p0}\right]$. In this chapter, the labor-supply ceiling is chosen sufficiently high so that the invariant distribution does not interact with full-employment constraints; the resulting measure therefore reflects only the intrinsic dynamics of the capital-adjustment mechanism without wage feedback. The simulation underlying the lower panel of Fig. 6.6 was performed using 10,000 iterations. The parameter values used in the simulation are reported in Table 6.2 of Appendix 6.H

window, stochastic disturbances allow occasional excursions beyond $[K_{q0}, K_{p0}]$. The thresholds therefore delimit a region of dominant drift rather than an absorbing range, and the invariant distribution reflects this by exhibiting thin but nonzero tails outside the core interval.

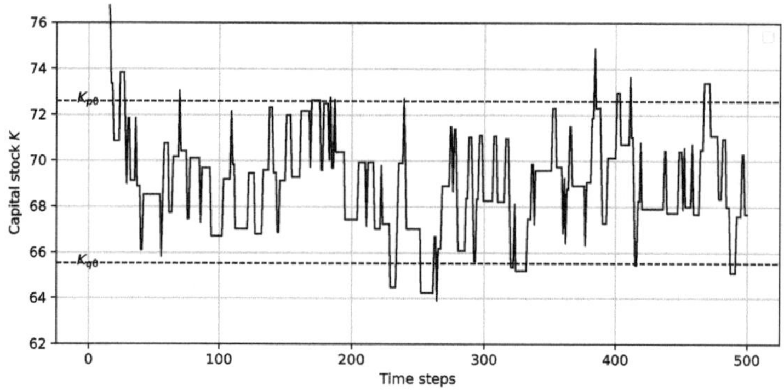


**Fig. 6.7** Sample trajectory of the capital stock under stochastic dynamics.
The figure shows a representative sample path of the capital stock $K_t$ generated by the stochastic transition dynamics. The trajectory fluctuates persistently around the neutral point $K^*$ and remains for most of the time within the admissible corridor $[K_{q0}, K_{p0}]$. In the presence of a unique invariant distribution, such sample paths are ergodic: Long-run time averages along a single realization coincide with the moments of the invariant distribution

At this stage of the analysis, labor supply is taken to be sufficiently high so that full-employment effects do not truncate the distribution. Consequently, the figure captures the intrinsic long-run behavior of the stochastic capital-adjustment mechanism before wage dynamics and price-level feedback are introduced in later chapters.

To complement the distributional results above, Fig. 6.7 illustrates a representative sample trajectory of the capital stock, providing a time-series counterpart to the invariant distribution and the *stability corridor* introduced in Chap. 1.

**Summary and Outlook**

In this chapter, we have analyzed the stochastic dynamics of capital adjustment under the simplifying assumption that the nominal wage $W$ remains fixed. With labor costs held constant, the investment mechanism operates independently of wage adjustment: The probabilities of upward and downward adjustment, the resulting drift structure, and the regeneration properties of the Markov chain jointly determine a unique invariant distribution. The equilibrium distribution obtained here therefore captures the *intrinsic* long-run fluctuations of capital when the cost of labor is rigid and does not respond to economic conditions. However, a fixed nominal wage is not a long-run equilibrium institution in a growing or contracting economy. As the economy repeatedly approaches full employment or experiences persistent slack, the discrepancy between the Wicksellian optimal wage $W^*$ and the prevailing nominal wage necessarily induces pressure for revision, setting in motion the mechanisms of inflation or deflation. The next chapter develops this wage-adjustment channel and embeds it into the present framework, enabling a unified analysis of output, employment, price-level movements, and the policy mix required to stabilize the macroeconomy.

# Appendix

## 6.A Summary of the Stochastic Transition Mechanism

This section summarizes the stochastic mechanism underlying the capital-stock dynamics analyzed in this chapter. The purpose is to clarify the economic structure of the transition process and its relation to the discrete Markov kernel used in the formal analysis:

1. **Exogenous draws per period.** Given the current aggregate capital stock $K_t$, each period is characterized by a collection of exogenous draws. First, a macrosentiment shock $\nu_t$ is drawn independently over time from a fixed distribution $F_\nu$. Second, at the industry level, idiosyncratic shocks and auxiliary random variables are drawn independently across firms. Because $\nu_t$ is i.i.d., it does not generate persistence directly; integrating over its distribution allows the aggregate transition probabilities to be expressed as functions of the current state $K_t$ alone.
2. **Branch probabilities (micro to macro).** Given $K_t$, the macrosentiment shock, and the cross section of idiosyncratic disturbances, firms form confidence indices and make threshold-based decisions to expand, contract, or maintain their capital stock. In the large-population limit, the law of large numbers implies that these micro decisions aggregate into state-dependent transition probabilities

$$p_+(K_t), \quad p_-(K_t), \quad p_0(K_t) = 1 - p_+(K_t) - p_-(K_t),$$

representing the fractions of industries that move up, move down, or remain inactive, respectively.

**Remark (Latent Industry-State Layer)** Between the macro sentiment shock and firm-level decisions lies an implicit cross-industry state variable capturing utilization and demand conditions at the industry level. While not modeled as a separate state variable, this layer affects how the macro shock translates into firm confidence and hence into the aggregate transition probabilities $p_\pm(K)$. In particular, when a large fraction of industries operate under weak conditions, upward adjustments tend to be attenuated even under favorable macro sentiment. Thus, the transition kernel reflects not only the aggregate state $K$, but also the latent distribution of industry conditions summarized through the aggregation.

3. **Multiplicative adjustment mechanism.** Conditional on the aggregate branch realization, capital evolves multiplicatively. On the upward branch, capital increases by a state-dependent factor; on the downward branch, it decreases analogously. Conceptually, these adjustments can be described using continuous random multipliers derived from underlying uniform draws. This representation highlights that, even conditional on a given branch, capital updates are dispersed rather than deterministic.

4. **Aggregation and expected drift.** With conditional independence across industries, the aggregate capital stock evolves according to the cross-sectional average of individual adjustments. The resulting expected one-step growth factor,

$$\mathbb{E}\left[\frac{K_{t+1}}{K_t}\,\middle|\, K_t = K\right],$$

summarizes the local drift of the process and provides intuition for the sign structure used in the Foster–Lyapunov analysis. This expectation motivates the identification of drift thresholds and the admissible region for long-run fluctuations.
5. **Discretized Markov kernel.** While the stochastic mechanism is naturally described in continuous terms, the formal analysis in this chapter proceeds by discretizing the state space for $K$ and representing the one-step law of motion as a time-homogeneous Markov kernel on a compact grid. The discretized kernel aggregates the branch probabilities and multiplicative adjustments into a finite transition matrix that preserves the Markov property. Existence, uniqueness, and ergodicity of the invariant distribution are established for this discrete Markov chain, which serves as the analytical object throughout the chapter.
6. **Definition**

**Definition 6.5 (Minimal Multiplicative Up-Factor and Threshold Index)** Let $\Gamma_{+,1} \le \Gamma_{+,2} \le \cdots \le \Gamma_{+,J}$ denote the discrete multiplicative factors associated with the interior SQSS configurations under the up-branch. Fix $\gamma_\uparrow > 1$ such that $\gamma_\uparrow^M K_{p0}^{(0.5)} \ge K_{q0} + \eta$ for the value of $M$ in Lemma 6.11. Define the minimal index

$$j_+^* := \min\{j : \Gamma_{+,j} \ge \gamma_\uparrow\}.$$

Then any selection of an up-configuration with index $j \ge j_+^*$ guarantees a proportional increase of size at least $\gamma_\uparrow$.

## 6.B Algebraic Identities

### 6.B.1 *Monotonicity of $\zeta(K)$*

**Lemma 6.23 (Monotonicity of $\zeta(K)$)** *Let $\kappa(r)$ be defined by*

$$\kappa(r) = \frac{\alpha}{1-\alpha}\,\theta_{\mathrm{PM}}\,(r+\delta)^{-\frac{1}{1-\alpha}}, \qquad \delta > 0, \quad 0 < \alpha < 1, \tag{6.B1}$$

*and suppose that the capital-demand schedule satisfies*

$$K^d(K) = \kappa(r(K))^{1-\alpha}\,(c+\delta)\,K, \qquad c+\delta > 0, \tag{6.B2}$$

*where $r(K)$ is continuous and strictly increasing in $K$. Recall that*

$$\zeta(K) := \frac{1}{\delta}\Big(\frac{K^d(K)}{K} - 1\Big).$$

*Then $\zeta(K)$ is continuous and strictly decreasing in $K$, and in particular*

$$\zeta'(K) < 0 \qquad \textit{for all } K > 0. \tag{6.B3}$$

***Proof*** From (6.B2),

$$\frac{K^d(K)}{K} = (c+\delta)\,\kappa(r(K))^{1-\alpha}, \qquad \Rightarrow \qquad \zeta(K) = \frac{(c+\delta)\,\kappa(r(K))^{1-\alpha} - 1}{\delta}.$$

Differentiate $\kappa(r)^{1-\alpha}$ using (6.B1). Since $\kappa(r)^{1-\alpha} = \left(\frac{\alpha}{1-\alpha}\theta_{\mathrm{PM}}\right)^{1-\alpha}(r+\delta)^{-1}$, we have

$$\frac{d}{dr}\kappa(r)^{1-\alpha} = -\frac{\kappa(r)^{1-\alpha}}{r+\delta}.$$

Hence, by the chain rule,

$$\zeta'(K) = \frac{c+\delta}{\delta}\frac{d}{dr}\kappa(r(K))^{1-\alpha}\,r'(K) = -\frac{c+\delta}{\delta}\frac{\kappa(r(K))^{1-\alpha}}{r(K)+\delta}\,r'(K).$$

Because $c+\delta > 0$, $\delta > 0$, $\kappa(r(K))^{1-\alpha} > 0$, $r(K)+\delta > 0$, and $r'(K) > 0$ by assumption, it follows that $\zeta'(K) < 0$. Continuity is immediate from continuity of $r(\cdot)$. □

## 6.B.2 *Proof of Lemma 6.1 (Clipping Thresholds Implied by Monotone Investment Sentiment)*

***Proof*** Since $\zeta(K)$ is continuous and strictly decreasing in $K$ (Lemma 6.23), it admits a continuous inverse on its range. Each threshold in Table 6.2 is defined as the unique solution to a level condition $\zeta(K) = \bar{\zeta}$ associated with a clipping boundary (e.g., raw probability equal to 0 or 0.5). The ordering of these level values in Table 6.2 therefore implies the stated ordering of the corresponding $K$-thresholds: $K_{p0}^{(0.5)} < K_{q0} < K_{p0} < K_{q0}^{(0.5)}$.

By construction of the clipped probabilities (6.3.3), the raw expressions fall below zero beyond the clipping thresholds, hence $p_+(K) = 0$ for all $K \geq K_{p0}$ and $p_-(K) = 0$ for all $K \leq K_{q0}$. The forward invariance of $S$ and the support restriction then follow from Lemma 6.26. □

## 6.C Continuity of Clipped Transition Primitives

This section collects proofs of the continuity results stated in Lemmas 6.4 and 6.5.

***Proof of Lemma 6.4*** Define $\Pi(x) := \operatorname{clip}[x; \underline{c}, \bar{c}] = \min\{\max\{x, \underline{c}\}, \bar{c}\}$. It is standard that $\Pi$ is continuous and 1-Lipschitz on $\mathbb{R}$, i.e., $|\Pi(x) - \Pi(y)| \le |x - y|$ for all $x, y$ (nonexpansiveness of the projection onto a closed interval).

Set $u(K) := a + b\, g(K)$. If $g$ is continuous, then $u$ is continuous; hence, $h = \Pi \circ u$ is continuous as a composition of continuous maps.

If $g$ is $L_g$-Lipschitz, then for any $K_1, K_2$,

$$|h(K_1) - h(K_2)| = |\Pi(u(K_1)) - \Pi(u(K_2))| \le |u(K_1) - u(K_2)| \tag{6.C1}$$

$$= |b|\, |g(K_1) - g(K_2)| \le |b| L_g\, |K_1 - K_2|. \tag{6.C2}$$

Thus $h$ is $|b| L_g$-Lipschitz. □

***Proof of Lemma 6.5*** Continuity of $\zeta(K)$ implies continuity of the affine expressions inside the clip operators; Lemma 6.4 then yields continuity of $p_\pm$ and hence of $\vartheta_\pm = 1 - p_\pm$. The explicit formulas Eqs. (6.4.2) express $\Gamma_\pm$ as products and affine transformations of continuous functions, so they are continuous as well. □

## 6.D Technical Proofs for the Stochastic Dynamics

### *6.D.1 Proof of Lemma 6.10 (Interior-Minorization: Discrete Kernel)*

***Sketch*** Fix $K \in D_\eta$ and a Borel set $A \subseteq \mathcal{K}$. Since the atoms $\{\xi_i\}$ are disjoint,

$$P(K, A) \ge \sum_{i=1}^{N} P(K, \{\xi_i\})\, \mathbf{1}_A(\xi_i) \ \ge\ c \sum_{i=1}^{N} \mathbf{1}_A(\xi_i).$$

On the other hand, by the definition of $\nu_G$,

$$\nu_G(A) = \frac{1}{N} \sum_{i=1}^{N} \mathbf{1}_A(\xi_i).$$

Combining the two displays gives

$$P(K, A) \ \ge\ c \sum_{i=1}^{N} \mathbf{1}_A(\xi_i) = cN \frac{1}{N} \sum_{i=1}^{N} \mathbf{1}_A(\xi_i) = cN\, \nu_G(A).$$

Setting $\varepsilon := cN$ yields

$$P(K, A) \geq \varepsilon\, \nu_G(A),$$

for all $K \in D_\eta$ and all Borel sets $A$, as claimed. □

### 6.D.2 *Proof of Lemma 6.11 (M-Step Push-up from the Lower Block)*

***Sketch*** Fix $K \in [K_{p0}^{(0.5)}, K_{q0}]$ and consider the event that the process realizes $M$ consecutive up-moves satisfying

$$K_{t+1} \geq \gamma K_t, \qquad t = 0, \ldots, M-1.$$

As long as the process has not yet reached $[K_{q0} + \eta, \infty)$, we have $K_t \in [K_{p0}^{(0.5)}, K_{q0} + \eta]$, so the uniform bound (6.5.2) applies at each step. Hence, by the Markov property, the probability of this consecutive-up event is at least $c_*^M$.

On this event,

$$K_M \geq \gamma^M K \geq \gamma^M K_{p0}^{(0.5)} \geq K_{q0} + \eta,$$

where the last inequality follows from the definition of $M$ in (6.5.3)[5]. Therefore

$$\mathbb{P}_K\{\tau_{[K_{q0}+\eta,\infty)} \leq M\} \geq c_*^M.$$

Since the bound is uniform in $K \in [K_{p0}^{(0.5)}, K_{q0}]$, the result follows. □

### 6.D.3 *Proof of Lemma 6.12 (Bridge Entry from the Upper Block)*

***Sketch*** Consider the path consisting of $M_U$ consecutive down-jumps at level $j_-^*$. For any starting point $K \in U = (K_{p0}, \bar{K}]$, the state after $m$ such jumps satisfies

$$K_m \leq K\, \gamma_\downarrow^m,$$

[5] Equation (6.5.3) follows immediately from taking the natural logarithm of Eq. (6.D1), yielding the requirement on $M$: $M \geq \left(\frac{\log(K_{q0}+\eta)}{\log \gamma}\right)$.

since selection of $j_-^*$ guarantees a multiplicative decrease of at least $\gamma_\downarrow < 1$. Because $K \le \bar{K}$ and $\gamma_\downarrow < 1$, we may choose $M_U$ large enough that the resulting state lies inside $D$ for every $K \in U$; it is sufficient to ensure

$$\bar{K}\,\gamma_\downarrow^{M_U} \le K_{p0}.$$

Let $D_U \subset D_\eta$ to be a compact interval containing all points of the form

$$K \mapsto \mathrm{clip}_{[1,\bar{K}]}(K\gamma_\downarrow^{M_U}) \text{ with } K \in U.$$

***Sketch*** Fix $K \in [K_{p0}^{(0.5)}, K_{q0}]$ and consider the event that the process realizes $M$ consecutive up-moves satisfying

$$K_{t+1} \ge \gamma K_t, \qquad t = 0, \dots, M-1.$$

As long as the process has not yet reached $[K_{q0} + \eta, \infty)$, we have $K_t \in [K_{p0}^{(0.5)}, K_{q0} + \eta]$, so the uniform bound (6.5.2) applies at each step. Hence, by the Markov property, the probability of this consecutive-up event is at least $c_*^M$.

On this event,

$$K_M \ge \gamma^M K \ge \gamma^M K_{p0}^{(0.5)} \ge K_{q0} + \eta. \tag{6.D1}$$

where the last inequality follows from the definition of $M$ in (6.5.3). Therefore

$$\mathbb{P}_K\{\tau_{[K_{q0}+\eta,\infty)} \le M\} \ge c_*^M.$$

Since the bound is uniform in $K \in [K_{p0}^{(0.5)}, K_{q0}]$, the result follows. □

By Assumptions 6.2 and 6.3, the probability of choosing the down-branch is uniformly bounded below on $U$, say by $\underline{p}_- > 0$, and conditional on the down-branch the probability of selecting the configuration $j_-^*$ is also bounded below, say by $\underline{\delta}_- > 0$. Therefore, the probability of realizing the specific path of $M_U$ consecutive down-jumps at this level is at least

$$\delta_U := (\underline{p}_-\,\underline{\delta}_-)^{M_U} > 0.$$

This implies

$$P^{M_U}(K, D_U) \ge \delta_U, \qquad \text{for all } K \in U,$$

and the claim follows. □

### 6.D.4 Proof of Proposition 6.13 (Two-Stage Regeneration and Small Set)

***Proof Sketch*** Let $D_\eta \subset D$ and let $\nu_G$ be the reference measure in Lemma 6.10. By Lemmas 6.11 and 6.12, there exist integers $M_L, M_U \geq 1$, constants $\delta_L, \delta_U > 0$, and compact intervals $D_L, D_U \subset D_\eta$ such that

$$P^{M_L}(K, D_L) \geq \delta_L \quad (K \in L), \qquad P^{M_U}(K, D_U) \geq \delta_U \quad (K \in U).$$

Let

$$C := J \cup D_L \cup D_U \subset D_\eta,$$

where $J \subset D_\eta$ is the interior interval used in the local minorization argument. Since $C \subset D_\eta$, Lemma 6.10 implies the one-step minorization

$$P(K, \cdot) \geq \varepsilon\, \nu_G(\cdot) \qquad (K \in C).$$

Together with the uniform entry bounds from $L$ and $U$ into $D_L$ and $D_U$, this gives a two-stage regeneration structure: first the chain enters the interior set $C$ with uniformly positive probability, and then the one-step minorization applies on $C$. Hence $C$ is a nonempty small set, and the regeneration structure follows. □

### 6.D.5 Proof of Lemma 6.14 (Smallness of the Communicating Set C)

***Proof Sketch*** Fix $K \in C$ and $A \in \mathcal{B}(\mathcal{K})$. By the Chapman–Kolmogorov equation,

$$P^{M+n_0}(K, A) = \int_{\mathcal{K}} P^{n_0}(K', A)\, P^M(K, \mathrm{d}K') \geq \int_J P^{n_0}(K', A)\, P^M(K, \mathrm{d}K').$$

Using the minorization on $J$ and the uniform entry bound into $J$,

$$P^{M+n_0}(K, A) \geq c\, \nu_G(A)\, P^M(K, J) \geq c\delta\, \nu_G(A).$$

Thus $C$ is a small set with $m = M + n_0$ and $\delta_C = c\delta$. □

### 6.D.6 Proof of Lemma 6.15 (Local Lebesgue–Minorization on an Interior Interval)

***Proof (Sketch)*** By Assumption 6.2 and the continuous multiplier construction on the interior region, we may select an integer $n_0 \geq 1$, a specific sequence of branch choices $(\sigma_1, \dots, \sigma_{n_0}) \in \{+, -\}^{n_0}$, and an open interval $J \subset D_\eta$ with $\overline{J} \subset D_\eta$ such that the following holds: for every initial state $K \in J$, conditional on the event that these $n_0$ branch selections occur, the resulting state $K_{n_0}$ admits a density $g_K$ with respect to Lebesgue measure on $J$. Moreover, this density is uniformly bounded away from zero: there exists $\underline{g} > 0$ such that

$$g_K(x) \geq \underline{g} \qquad \text{for all } x \in J, \text{ and all } K \in J.$$

Let $E$ denote the event that this branch sequence is followed during the first $n_0$ steps. By Assumption 6.2, the probability of $E$ is uniformly bounded below on $J$; that is, there exists $\underline{\pi} > 0$ such that

$$\mathbb{P}(E \mid K_0 = K) \geq \underline{\pi} \qquad \text{for all } K \in J.$$

Hence, for any $K \in J$ and any $A \in \mathcal{B}(\mathcal{K})$,

$$P^{n_0}(K, A) \geq \mathbb{P}(E \mid K_0 = K) \int_{A \cap J} g_K(x)\, \mathrm{d}x \geq \underline{\pi}\, \underline{g}\, \lambda(A \cap J).$$

Setting $c_0 := \underline{\pi}\, \underline{g} > 0$ yields the claimed minorization. □

### 6.D.7 *Proof of Corollary 6.1 (Uniform Entry into J)*

***Sketch*** Write $C = L \cup D \cup U$ for the decomposition into lower block, interior, and upper block.

*Step 1: From L and U into D.* By Lemma 6.11 and Lemma 6.12, there exist $M_L, M_U \geq 1$, $\delta_L, \delta_U > 0$, and intervals $D_L, D_U \subset D$ such that

$$P^{M_L}(K, D_L) \geq \delta_L \quad (K \in L), \qquad P^{M_U}(K, D_U) \geq \delta_U \quad (K \in U).$$

*Step 2: From $D_L$ and $D_U$ into J.* Using the interior structure on $D$ (Assumption 6.2) and the one-step entry lemma, we can find an integer $N \geq 1$, a constant $\eta > 0$, and an interval $J \subset D$ with $\overline{J} \subset D$ such that

$$P^N(K, J) \geq \eta \qquad \text{for all } K \in D_L \cup D_U \cup J.$$

*Step 3: Combine the bounds.* Let $M := M_L + N$ and $\delta := \min\{\delta_L, \delta_U, 1\}\, \eta$. Then for $K \in L$,

$$P^M(K, J) \geq \int_{D_L} P^N(K', J)\, P^{M_L}(K, \mathrm{d}K') \geq \eta\, \delta_L \geq \delta.$$

A symmetric argument gives $P^M(K, J) \geq \eta\, \delta_U \geq \delta$ for $K \in U$, while for $K \in J$ we have $P^M(K, J) \geq P^N(K, J) \geq \eta$. This yields the claimed uniform bound on $C$.

□

### *6.D.8 Proof of Lemma 6.22 (Bounds on the Support of the Invariant Distribution)*

***Proof Sketch*** Near the upper clipping point $K_{p0}$, the expression

$$p_+(K) = \max\{\, 0,\ 0.5 - b(\varphi - \zeta(K))\,\}$$

remains equal to zero on a right neighborhood of $K_{p0}$, extending up to approximately

$$K_{p0}\big(1 + \tfrac{\delta}{2b}\big).$$

Over this region, the chain cannot move upward and therefore has a strictly positive probability of executing a downward adjustment whenever the state lies sufficiently close to $K_{p0}$. Consequently, no invariant distribution can place mass on values exceeding $\big(1 + \frac{\delta}{2b}\big)K_{p0}$.

A symmetric argument applies at the lower clipping point $K_{q0}$. The condition $p_-(K_{q0}) = 0$ implies the existence of a left neighborhood of $K_{q0}$ over which the chain cannot move further downward, ensuring recurrent return from below and preventing the stationary distribution from placing mass below $\big(1 - \frac{\delta}{2b}\big)K_{q0}$.

The preceding arguments show that the support of any invariant probability measure is contained in the displayed interval. Finally, by the definition of support, every open subinterval that intersects $\operatorname{supp}(\pi)$ has positive $\pi$-measure. □

## 6.E Foster–Lyapunov Stability: Push-Up and Entry Mechanisms

**Lemma 6.24 (Multi-Step Push-Up under Consecutive Good Up-Steps)** *Fix $K_{q0} > 0$ and $\eta > 0$ and define*

$$I := (K_{q0}, K_{q0} + \eta).$$

*Suppose there exists $\alpha_1 > 0$ such that, whenever the up-branch is chosen and a good up-shock is realized,*

$$K_{t+1} = (1 + \varepsilon_t^+)K_t \quad \text{with} \quad \varepsilon_t^+ \geq \alpha_1.$$

*Let* $\gamma := 1 + \alpha_1 > 1$ *and set*

$$M := \left\lceil \frac{\log(K_{q0} + \eta)}{\log\gamma} \right\rceil. \tag{6.E1}$$

*Then for any* $K_0 \geq 1$, *any sequence of M consecutive good up-steps satisfies*

$$K_M \geq \gamma^M K_0 \geq \gamma^M \geq K_{q0} + \eta.$$

***Proof*** Each good up-step multiplies the current state by at least $\gamma > 1$; hence, $K_j \geq \gamma^j K_0$ for all $j$. For $K_0 \geq 1$, this implies $K_M \geq \gamma^M$. By the definition of $M$, we have $\gamma^M \geq K_{q0} + \eta$, and therefore $K_M \geq K_{q0} + \eta$. □

**Corollary 6.25 (Uniform $M$-Step Entry Bound Above $K_{q0}$)** *In addition to the assumptions of Lemma 6.24, suppose that on the compact interval* $[1, K_{q0} + \eta]$, *the following hold:*

1. *The probability of choosing the up-branch satisfies* $p_+(K) \geq p_{\min} > 0$.
2. *Conditional on choosing the up-branch, the probability of a good up-shock with* $\varepsilon^+ \geq \alpha_1$ *satisfies* $q(K) \geq q_{\min} > 0$.

*Let M be as in Lemma 6.24 and set* $\delta := (p_{\min}q_{\min})^M$. *Then for every* $K_0 \in [1, K_{q0}]$,

$$P^M\big(K_0, [K_{q0} + \eta, \infty)\big) \geq \delta.$$

***Proof*** For any state $K \in [1, K_{q0} + \eta]$, the conditional probability of a good up-step at time $t$ satisfies

$$\mathbb{P}(\text{good up-step at time } t \mid \mathcal{F}_t) \geq p_{\min}q_{\min} =: \rho.$$

By iterated conditioning,

$$\mathbb{P}(M \text{ consecutive good up-steps starting from} K_0) \geq \rho^M = \delta,$$

uniformly for $K_0 \in [1, K_{q0}]$. On this event, Lemma 6.24 yields $K_M \geq K_{q0} + \eta$, proving the claim. □

This corollary translates the multi-step growth mechanism of Lemma 6.24 into a uniform entry bound, thereby providing the probabilistic foundation for the push-up result stated in Lemma 6.11.

**Remark 6.14** If the natural lower bound of the state space is $K_{p0}^{(0.5)} > 0$ rather than 1, apply the argument to the rescaled process $\tilde{K}_t := K_t / K_{p0}^{(0.5)}$, so that $\tilde{K}_0 \geq 1$. The required number of good up-steps becomes

$$M = \left\lceil \frac{\log\big((K_{q0}+\eta)/K_{p0}^{(0.5)}\big)}{\log \gamma} \right\rceil,$$

with $\gamma = 1 + \alpha_1$ as before.

## 6.F Forward Invariance and Support of the Invariant Distribution

This appendix collects the confinement results that complement the push-up and entry arguments in Appendix 6.E, linking forward invariance of trajectories to restrictions on the support of invariant distributions.

**Lemma 6.26 (Forward-Invariant Support Under Bounded Multiplicative Jumps)** *Consider the Markov chain $\{K_t\}$ initially defined on $(0,\infty)$ with transition kernel (6.5.1), where $\Gamma_+(K,j) = K\gamma_+^j$ and $\Gamma_-(K,j) = K\gamma_-^j$ with $\gamma_+ > 1 > \gamma_- > 0$ and $j \in \{0,1,\ldots,J_{\max}\}$. Assume there exist two clipping thresholds $K_{q0} < K_{p0}$ such that*

$$p_+(K) = 0 \qquad \text{for all } K \geq K_{p0}, \qquad p_-(K) = 0 \qquad \text{for all } K \leq K_{q0}.$$

*Define*

$$\underline{K} := K_{q0}\,\gamma_-^{J_{\max}}, \qquad \overline{K} := K_{p0}\,\gamma_+^{J_{\max}}, \qquad \bar{S} := [\underline{K}, \overline{K}].$$

*Then $\bar{S}$ is forward invariant: If $K_t \in \bar{S}$, then $K_{t+1} \in \bar{S}$ almost surely. Consequently, for any invariant probability measure $\pi$ of $\{K_t\}$,*

$$\operatorname{supp}(\pi) \subseteq \bar{S}.$$

*If the chain is additionally constrained by a hard ceiling $\bar{K}$ (i.e., $K_{t+1} \leq \bar{K}$ a.s.), then*

$$\operatorname{supp}(\pi) \subseteq \big[\underline{K},\ \min\{\overline{K}, \bar{K}\}\big].$$

***Sketch of Proof*** Let $K \in \bar{S} = [\underline{K}, \overline{K}]$.

*Upper bound.* If $K \geq K_{p0}$, then $p_+(K) = 0$, so no up-move can occur and hence $K_{t+1} \leq K \leq \overline{K}$. If instead $K < K_{p0}$, any up-move has the form $K_{t+1} = K\gamma_+^j \leq K\gamma_+^{J_{\max}} \leq K_{p0}\gamma_+^{J_{\max}} = \overline{K}$.

*Lower bound.* If $K \leq K_{q0}$, then $p_-(K) = 0$, so no down-move can occur, and hence $K_{t+1} \geq K \geq \underline{K}$. If instead $K > K_{q0}$, any down-move has the form $K_{t+1} = K\gamma_-^j \geq K\gamma_-^{J_{\max}} \geq K_{q0}\gamma_-^{J_{\max}} = \underline{K}$.

Combining the two bounds yields $K_{t+1} \in \bar{S}$ almost surely whenever $K_t \in \bar{S}$, so $\bar{S}$ is forward invariant. If $\pi$ is invariant, then $\pi(\bar{S}) = \int P(K, \bar{S})\,\pi(dK) = 1$, which implies $\mathrm{supp}(\pi) \subseteq \bar{S}$. The final claim with a hard ceiling $\bar{K}$ follows by intersecting $\bar{S}$ with $(-\infty, \bar{K}]$. □

The following corollary translates the forward-invariance property established in the previous section into a restriction on the support of any invariant probability measure.

**Corollary 6.27 (Support of the Invariant Distribution)** *Under the clipped transition probabilities* (6.3.2)–(6.3.3) *and the strict monotonicity of* $\zeta(K)$ *established in Lemma 6.23, the clipping thresholds* $K_{q0} < K_{p0}$ *are well defined and satisfy* $p_+(K) = 0$ *for all* $K \geq K_{p0}$ *and* $p_-(K) = 0$ *for all* $K \leq K_{q0}$. *Hence, by Lemma 6.26, the admissible interval* $[K_{q0}\gamma_-^{J_{\max}}, K_{p0}\gamma_+^{J_{\max}}]$ *is forward invariant, and any invariant probability measure* $\pi$ *of* $\{K_t\}$ *satisfies*

$$\mathrm{supp}(\pi) \subseteq \left[K_{q0}\gamma_-^{J_{\max}},\ K_{p0}\gamma_+^{J_{\max}}\right].$$

## 6.G Auxiliary Continuous-Jump Representation for Drift Analysis

Although the core model in the main text is formulated with discrete multiplicative jumps, this appendix introduces a continuous-jump representation that shares the same first-moment structure and is used solely to establish Foster–Lyapunov drift conditions.

This appendix provides a self-contained derivation of the jump-size expectations and their connection to the Markov transition operator used in the Foster–Lyapunov drift analysis in Sect. 6.6. The computation begins from the simulation-level sampling mechanism and leads to an explicit representation of the transition operator acting on bounded measurable functions.

### *6.G.1 Upward and Downward Jumps from Simulation Rules*

Let the state variable $K_t \in [0, \bar{K}]$ denote the capital stock at time $t$. At each period, the following procedure updates $K_t$:

1. Draw $U \sim \mathrm{Unif}[0, 1]$. If $U < p_+(K_t)$, an *upward* move occurs,

$$K_{t+1} = K_t\big(1 + \varepsilon^+\big), \qquad \varepsilon^+ = \frac{1-U}{2b}\,\delta,$$

where $p_+(K)$ is the probability of an upward innovation and $\delta > 0$ is the maximal proportional adjustment size.

2. Independently, draw $D \sim \mathrm{Unif}\,[0, 1]$. If $D < p_-(K_t)$, a *downward* move occurs,

$$K_{t+1} = K_t\bigl(1 - \varepsilon^-\bigr), \qquad \varepsilon^- = \frac{1-D}{2b}\,\delta,$$

   where $p_-(K)$ is the probability of a downward movement.
3. If neither event is triggered, the process remains at $K_{t+1} = K_t$.

We assume $p_+(K)+p_-(K) \le 1$ for all $K$, so that "no jump" occurs with residual probability $1 - p_+(K) - p_-(K)$.

### 6.G.2 Conditional Expectations of Jump Sizes

We now compute the conditional expectations of the jump magnitudes, given that a jump occurs. The derivation for $\varepsilon^+$ is shown; the same steps apply to $\varepsilon^-$.

**Upward Move** Conditional on $U < p_+(K)$, the variable $U$ is uniformly distributed on $[0, p_+(K)]$ with density $1/p_+(K)$. Hence

$$\varepsilon^+ = \frac{(1-U)\delta}{2b} \qquad \Longrightarrow \qquad \mathbb{E}\bigl[\varepsilon^+ \,\big|\, U < p_+(K)\bigr] = \frac{\delta}{2b}\,\mathbb{E}[1 - U \mid U < p_+(K)].$$

Since $U \mid U < p_+(K) \sim \mathrm{Unif}\,[0, p_+(K)]$,

$$\mathbb{E}[1 - U \mid U < p_+(K)] = \int_0^{p_+(K)} (1-u)\,\frac{du}{p_+(K)} = 1 - \frac{p_+(K)}{2}.$$

Therefore,

$$\mathbb{E}\bigl[\varepsilon^+ \,\big|\, \mathrm{up}\bigr] = \frac{\delta}{2b}\left(1 - \frac{p_+(K)}{2}\right) =: \mu_+(K). \tag{6.G1}$$

**Downward Move** An identical calculation yields for the downward branch,

$$\mathbb{E}\bigl[\varepsilon^- \,\big|\, \mathrm{down}\bigr] = \frac{\delta}{2b}\left(1 - \frac{p_-(K)}{2}\right) =: \mu_-(K). \tag{6.G2}$$

### 6.G.3 Transition Operator Representation

Let $f\colon [0, \bar{K}] \to \mathbb{R}$ be bounded and measurable. The transition operator $P$ associated with the Markov process $\{K_t\}_{t\ge 0}$ acts by

$$(Pf)(K) = \mathbb{E}\bigl[f(K_{t+1}) \,\big|\, K_t = K\bigr].$$

To express $(Pf)(K)$ explicitly, notice that the upward jump can be parameterized by $j := 1 - U$, which is conditionally uniform on $[1 - p_+(K), 1]$ when an upward move occurs. In that event,

$$K_{t+1} = \Gamma_+(K, j) \;:=\; K\left(1 + \frac{\delta}{2b}\, j\right), \qquad j \in [1 - p_+(K), 1].$$

Similarly, the downward jump maps

$$K_{t+1} = \Gamma_-(K, j') \;:=\; K\left(1 - \frac{\delta}{2b}\, j'\right), \qquad j' \in [1 - p_-(K), 1].$$

Thus the operator decomposes into three branches:

$$(Pf)(K) = \big(1 - p_+(K) - p_-(K)\big) f(K) + p_+(K) \int_{1-p_+(K)}^{1} f(\Gamma_+(K, j))\, \frac{dj}{p_+(K)} + p_-(K) \int_{1-p_-(K)}^{1} f(\Gamma_-(K, j'))\, \frac{dj'}{p_-(K)}. \tag{6.G3}$$

This integral form links naturally with the simulation procedure and is convenient for applying drift criteria.

### 6.G.4 Foster–Lyapunov Drift for $V(K) = K$

The expression below is quoted in Sect. 6.6. For the linear Lyapunov function $V(K) = K$, the action of $P$ can be read off from (6.G3):

$$(PV)(K) = \big(1 - p_+ - p_-\big)K + p_+\,\mathbb{E}\big[K\big(1 + \varepsilon^+\big)\big|\text{up}\big] + p_-\,\mathbb{E}\big[K\big(1 - \varepsilon^-\big)\big|\text{down}\big] \tag{6.G4}$$

$$= K\Big[(1 - p_+ - p_-) + p_+\big(1 + \mu_+(K)\big) + p_-\big(1 - \mu_-(K)\big)\Big] \tag{6.G5}$$

$$= K\Big[1 + p_+(K)\mu_+(K) - p_-(K)\mu_-(K)\Big], \tag{6.G6}$$

with $\mu_+$ and $\mu_-$ given by (6.G1)–(6.G2). Hence, the drift is

$$\Delta V(K) \;:=\; (PV)(K) - V(K) \;=\; K\Big[p_+(K)\mu_+(K) - p_-(K)\mu_-(K)\Big]. \tag{6.G7}$$

A sufficient condition for the existence of an invariant measure is that $\Delta V(K) \leq -\varepsilon$ for all $K$ outside a compact interval, for some $\varepsilon > 0$. In economic terms, this requires the expected proportional downward pull to dominate the upward drift in the tails, a property enforced by the threshold structure of $p_+(K)$ and $p_-(K)$ in the main text.

## 6.H Illustrative Python Excerpts for Generating Figure 6.6

This section provides a minimal Python illustration of how the stochastic transition mechanism developed in this chapter is implemented computationally. The excerpts are intended to clarify the mapping between the analytical definitions and their numerical realization.

The code mirrors the analytical definitions of the transition probabilities and the Foster–Lyapunov drift for $V(K) = K$. For analytical transparency, we also recall a continuous-jump representation of the adjustment mechanism. This auxiliary representation shares the same first-moment structure as the discrete simulation model and is used solely to derive the Foster–Lyapunov drift conditions; all simulations reported in the book are based on the discrete jump process.

The simulations are designed to illustrate the qualitative dynamics implied by the analytical framework rather than to provide a fully calibrated quantitative exercise. Parameter values are either standard in the DSGE literature or pinned down by model identities. A full computational implementation of the model is not pursued here and is left to future work, including a planned graduate textbook.

The baseline values $\alpha = 1/3$ and $\beta = 1/2$ imply $a := \alpha/(1-\alpha) = \beta$, consistent with the exponent-matching assumption maintained throughout the book.

The complete Python implementation—including the discrete jump mechanism, transition matrix construction, and numerical computation of invariant distributions—is available at: https://github.com/ABMecon/micro-keynes-evolutionary-model.

**Table 6.2** Baseline parameter values used in the simulations

| Parameter | Description | Baseline value |
|---|---|---|
| $\alpha$ | Capital income share | 1/3 |
| $\beta$ | Consumption–money substitution elasticity | 1/2 |
| $a$ | $\alpha/(1-\alpha)$ | 1/2 |
| $c$ | Investment confidence parameter | 0.175 |
| $b$[a] | Stickiness in revising transition probabilities | 0.7 |
| $M$ | Quantity of money | 1000 |
| $W$ | Nominal wage | 25.0 |
| $\lambda$ | Transactions demand coefficient | 0.1 |
| $\bar{N}$ | Labor endowment | 12.0 |

[a] The parameter $b$ is not a deep parameter but is defined as $b = c_b/V$

### *6.H.1 Transition Primitives (`transition_kernel.py`)*

```
def hat_r_K(K, sigma_0, delta):
    """
    Implied interest rate r(K) (the r(K) mapping used in the
    ↪  simulations).
    """
    K = np.asarray(K, dtype=float)
    safe = np.where(sigma_0 * K < 1e-12, 1e-12, sigma_0 * K)
    disc = ((1.0 / safe) ** 2 - (1.0 + delta)) ** 2 - 4.0 *
    ↪  delta
    disc = np.maximum(disc, 0.0)
    return ((1.0 / safe) ** 2 - (1.0 + delta) - np.sqrt(disc))
    ↪  / 2.0


def Kd_K(alpha, eta, delta, rK, c, K):
    """Capital demand K^d(K) as implemented in the code."""
    return alpha * (eta - 1.0) / eta / (rK + delta) * (c +
    ↪  delta) * K


def compute_zeta(K, sigma_0, c, alpha, eta, delta):
    """Compute zeta(K) = (K^d(K)/K - 1)/delta."""
    rK = hat_r_K(K, sigma_0, delta)
    Kd = Kd_K(alpha, eta, delta, rK, c, K)
    util = Kd / K
    return (util - 1.0) / delta


def p_func(b, phi, zeta):
    """Up probability p_+(K) with clipping to [0, 0.5]."""
    return np.clip(0.5 - b * np.maximum(phi - zeta, 0.0), 0.0,
    ↪  0.5)

def q_func(b, phi, cb, zeta):
    """Down probability p_-(K) with clipping to [0, 0.5]."""
    return np.clip(0.5 - b * np.maximum(zeta + 1.0 + cb - phi,
    ↪  0.0), 0.0, 0.5)


def s_func(b, phi, cb, zeta):
    """Stay probability p_0(K) = 1 - p_+(K) - p_-(K)."""
```

```
    return 1.0 - p_func(b, phi, zeta) - q_func(b, phi, cb,
    ↪  zeta)

def P_of_f(K, f, p_plus, p_minus, delta_plus, delta_minus,
↪  gamma_plus, gamma_minus):
    """
    Markov transition operator: (Pf)(K) = E[f(K_{t+1}) | K_t =
    ↪  K].
    """
    p0 = 1.0 - p_plus(K) - p_minus(K)

    val = p0 * f(K)

    # Up branch
    if p_plus(K) > 0:
        for j in range(J_max + 1):
            val += p_plus(K) * delta_plus(K, j) * f(K *
            ↪  (gamma_plus ** j))

    # Down branch
    if p_minus(K) > 0:
        for j in range(J_max + 1):
            val += p_minus(K) * delta_minus(K, j) * f(K *
            ↪  (gamma_minus ** j))

    return val
```

## 6.H.2 *Drift and Diagnostic Checks* (**diagnostics.py**)

```
def drift_function(K, delta, b, p_plus, p_minus):
    """
    Foster--Lyapunov drift Delta-V(K) for V(K)=K
    in the auxiliary continuous-jump representation.
    See eq. (eq-delta-VK) in the Appendix.
    """
    return K * (delta / (2.0 * b)) * (
        p_plus * (1.0 - p_plus / 2.0)
      - p_minus * (1.0 - p_minus / 2.0)
    )


def check_row_sums(P, tol=1e-12):
```

```
    """Verify that rows of the transition matrix sum to one (up
    ↪  to rounding)."""
    row_sums = P.sum(axis=1)
    max_err = abs(row_sums - 1.0).max()
    print(f"[diagnostic] max |row sum - 1| = {max_err:.3e}")
    if max_err > tol:
        print(f"[warning] exceeds tol={tol:.1e}")
```

### 6.H.3 Minimal Runnable Entry Point (`RUN_DIAGNOSTICS.py`)

**Notation** The symbol `phi` in the code denotes the same threshold parameter as $\varphi$ in the main text; the change in notation reflects Python naming conventions.

```
def main():
    # Parameter container (see GitHub for full script)
    p = Params()

    # Grid for capital stock
    K_grid = np.linspace(p.K_min, p.K_max, p.nK)

    # Build transition matrix P(K'|K) for the discrete kernel
    P = tk.build_transition_matrix(
        K_grid=K_grid,
        sigma_0=p.sigma_0, b=p.b, phi=p.phi, c=p.c,
        alpha=p.alpha, eta=p.eta, delta=p.delta, cb=p.cb,
        J=p.J, u_mode=p.u_mode, normalize_rows=False
    )

    # Basic diagnostic: rows should sum to one (up to rounding)
    check_row_sums(P)

    # Auxiliary continuous-jump drift diagnostic (Appendix eq.
    ↪  (6.G5))
    zeta = tk.compute_zeta(K_grid, p.sigma_0, p.c, p.alpha,
    ↪  p.eta, p.delta)
    p_plus  = tk.p_func(p.b, p.phi, zeta)
    p_minus = tk.q_func(p.b, p.phi, p.cb, zeta)

    drift_vals = [drift_function(K, p.delta, p.b, pp, pm)
                  for (K, pp, pm) in zip(K_grid, p_plus,
                  ↪  p_minus)]
```

```
    print("Drift summary for V(K)=K (auxiliary continuous-jump
     ↪  representation)")
    print("min Delt-V:", min(drift_vals))
    print("max Delta-V:", max(drift_vals))

if __name__ == "__main__":
    main()
```

## References

Davis, M. H. A. (1984). Piecewise-deterministic Markov processes: A general class of non-diffusion stochastic models (with discussion). *Journal of the Royal Statistical Society Series B (Methodological), 46*(3), 353–388. https://doi.org/10.1111/j.2517-6161.1984.tb01308.x.

Debreu, G. (1959). *Theory of value: An axiomatic analysis of economic equilibrium*. Monograph (Vol. 17), Yale University Press.

Einsiedler, M., & Ward, T. (2011). *Ergodic theory: With a view towards number theory*. Graduate Texts in Mathematics (Vol. 259). Springer. https://doi.org/10.1007/978-0-85729-021-2.

Ethier, S. N., & Kurtz, T. G. (1986). *Markov processes: Characterization and convergence*. John Wiley & Sons. https://doi.org/10.1002/9780470316658.

Hairer, M. (2018). Ergodic properties of Markov processes. https://www.hairer.org/notes/Markov.pdf. Lecture notes, July 29, 2018

Kandori, M., Mailath, G. J., & Rob, R. (1993). Learning, mutation, and long run equilibria in games. *Econometrica: Journal of the Econometric Society*, 29–56. https://doi.org/10.2307/2951777

Meyn, S. P., & Tweedie, R. L. (1993). *Markov chains and stochastic stability*. Communications and Control Engineering. Springer-Verlag.

Young, H. P. (1993). The evolution of conventions. *Econometrica: Journal of the Econometric Society*, 57–84. https://doi.org/10.2307/2951778

# Chapter 7
# Wage Setting, Fairness, and Firm Performance

*It is necessary for efficiency that the wage contract should be felt—by both parties, but especially by the worker—to be fair.*

— John Hicks, *The Crisis in Keynesian Economics,* Ch. 3

## 7.1 Introduction

This chapter examines how firms revise nominal wages in response to economic conditions, and how such revisions are constrained by institutional and behavioral considerations. Wage setting plays a central role in the macroeconomic adjustment process developed in this book, because nominal wages serve as a key anchor for prices, employment, and investment dynamics. Rather than assuming fully rational wage setters, we model firms as boundedly rational agents operating in an environment where nominal wages display substantial rigidity, especially in the downward direction.

A large theoretical literature seeks to explain the pronounced asymmetry between wage increases and wage cuts. A central element underlying nominal wage rigidity is workers' perception of fairness, which links wage outcomes to firm performance rather than to job characteristics alone. We incorporate fairness considerations to capture a well-documented empirical regularity in wage setting: Workers performing similar tasks with comparable productivity often receive substantially different wages across firms within the same industry. These wage differentials are driven primarily by firm-specific economic conditions rather than individual performance. In particular, more successful firms tend to pay systematically higher wages even for identical jobs, while less successful firms pay less. This evidence suggests that wages are shaped not only by market-clearing conditions, but also by workers' perceptions of fairness relative to their firm's economic performance. Such perceptions are widely argued to play a central role in determining effort provision, resistance to wage reductions, and the degree of nominal wage rigidity (see Abowd et al., 1999).

 
I. Takahashi, *Microfoundations of Keynesian Economics*, Evolutionary Economics and Social Complexity Science 37, https://doi.org/10.1007/978-981-95-9390-3_7

While the present analysis emphasizes fairness and its impact on effort and firm performance, the broader literature has long recognized that wage determination departs from the competitive benchmark. In particular, McDonald and Solow (1981) develop a framework in which wages and employment are jointly determined through bargaining between firms and workers, reflecting institutional and strategic considerations rather than pure market clearing. Although the mechanism differs, the fairness-based approach developed here and the bargaining framework of McDonald and Solow (1981) both underscore that wages are shaped by forces beyond marginal productivity—whether through perceptions of fairness, as emphasized here, or through negotiated outcomes, as in the bargaining framework.

Unlike goods or money markets, the workplace is an experience good: Workers gradually learn the true quality of their working environment through on-the-job experience. At the same time, workers dislike fluctuations in income and place a premium on stability[1]. As John Maynard Keynes emphasized, employees strongly resist cuts in money wages (see Keynes (1936, Chs. 2 and 9). A large subsequent literature has sought to account for this asymmetry, often appealing to loss aversion and reference-dependent preferences, whereby wage cuts are perceived as disproportionately costly relative to equivalent gains (see (Bhaskar, 1990; Lye and McDonald, 2021)). Because downward rigidity discourages wage reductions, any wage increase is difficult to reverse, which naturally gives rise to upward rigidity as well. Consequently, firms rarely adjust nominal wages; instead, they maintain the prevailing wage for long periods and revise it only when economic conditions change sufficiently. Iwai captures this idea by proposing that wage adjustment occurs only when the prevailing real wage deviates substantially from the firm's target real wage.

Work environments differ across firms, and in this sense, each firm possesses a degree of monopsonistic power in wage setting. However, firms act as wage setters only intermittently. Because it is neither feasible nor desirable to adjust wages continuously, firms do not engage in full optimization. Rather, they follow a satisficing principle: Nominal wages are set so as to maintain a sufficiently stable and adequate workforce in most periods.

Although the broader question of how policies can prevent inflation or deflation is important, a full dynamic theory of price and wage movements lies beyond the scope of this book. The purpose of this chapter is more modest: to identify the threshold conditions that trigger upward or downward wage adjustments, without attempting to model the complete dynamics of inflation and deflation.

The wage-setting framework adopted here is fully consistent with Keynes's original position on the labor market. Keynes accepted the neoclassical principle that firms' demand for labor is governed by profitability and marginal productivity, but rejected the auxiliary assumption that real wages adjust to clear the labor market. The present model preserves the former, while replacing the latter with

---

[1] This idea dates back to the implicit contracts literature, which emphasizes workers' relative risk aversion and firms' ability to smooth income over time (e.g., Hall, 1975; Azariadis, 1975).

a fairness-sensitive wage-setting mechanism in which nominal wages are institutionally anchored and effort responds asymmetrically to wage revisions. As a result, employment outcomes are demand determined even though firms behave competitively on the production side.

This chapter proceeds as follows. Section 7.2 formulates a fairness-based model of wage setting under bounded rationality. Building on this framework, Sect. 7.3 characterizes the inflationary threshold $K_F$ at which upward wage adjustment is triggered under tight labor-market conditions. Section 7.4 then analyzes wage determination under unemployment and identifies the deflationary threshold $K_D$.

## 7.2 Wage-Setting Model

### 7.2.1 Environment and Notation

We study wage adjustment in a quasi-steady-state environment (SQSS), taking aggregate variables and product-market conditions as given. Each firm sets a money wage $w$ and then chooses employment $n$ under the right-to-manage convention. Within a period, the capital stock $k$ is predetermined, and product-market conditions are summarized by a demand shifter $F > 0$ and an elasticity composite $g \in (0, 1)$ defined below.

Table 7.1 summarizes the notation used in the wage-setting framework developed in this chapter.

**Table 7.1** Notation for wage setting and fairness

| Symbol | Description | Note |
|---|---|---|
| $e(\varpi)$ | effort (fairness-sensitive) | kinked at $\varpi = 1$ |
| $w$ | firm-level money wage | |
| $W$ | prevailing money wage norm | in the SQSS |
| $W_0$ | external reference (market) wage | external fairness benchmark |
| $\varpi = w/W$ | wage relative to internal norm | Wage Fairness Ratio (WFR)[a] |
| $\omega = w/P$ | real wage | |
| $\varpi^* = w/W_0$ | wage relative to external benchmark | Relative Wage Index (RWI)[b] |
| $P$ | price level | |
| $\Pi(w, W)$ | short-run profit | internal fairness component |
| $U(w; W_0)$ | external payoff (outside option) | external fairness component |
| $\mathrm{NP}(w)$ | Nash product | wage-setting objective |
| $b_-, b_+$ | effort elasticities | $0 < b_+ < b_- < 1$ |
| $\rho \in (0, 1)$ | discount on outside option | |
| $\theta$ | bargaining weight | firm's weight in NP |

[a] WFR focuses on the perception of fairness or equity, which ties directly into the motivational effects discussed in Sirota's work

[b] RWI emphasizes the idea of comparison between an individual's wage and the equilibrium wage

### 7.2.2 External Reference Wage

We begin by characterizing the external wage benchmark, then embed it into the firm's wage-setting problem.

We treat $W_0$ as the market-wide fairness benchmark. It coincides with the SQSS marginal-product wage when employment is normal, but rises with aggregate labor-market tightness as $N^d(K)/\bar{N}$ increases.

**Lemma 7.1 (External Reference Wage and Labor-Market Tightness)** *Let $W$ denote the prevailing money wage in the SQSS and define the external reference wage $W_0(K)$ by*

$$\frac{W_0}{P} := \left(\frac{\eta - 1}{\eta}\right)(1 - \alpha)\kappa_t^{\alpha}\left(\frac{N^d(K)}{\bar{N}}\right)^{\gamma_0}, \qquad \gamma_0 > 0.$$

*Using Eq.* (4.B10), *it follows that*

$$\frac{W_0}{W} = \left(\frac{N^d(K)}{\bar{N}}\right)^{\gamma_0}.$$

*Since $N^d(K)$ is increasing in $K$, the ratio $W_0/W$ is increasing in $K$. Moreover, if $K_F$ is defined by $N^d(K_F) = \bar{N}$, then $W_0/W = 1$ at $K = K_F$.*

This external reference wage will play a central role in the firm's wage-setting problem below, where firms balance internal motivation against external competitiveness.

**Remark 7.1 (Interpretation of $\gamma_0$)** The parameter $\gamma_0 > 0$ governs the sensitivity of the external benchmark $W_0$ to labor-market tightness, and thus the speed with which perceived fairness norms respond to changes in employment conditions.

**Loss Aversion (Internal Norm)**
A central behavioral ingredient behind wage rigidity is reference dependence around the internal wage norm $W$. Workers react more strongly to wage cuts than to comparable wage increases. We capture this asymmetry through a kinked effort response to the relative wage $\varpi = w/W$, following the prospect-style intuition of Kahneman and Tversky (1979) and Tversky and Kahneman (1991).

### 7.2.3 Fairness-Sensitive Effort

To formalize how wage setting affects firm performance, we first characterize the firm's short-run profit as a function of wages and effort. This provides the payoff component that enters the Nash product, linking wage decisions to both internal motivation and external competitiveness.

**Short-Run Profit**
Replacing employment $n$ with effective labor input $en$ and neglecting sunk capital, Eq. (4.B4) in Appendix 4.B implies

$$\Pi(w, n; W) = X^{1/\eta} P \left(k^{\alpha} (en)^{1-\alpha}\right)^{\frac{\eta-1}{\eta}} - wn.$$

Define

$$F := X^{1/\eta} P k^{\frac{\alpha(\eta-1)}{\eta}}, \qquad g := \frac{(1-\alpha)(\eta-1)}{\eta} \in (0, 1),$$

so that $\Pi(w, n; W) = F\,(en)^g - wn$.

To capture how fairness perceptions translate into productivity, we model effort as a function of the wage paid relative to an internal reference level, allowing for asymmetric responses to wage gains and losses.

**Effort and the Internal Reference**
Workers assess wage fairness at two levels. The external benchmark $W_0$ reflects economy-wide labor-market conditions and governs when wage-revision pressure arises, while effort provision responds to the firm-level comparison between the paid wage and the prevailing nominal norm $W$. Workers evaluate their compensation relative to an internal reference wage level prevailing within the labor market.

Let $\varpi \equiv w/W$ denote the wage fairness ratio, where $w$ is the firm-specific nominal wage and $W$ is the average nominal wage. Effort is assumed to depend directly on this fairness ratio through a monotone response function capturing morale and perceived fairness:

$$e(\varpi) = \begin{cases} \varpi^{b_-}, & \varpi < 1, \\ 1, & \varpi = 1, \\ \varpi^{b_+}, & \varpi > 1. \end{cases}$$

The parameters $0 < b_+ < b_- < 1$ capture asymmetric behavioral responses to underpayment and overpayment relative to the reference wage. Thus, wage increases above $W$ raise effort only weakly, whereas wage cuts below $W$ depress effort more strongly.

Substituting the fairness-sensitive effort function into the profit maximization problem yields a reduced-form expression for profits as a function of wages alone.

**Reduced-Form Profit in $w$**
The first-order condition $\partial\Pi/\partial n = 0$ yields $en = (Fge/w)^{1/(1-g)}$. Substituting back,

$$\Pi(w, W) = (1 - g)\, g^{\frac{g}{1-g}}\, F^{\frac{1}{1-g}} \left( \frac{e(\varpi)}{w} \right)^{\frac{g}{1-g}}.$$

On each side $m \in \{-, +\}$, where $e(\varpi) = (w/W)^{b_m}$,

$$\frac{d}{dw} \log \Pi_m(w) = -\frac{\gamma_m}{w}, \qquad \gamma_m := \frac{(1 - b_m)g}{1 - g} > 0. \tag{7.2.1}$$

Because the elasticity of effort differs across the two regions, the marginal effect of wages on profit changes discretely at $w = W$, generating a kink that reflects asymmetric behavioral responses.

**External Reference and Worker Outside Option**
Wage adjustment is inherently asymmetric. During periods of labor-market tightness, firms must raise wages to attract and retain workers, and such adjustments are generally accommodated without conflict. By contrast, when demand weakens, firms have an incentive to reduce wages, but workers resist such cuts. Wage determination in these circumstances reflects not only the firm's short-run profitability, but also the worker's outside option or fallback position. To capture this external dimension, we introduce a reduced-form welfare component associated with the worker's position relative to the broader labor market.

We capture the worker's outside option relative to the external benchmark $W_0$, incorporating an endowment (status-quo) effect that makes wage reductions particularly costly.

$$U(w; W_0) := (w - \rho W_0)^{\upsilon}, \qquad \upsilon > 0, \qquad w > \rho W_0. \tag{7.2.2}$$

Within the Nash-product formulation introduced below, the external component $U(w; W_0)$ may be viewed as representing the worker's fallback position in wage negotiations or, more generally, the firm's long-run constraint associated with employee retention and reputational concerns, thereby complementing the short-run profit component $\Pi(w, W)$.

We focus on the empirically relevant region $w > \rho W_0$, where $U(w; W_0) > 0$.

### 7.2.4 Nash Wage-Setting Problem

As discussed above, wage setting reflects a trade-off between internal incentives and external constraints. Firms care about short-run profitability, which depends on effort and internal wage comparisons, while workers evaluate wages relative to external benchmarks reflecting outside opportunities. We summarize this interaction in a reduced-form bargaining framework, in which the wage is determined by maximizing a Nash product that balances these two considerations.

The firm and workers jointly determine the money wage $w$ by maximizing the Nash product

$$\mathrm{NP}(w) = \Pi_m(w)^{\theta} U(w; W_0)^{1-\theta},$$

where $\Pi_m(w)$ denotes short-run profit on side $m$ of the internal reference and $U(w; W_0)$ captures valuation relative to the external benchmark $\rho W_0$. The parameter $\theta$ captures the relative weight placed on firm profitability versus external valuation in wage determination. The economic interpretation of these two components in terms of internal and external fairness is developed in detail in Sect. 7.4.

**Remark 7.2 (Interpretation of $\theta$)** The parameter $\theta$ governs the relative weight placed on short-run profitability versus external fairness. At this stage, $\theta$ is best interpreted as a reduced-form priority parameter, reflecting bargaining power, governance norms, or institutional constraints.

The Nash product is piecewise smooth in the wage $w$, reflecting the kink in the effort function at the internal reference point $w = W$. As a result, the marginal profit component and hence the marginal contributions—$\Pi_m(w)$ $(m \in \{-, +\})$—are discontinuous at this point. The marginal contribution of the external component $U(w; W)$, by contrast, is smooth on its domain $w > \rho W_o$. Accordingly, the first-order condition must be evaluated separately on each side of the internal reference, treating the problem as piecewise differentiable. In what follows, we characterize interior solutions within each region; the non-differentiability at $w = W$ can be handled by comparing the one-sided conditions and therefore does not affect the subsequent analysis.

**Nash First-Order Condition**

The optimal wage, therefore, solves a trade-off between short-run profitability and external valuation, as summarized by the Nash product; the first-order condition characterizes this balance.

Taking logs and differentiating,

$$\frac{d}{dw}\log \mathrm{NP}(w) = \theta \frac{d}{dw}\log \Pi_m(w) + (1-\theta)\frac{d}{dw}\log U(w; W_0).$$

Using (7.2.1) and (7.2.2), the first-order condition is given by

$$-\theta \frac{\gamma_m}{w} + (1-\theta)\frac{v}{w - \rho W_0} = 0. \tag{7.2.3}$$

The existence and location of an interior solution depend on whether this condition holds on each side of the internal reference point $w = W$.

**Marginal Profit and Marginal Fairness**

The first-order condition (7.2.3) equates two opposing forces in wage determination: the marginal effect of wages on firm profitability and the marginal valuation associated with external fairness. To make this trade-off transparent, it is useful to decompose the condition into these two components.

Equation (7.2.3) decomposes wage setting into a *marginal-profit effect* (MP),

$$-\theta\,\frac{\gamma_m}{w},$$

and a *marginal fairness effect* (MF),

$$(1-\theta)\frac{v}{w-\rho W_0}.$$

While this condition is already informative, the two components are expressed in different functional forms. To facilitate comparison, we transform the first-order condition into a normalized representation in which the marginal profit and marginal fairness effects can be directly compared.

**MP and MF Indices (Normalized)**

Multiplying (7.2.3) by $w(w-\rho W_0) > 0$ clears denominators and yields an equivalent condition:

$$\theta\gamma_m\Big(1-\rho\frac{W_0}{w}\Big) = (1-\theta)v.$$

Define the marginal-profit index

$$\widehat{\mathrm{MP}}_m(w;\,W_0) := \theta\gamma_m\Big(1-\rho\frac{W_0}{w}\Big), \tag{7.2.4}$$

and the marginal fairness index

$$\widehat{\mathrm{MF}} := (1-\theta)v. \tag{7.2.5}$$

In this representation, $\widehat{\mathrm{MP}}_m(w;\,W_0)$ measures the effective marginal gain from wage adjustment after accounting for the external benchmark, while $\widehat{\mathrm{MF}}$ is a constant reflecting the strength of external fairness considerations. Wage determination therefore reduces to equating a state-dependent marginal profit index with a fixed marginal fairness benchmark.

The wage-setting condition is characterized by $\widehat{\mathrm{MP}}_m(w;\,W_0) = \widehat{\mathrm{MF}}$. Because $\gamma_m$ depends on the side of the internal reference, $\widehat{\mathrm{MP}}_m$ inherits a kink at $w = W$, whereas the marginal-fairness index remains smooth.

This normalized representation will prove particularly useful in characterizing how wage-setting behavior varies across economic conditions. In particular, it provides a convenient basis for identifying the deflationary threshold $K_D$ at which the balance between marginal profit and external fairness shifts and downward wage pressures begin to bind.

## 7.3 Inflationary Threshold: $K_F$

This section investigates the level of the capital stock at which upward wage adjustment—corresponding to inflationary pressure—is triggered. In a fully rational monopsonistic model of wage setting, firms would adjust wages continuously so as to maintain an optimal vacancy rate or a "normal" degree of labor shortage, as in Iwai (1987). In such a framework, the wage-adjustment threshold would generally lie slightly below the full-employment capital stock $K_F$. In the present model, however, firms are boundedly rational and do not continuously optimize vacancy conditions. Instead, they revise wages only when labor shortages become sufficiently severe.

For expositional clarity, and in line with the state space restriction $K \in [1, K_F]$, we therefore identify the inflationary wage-adjustment threshold with the full-employment level and set

$$K_I = K_F.$$

This specification captures the idea that firms tolerate moderate labor shortages but respond once tightness becomes extreme. A microfounded analysis of the fully rational wage setter is provided in Appendix 7.A.

**Remark 7.3 (Rational Benchmark)** If wages were chosen to maximize profits subject to an explicit vacancy technology, the implied threshold $K_I$ would generally differ from $K_F$, reflecting the firm's optimal "normal" labor shortage rate. The present analysis can be viewed as a bounded-rational approximation in which wage adjustment is triggered only once the economy enters the full-employment region. A more elaborate vacancy model would refine the precise location of $K_I$ without altering the qualitative mechanism developed below.

To formalize how labor-market tightness translates into wage adjustment, we retain Iwai's Wicksellian target real wage $\omega_t^*$ derived in Appendix 7.A; see (7.A1). This target represents the real wage a fully rational wage setter would choose in period $t$ given the prevailing capital stock and labor-market conditions. In particular, a Wicksellian firm exploits its wage-setting power to maintain a "normal" degree of labor shortage, and hence, $\omega_t^*$ need not coincide with the inherited real wage $\omega_{t-1}$.

In the present model, firms do not continuously compute or implement $\omega_t^*$. Instead, the target wage serves as a diagnostic reference indicating whether current labor-market conditions have deviated sufficiently from the status quo to justify wage revision. Nominal wages are revised only intermittently, according to a satisficing rule reflecting adjustment costs and the downward rigidity of money wages.

Because the number of workers effectively available to firms is typically smaller than the measured labor endowment, we write effective labor supply as $L_{\text{eff}} = \bar{\upsilon}\bar{N}$ with $\bar{\upsilon} < 1$. Under this specification, the target real wage $\omega_t^*$ exceeds the inherited wage $\omega_{t-1}$ when labor-market tightness becomes sufficiently high. Under the condition $f^*\bar{\upsilon} < 1$ derived in Appendix 7.A, this occurs in a neighborhood of the full-employment boundary $K_F$.

Consequently, as the capital stock approaches $K_F$, the deviation between $\omega_t^*$ and $\omega_{t-1}$ eventually exceeds the inaction threshold $\theta_w$, activating wage adjustment. In this bounded-rational sense, sufficiently tight labor-market conditions near $K_F$ trigger *upward* revisions of the nominal wage, even though the notional Wicksellian target wage may lie below the Walrasian equilibrium wage in a fully rational setting.

This behavior is summarized by the following wage-revision rule:

$$\bar{W}_t = \begin{cases} \bar{W}_{t-1}, & \text{if } \left|\log \omega_t^* - \log \omega_{t-1}\right| \le \theta_w, \\ \omega_t^* P_{t-1}, & \text{otherwise,} \end{cases} \tag{7.3.1}$$

where $\theta_w > 0$ captures the costs and frictions associated with frequent wage adjustment.

Under $f^*\bar{\upsilon} < 1$ and for sufficiently small $\theta_w$, repeated episodes in which the economy approaches the neighborhood of $K_F$ eventually activate (7.3.1), resulting in an upward revision of the nominal wage. In this way, the inflationary threshold $K_F$ marks the point at which labor-market tightness becomes sufficiently pronounced to overcome nominal rigidities and trigger upward wage adjustment. The consequences of such wage increases for the price level, real balances, and the invariant distribution of the capital-stock process are analyzed in the next chapter.

One might worry that upward wage revision under tight labor-market conditions could lead to a self-sustaining inflationary process. In the present fixed money environment, however, such dynamics are ruled out. An increase in the nominal wage raises the price level, reduces real balances, and contracts aggregate demand. This feedback pushes the economy away from the region in which further wage increases would be triggered. Wage inflation is therefore self-limiting under a constant nominal money supply.

We now turn to the opposite case, in which aggregate conditions deteriorate, labor-market slack emerges, and downward pressure on wages may arise. This motivates the definition of the deflationary threshold $K_D$.

## 7.4 Deflationary Threshold: $K_D$

We now consider the opposite regime, in which aggregate conditions deteriorate, and labor-market slack emerges. In contrast to the inflationary case, where wage increases are accommodated relatively smoothly, downward wage adjustment gives rise to conflict between firms and workers. Firms have an incentive to reduce wages in response to weak demand, but workers resist such cuts due to fairness concerns and loss aversion. As a result, wage determination in this region cannot be described by a simple adjustment rule; instead, it reflects a balance between profitability and external constraints.

This asymmetry contrasts with the inflationary threshold $K_F$, where wage adjustment is triggered by labor-market tightness and implemented through a satisficing rule. In the deflationary region, by contrast, wage setting reflects a tension between internal incentives and external resistance.

To formalize this tension, we introduce a reduced-form bargaining framework in which the wage is determined by balancing short-run profitability and external valuation. This role is played by the Nash product introduced above, which summarizes the competing forces governing wage adjustment in the presence of downward rigidity.

The deflationary threshold $K_D$ is defined as the point at which this balance ceases to admit an interior solution, and downward wage pressure becomes binding.

A well-established empirical fact is that money wages exhibit strong downward rigidity during recessions. Firms hesitate to cut wages not only because reductions may harm worker morale, but also because such cuts are often perceived as unfair, thereby weakening effort and productivity.

Sirota et al. (2005) emphasize that workers are most motivated when treated fairly, especially when pay is perceived to be aligned with contribution. Their extensive fieldwork across leading firms shows that when employees feel underpaid relative to their efforts, they interpret the situation as unfair and disproportionately reduce effort.[2] Complementary theoretical foundations are provided by Akerlof (1982), Akerlof and Yellen (1990), who argue that effort depends on perceived fairness, and by Shapiro and Stiglitz (1984), whose efficiency-wage model shows that firms may pay above-market wages to deter shirking. Psychological and experimental evidence, such as Fehr et al. (1993), likewise documents strong reciprocity motives reinforcing downward rigidity. Consistent with these findings, Bewley (2004) reports, based on interviews with managers, that firms resist cutting wages for fear of damaging morale and productivity. Taken together, these insights motivate the asymmetric effort specification adopted here and provide the behavioral foundation for a state-dependent deflationary threshold, below which wage cuts are resisted despite declining labor demand.

These considerations alone, however, do not explain why wage cuts do occur when economic conditions deteriorate severely or when a firm's survival is at stake. In such circumstances, workers' outside options decline sharply, and wage reductions are perceived as less unfair. As employees evaluate wages relative to prevailing labor-market conditions rather than past compensation, resistance

---

[2] "[I]t is difficult for a company to obtain loyalty and commitment from its workers when it is seen as one of the lower-paying companies (especially in relation to its competition or its profitability)" (Sirota et al., 2005, p. 81).

weakens and wage cuts may be reluctantly accepted. This asymmetry between wage increases and wage cuts will be central to the definition of the deflationary threshold $K_D$ developed below.

These observations naturally lead to two distinct notions of fairness that shape wage setting. In what follows, $\Pi(w, W)$ denotes short-run profitability evaluated relative to the prevailing average wage $W$ in the symmetric quasi-steady state. It reflects internal fairness considerations governing effort, morale, and current profits. By contrast, $U(w, W_0)$ is benchmarked against a long-run reference wage $W_0$, representing outside opportunities and normative labor-market conditions. In this sense, $U$ captures external fairness and the long-run sustainability of employment relations. Although both components are evaluated within the same period, they reflect distinct reference standards associated with short-run and long-run considerations:

- *Internal fairness* concerns whether a worker's wage is appropriate relative to others within the same firm. This dimension primarily affects motivation and effort, and hence short-run profits $\Pi(w, W)$.
- *External fairness* evaluates wages relative to the broader labor market or an external reference level $W_0$. Higher wages improve retention and strengthen long-run profitability or worker welfare through $U(w, W_0)$.

This distinction between internal and external fairness mirrors the asymmetry in wage adjustment emphasized earlier.

Firms implicitly balance internal motivation against external competitiveness, the latter being summarized by the external reference wage introduced above. A convenient way to formalize this trade-off is through the Nash product

$$\mathrm{NP}(w; w^*) \equiv \Pi(w, W)^{\theta}\, U(w, W_0)^{1-\theta},$$

which captures the idea that internal motivation and external competitiveness jointly determine wage desirability. The key question, then, is whether this balance admits an interior solution under all economic conditions. The deflationary threshold $K_D$ is defined as the level of the capital stock at which this balance ceases to admit an interior solution, so that downward wage pressure becomes binding.

The use of the Nash product here is not derived from a fully specified bargaining model, but is motivated by a general result due to Yoshihara (2006). When competing fairness principles cannot be jointly satisfied under efficiency, the Nash solution emerges as the unique efficient and symmetric second-best compromise. This logic closely parallels the present wage-setting problem. Table 7.2 summarizes the correspondence between our framework and Yoshihara's axiomatic analysis of solidarity in cooperative bargaining.

In the present context, internal fairness plays a role analogous to responsibility-sensitive incentives: Wages must sustain effort and motivation within the firm in order to maintain short-run profitability. External fairness, by contrast, reflects a compensation motive, as wages are evaluated relative to outside opportunities and broader labor-market conditions that affect workers' welfare and the firm's long-run

**Table 7.2** Correspondence between the wage-setting framework and Yoshihara (2006)

| This model | Yoshihara (2006) |
|---|---|
| Internal fairness (effort, morale, short-run profits $\Pi$) | Responsibility principle (IUI) |
| External fairness (outside options, welfare $U$) | Compensation principle (SM/weakened SM) |
| Flex–fix wages | Second-best compromise under impossibility |
| Nash product $\Pi^{\theta} U^{1-\theta}$ | Unique efficient, symmetric second-best solution |

viability. The Nash solution provides a natural second-best compromise between these considerations without requiring full compensation or full insurance against adverse conditions.[3]

The function $U(w, W_0)$ admits two complementary interpretations. It may be viewed as capturing the firm's long-run benefit from maintaining a stable and loyal workforce, or alternatively as workers' welfare evaluated relative to an external reference wage. Both interpretations lead to the same reduced-form wage rule and differ only in their microeconomic narratives.

Firms, however, are not assumed to optimize this expression continuously. Wage setting in this environment is bounded rational and highly discrete. Downward rigidity implies that wage cuts occur only under sufficiently adverse conditions, while the difficulty of reversing such cuts induces reluctance to raise wages in response to temporary labor shortages. As a result, money wages exhibit *flex–fix* behavior (cf. (McDonald, 2019)): They remain fixed over extended periods and adjust only when economic conditions cross critical thresholds. The goal of this section is to formalize a wage-determination rule with these flex–fix properties and to identify the deflationary threshold $K_D$ at which firms are willing to implement wage reductions.

The mechanism can be visualized in terms of the interaction between the marginal-profit and marginal-fairness indices. In particular, the location of the intersection determines whether the current wage remains optimal or downward adjustment becomes desirable. Figures 7.1 and 7.2 illustrate how this interaction varies with aggregate conditions. Figure 7.1 depicts a *moderate downturn*, in which the deterioration in aggregate conditions is limited and the external reference wage remains sufficiently close to the status quo wage, so that downward wage rigidity continues to bind.

Figure 7.2 considers a *severe downturn*, in which the external reference wage has fallen substantially relative to the status quo wage. In this case, the fairness constraint weakens and downward wage rigidity breaks down, making nominal wage reductions admissible.

These two cases illustrate how the interaction between marginal profitability and external fairness governs wage adjustment. We now formalize this mechanism.

[3] Yoshihara (2006) shows that responsibility-sensitive fairness and full compensation for adverse skill conditions cannot be jointly satisfied in efficient bargaining outcomes.

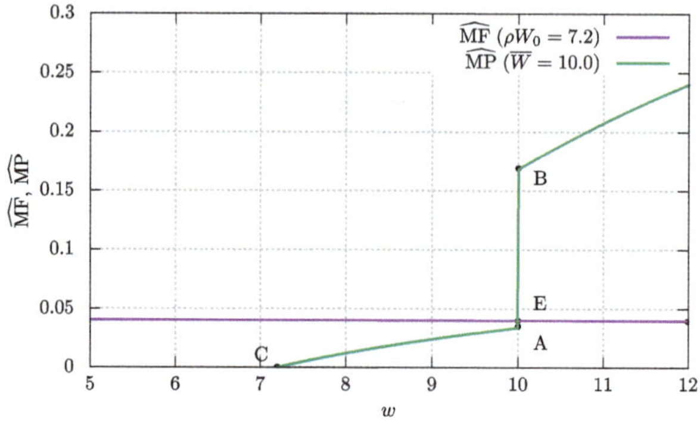


**Fig. 7.1** Moderate downturn: Downward wage rigidity remains binding.
At the status quo wage $w = W$, the marginal fairness index $\widehat{\text{MF}}(w, W_0)$ still intersects the kinked segment of the marginal-profit index $\widehat{\text{MP}}(w)$. Although the external reference wage $\rho W_0$ has declined, it remains sufficiently close to $W$ that the marginal fairness cost of a wage cut outweighs its marginal profitability. As a result, the firm optimally maintains the current money wage

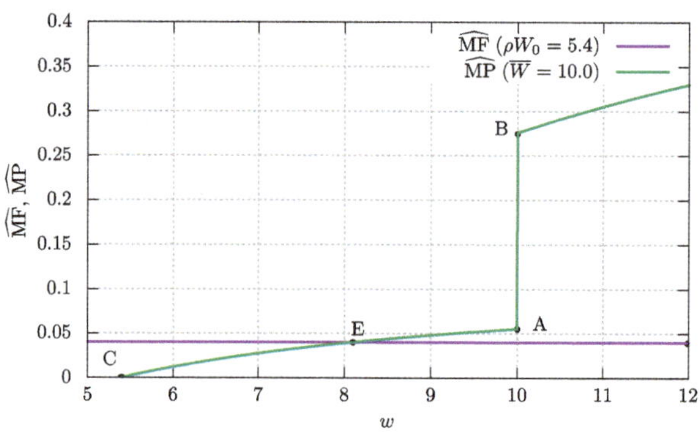


**Fig. 7.2** Severe downturn: breakdown of downward wage rigidity.
A further deterioration in aggregate conditions lowers the external reference wage $\rho W_0$ substantially relative to $W$. The marginal fairness index at $w = W$ now lies below the kinked segment of the marginal-profit index, eliminating the intersection at the status quo wage. In this case, the Nash criterion favors a downward revision of the money wage

### *7.4.1 Interior Optimum (on a Given Side)*

If the solution lies entirely on side $m \in \{-, +\}$, the first-order condition yields

$$w_m^* = \frac{\theta \gamma_m}{\theta \gamma_m - (1-\theta)v} \rho W_0, \qquad \text{valid if } \theta \gamma_m > (1-\theta)v.$$

If $\theta \gamma_m \leq (1-\theta)v$, the Nash objective prefers the boundary.

### 7.4.2 *Downward Rigidity and the Deflationary Threshold ("flex–fix")*

Downward nominal wage adjustment is disciplined by workers' asymmetric effort responses to relative wages. In the wage-bargaining formulation of the previous chapter, a wage cut below the status quo level reduces effort disproportionately, reflecting loss aversion and fairness considerations. As a result, small wage cuts are rejected by the Nash solution unless profitability conditions deteriorate sufficiently. This asymmetry gives rise to a deflationary threshold $K_D$, below which downward wage adjustment becomes admissible.

To characterize this threshold, suppose the firm currently pays the status quo wage $W$ and considers a small cut (just below $W$), which places the profit term on the left regime $m = -$. A cut is not adopted if the left-limit derivative of $\log \mathrm{NP}$ at $W$ is nonnegative, so that a marginal wage reduction lowers the Nash objective.

$$D_- := \frac{d}{dw} \log \mathrm{NP}(w)\bigg|_{w=W^-} = -\theta\,\frac{\gamma_-}{W} + (1-\theta)\,\frac{v}{W - \rho W_0} \geq 0.$$

Equivalently,

$$(1-\theta)\,v(1-g) \geq \theta\, g(1-b_-)\left(1 - \rho\frac{W_0}{W}\right). \tag{7.4.1}$$

In view of Eqs. (7.2.4) and (7.2.5), this condition reduces to

$$\widehat{\mathrm{MF}} > \widehat{\mathrm{MP}}_-(W; W_0).$$

That is, downward rigidity persists when the marginal fairness cost of a wage cut exceeds its marginal profitability. This condition characterizes the region in which the status-quo wage remains optimal; the deflationary threshold $K_D$ is reached when this inequality is reversed.

**Graphical Interpretation**

For a given status quo wage $W$, a deterioration in aggregate conditions lowers the external reference wage $W_0$, reflecting weaker labor-market tightness. This shifts the marginal fairness index $\widehat{\mathrm{MF}}$ downward relative to the marginal-profit schedule. As illustrated in Fig. 7.1, as long as the horizontal marginal fairness line $\widehat{\mathrm{MF}}$ lies above the left-side marginal-profit index $\widehat{\mathrm{MP}}_-(W)$ at the kink, the firm has no incentive to cut the current money wage. Downward wage adjustment is triggered only once the decline in $W_0$ is sufficiently large for $\widehat{\mathrm{MF}}$ to fall below $\widehat{\mathrm{MP}}_-(W)$.

Define the *floor* external reference wage $\underline{W}^0$ as the lowest value of $W_0$ for which maintaining the status quo wage $W$ remains optimal. It is implicitly defined by (7.4.1) at equality:

$$\frac{\underline{W}^0}{W} = \frac{1}{\rho}\left[1 - \frac{(1-\theta)v(1-g)}{\theta g(1-b_-)}\right].$$

A lower value of $\underline{W}^0$ corresponds to stronger downward rigidity.

**Deflationary Threshold**

Let the ratio of reference wage $W_0$ and $W$ in an SQSS equilibrium be defined as $f_w(K) = \frac{W_0(K)}{W(K)}$. Since $W_0/W$ is increasing in $K$ by Lemma 7.1, the deflationary threshold $K_D$ is characterized as the unique solution to

$$f_w(K_D) = \frac{1}{\rho}\left[1 - \frac{(1-\theta)v(1-g)}{\theta g(1-b_-)}\right],$$

provided $\theta\gamma_- > (1-\theta)v$ (interior feasibility on the left side). This condition identifies the point at which the marginal fairness constraint ceases to bind, and downward wage adjustment becomes optimal.

The monotonicity of $f_w(K)$ then allows this wage-setting condition to be mapped directly into a capital-stock threshold, thereby completing the characterization of $K_D$.

**Proposition 7.2 (Deflationary Threshold $K_D$)** *Suppose $f_w(K) \equiv \left(\frac{N^d(K)}{\bar{N}}\right)^{\gamma_0}$ is strictly increasing in $K$ and define*

$$\Xi(\theta, b_-, v, \rho) := \frac{1}{\rho}\left[1 - \frac{(1-\theta)v(1-g)}{\theta g(1-b_-)}\right].$$

*Assume $\Xi(\theta, b_-, v, \rho) \in (0, \infty)$. Then there exists a unique $K_D$ satisfying $f_w(K_D) = \Xi(\theta, b_-, v, \rho)$. Moreover,*

$$\frac{\partial K_D}{\partial \theta} > 0, \qquad \frac{\partial K_D}{\partial b_-} < 0, \qquad \frac{\partial K_D}{\partial v} < 0, \qquad \frac{\partial K_D}{\partial \rho} < 0, \qquad \frac{\partial K_D}{\partial \bar{N}} > 0.$$

***Proof (Sketch)*** Because $f_w$ is strictly increasing, it admits an increasing inverse $f_w^{-1}$ on its range, so that

$$K_D = f_w^{-1}(\Xi(\theta, b_-, v, \rho)).$$

The comparative statics with respect to $\theta, b_-, v, \rho$ follow from the corresponding monotonicities of $\Xi$.

For $\bar{N}$, note that for each fixed $K$, $f_w(K) = (N^d(K)/\bar{N})^{\gamma_0}$ is strictly decreasing in $\bar{N}$. Hence, to maintain the equality $f_w(K_D) = \varXi$ when $\bar{N}$ increases, $K_D$ must increase. This yields $\partial K_D/\partial \bar{N} > 0$. □

**Intuition**
The mechanism underlying the deflationary threshold $K_D$ is straightforward. At the status quo wage $w = W$, firms compare the marginal profitability of maintaining the current wage with the marginal loss in worker welfare relative to the external reference $\rho W_0$. As long as the outside option is sufficiently attractive, the marginal fairness effect remains strong enough to deter wage cuts, even in the presence of unemployment. However, when aggregate conditions deteriorate, the reference wage $W_0$ declines relative to $W$, weakening workers' marginal valuation of the existing wage. Once this decline is large enough, the marginal fairness effect can no longer offset the rising marginal profitability of wage reductions at the internal reference point. At this point, firms optimally revise money wages downward, marking the deflationary threshold $K_D$.

**Summary and Transition**
The analysis above identifies two state-dependent boundaries for nominal wage adjustment: an inflationary threshold $K_F$ and a deflationary threshold $K_D$. While $K_F$ marks the point at which labor-market tightness triggers upward wage revision, $K_D$ identifies the point at which the fairness constraint ceases to bind and downward wage adjustment becomes admissible. Together, these thresholds define an inaction region within which nominal wages remain fixed despite fluctuations in economic conditions.

The next chapter embeds these thresholds within a stochastic long-run framework, where repeated investment shocks determine how frequently the economy approaches, crosses, or retreats from each boundary, thereby shaping the invariant distribution of macroeconomic outcomes.

# Appendix

## 7.A Fully Rational Wage Setter (Iwai Style)

Following Iwai's formulation, the firm perceives a subjective labor-demand schedule of the form

$$l(w) = \left(\frac{w}{W}\right)^{\xi} \bar{N} = W^{-\xi}\bar{N} \cdot w^{\xi} = b_N^{-1} w^{\xi}, \qquad b_N \equiv \frac{W^{\xi}}{\bar{N}},$$

where the parameter $b_N$ summarizes the tightness of the labor market.

Profit maximization yields the optimal (target) real wage[4]

$$\omega^* = \frac{(\eta - 1)\xi}{\eta(1+\xi)}(1-\alpha)k^{\alpha}\left(\frac{N^d}{\bar{N}}\right)^{\frac{1+\alpha(\eta-1)}{\eta}}, \tag{7.A1}$$

[4] This equation is derived by applying a straightforward computation to Eq. (4.B5) in Appendix 4.B.

while the SQSS real wage is given by

$$\omega = \frac{\eta - 1}{\eta}(1 - \alpha)k^{\alpha}. \tag{7.A2}$$

Taking the ratio of (7.A1) and (7.A2) yields

$$\frac{\omega^*}{\omega} = \frac{\xi}{1 + \xi}\left(\frac{N^d}{\bar{N}}\right)^{\frac{1+\alpha(\eta-1)}{\eta}}. \tag{7.A3}$$

Let $f^* > 1$ denote the "normal shortage factor" implicit in Iwai's wage-setting rule, defined implicitly by

$$\frac{\xi}{1 + \xi} = \left(\frac{1}{f^*}\right)^{\frac{1+\alpha(\eta-1)}{\eta}}.$$

Then (7.A3) may be rewritten as

$$\frac{\omega^*}{\omega} = \left(\frac{N^d}{f^*\bar{N}}\right)^{\frac{1+\alpha(\eta-1)}{\eta}}.$$

The firm must revise its nominal wage upward whenever

$$N^d > f^*\bar{N}.$$

Since $f^* > 1$, such a threshold might never be attained. However, effective labor supply is typically smaller than $\bar{N}$ due to frictional labor turnover and the normal allowance required for smooth operation. Let

$$L_{\text{eff}} = \bar{\upsilon}\,\bar{N}, \qquad 0 < \bar{\upsilon} < 1,$$

denote the effective labor endowment. Then, the wage-revision condition becomes

$$N^d > f^* L_{\text{eff}} = f^* \bar{\upsilon}\,\bar{N}.$$

If the normal allowance satisfies

$$\bar{\upsilon} < \frac{1}{f^*},$$

then the wage-revision condition is already met when $N^d = \bar{N}$, that is, at the full-employment point for the given nominal wage. Since the state space of the capital-stock process is confined to $[1, K_F]$, the threshold is reached precisely at $K_F$, and the inflation-trigger level of the capital stock coincides with the full-employment level:

$$K_I = K_F.$$

In the foregoing analysis, $\omega_t^*$ denotes the Wicksellian target real wage implied by Iwai's wage-setting rule, while $\omega_{t-1} = \bar{W}_{t-1}/P_{t-1}$ is the real wage inherited from the previous period. Under the condition $f^*\bar{\upsilon} < 1$, the target real wage exceeds the inherited wage at full employment, as shown above.

At the same time, nominal wage adjustment is subject to various frictions and costs (administrative costs, implicit contracts, and internal norms). We represent these frictions by an *inaction band* in the log real wage: As long as the deviation between the target and the inherited real wage remains within a tolerance level $\theta_w > 0$, the firm finds it optimal to refrain from changing the nominal wage. Once the deviation exceeds this band, the firm revises its nominal wage so as to attain the target real wage at the prevailing price level.[5]

Formally, the revision rule for the representative firm's nominal wage is given in Eq. (7.3.1). When the normal allowance satisfies $\bar{\upsilon} < 1/f^*$, the economy eventually reaches a region in which $\omega_t^*$ lies above $\omega_{t-1}$ by more than the tolerance $\theta_w$, so that the upper boundary $K_F$ can be interpreted as the point at which the nominal wage starts to move upward in response to labor-market tightness.

Using $L_{\text{eff}} = \bar{\upsilon}\bar{N}$, we can write the ratio of the target and inherited real wages as

$$\frac{\omega_t^*}{\omega_{t-1}} = \left(\frac{N_t^d}{f^* L_{\text{eff}}}\right)^{\gamma}, \qquad \gamma := \frac{1+\alpha(\eta-1)}{\eta},$$

so that

$$\log\frac{\omega_t^*}{\omega_{t-1}} = \gamma\Big[\log N_t^d - \log\big(f^*\bar{\upsilon}\bar{N}\big)\Big].$$

Evaluating this expression at the full-employment boundary $N_t^d = \bar{N}$ yields

$$\Delta_F := \log\frac{\omega_t^*}{\omega_{t-1}}\bigg|_{N_t^d=\bar{N}} = \gamma\log\frac{1}{f^*\bar{\upsilon}}.$$

Under the condition $f^*\bar{\upsilon} < 1$, we have $\Delta_F > 0$, so the target real wage exceeds the inherited wage in the full-employment region. The inaction band in (7.3.1) then implies that an upward revision of the nominal wage is guaranteed to occur by the time the economy reaches $K_F$ whenever

$$0 < \theta_w < \Delta_F = \gamma\log\frac{1}{f^*\bar{\upsilon}}.$$

Thus $\bar{\upsilon}$ and $f^*$ do not pin down $\theta_w$ uniquely, but they impose a natural upper bound on the size of the inaction band consistent with wage adjustment in tight labor markets.

[5] A fixed or convex cost of wage adjustment in a dynamic optimization problem typically generates $(S, s)$-type or band-type adjustment rules. Here we take the piecewise rule below as a reduced-form representation of such behavior.

## References

Abowd, J. M., Kramarz, F., & Margolis, D. N. (1999). High wage workers and high wage firms. *Econometrica, 67*(2), 251–333. https://doi.org/10.1111/1468-0262.00020

Akerlof, G. A. (1982). Labor contracts as partial gift exchange. *The Quarterly Journal of Economics, 97*(4), 543–569. https://doi.org/10.2307/1885099

Akerlof, G. A., & Yellen, J. L. (1990). The fair wage-effort hypothesis and unemployment. *The Quarterly Journal of Economics, 105*(2), 255–283. https://doi.org/10.2307/2937787

Azariadis, C. (1975). Implicit contracts and underemployment equilibria. *Journal of Political Economy, 83*(6), 1183–1202. https://doi.org/10.1086/260388

Bewley, T. F. (2004). Fairness, reciprocity, and wage rigidity. Discussion Paper 1137. IZA – Institute of Labor Economics. https://doi.org/10.2139/ssrn.348140

Bhaskar, V. (1990). Wage relativities and the natural range of unemployment. *The Economic Journal, 100*(400), 60–66. htpps://doi.org/10.2307/2234184

Fehr, E., Kirchsteiger, G., & Riedl, A. (1993). Does fairness prevent market clearing? An experimental investigation. *The Quarterly Journal of Economics, 108*(2), 437–459. https://doi.org/10.2307/2118338

Hall, R. E. (1975). The rigidity of wages and the persistence of unemployment. *Brookings Papers on Economic Activity, 6*(2), 301–350. https://doi.org/10.2307/2534105

Iwai, K. (1987). *Hukinko Dogaku-no Riron, A theory of disequilibrium dynamics*. Iwanami Shoten.

Kahneman, D., & Tversky, A. (1979). Prospect theory: An analysis of decision under risk. *Econometrica, 47*(2), 263–292. https://doi.org/10.2307/1914185

Lye, J. N., & McDonald, I. M. (2021). Can loss aversion shed light on the deflation puzzle? *Review of Keynesian Economics, 9*(1), 11–42, https://doi.org/10.4337/roke.2021.01.02

McDonald, I. M. (2019). John Maynard Keynes, Joan Robinson and the prospect theory approach to money wage determination. *Metroeconomica, 70*(1), 45–67. https://doi.org/10.1111/meca.12226

McDonald, I. M., & Solow, R. M. (1981). Wage bargaining and employment. *American Economic Review, 71*(5), 896–908.

Shapiro, C., & Stiglitz, J. E. (1984). Equilibrium unemployment as a worker discipline device. *The American Economic Review, 74*(3), 433–444. https://doi.org/10.1017/cbo9780511559594.004

Sirota, D., Mischkind, L. A., & Meltzer, M. I. (2005). *Enthusiastic employee: How companies profit by giving workers what they want*. Wharton School Publishing.

Tversky, A., & Kahneman, D. (1991). Loss aversion in riskless choice: A reference-dependent model. *The Quarterly Journal of Economics, 106*(4), 1039–1061. https://doi.org/10.2307/2937956

Yoshihara, N. (2006). Solidarity and cooperative bargaining solutions. In *Inequality and Economic Integration, Banach Center Publications* (Vol. 71, pp. 289–312). Polish Academy of Sciences, Institute of Mathematics. https://doi.org/10.4064/bc71-0-26

# Chapter 8
# Stochastic Long-Run Equilibrium under Threshold-Based Wage Adjustment: Neutrality of Money

*The theory of money which I have been propounding is not neutral in the sense that changes in the quantity of money have no effect on real magnitudes. But it is neutral in the long run, in the sense that it does not determine the ultimate real position of the system.*

— John Maynard Keynes, *The General Theory*, Ch. 18

## 8.1 Introduction

The preceding chapter identified two state-dependent thresholds governing nominal wage adjustment: an inflationary threshold $K_F$, at which labor-market tightness triggers upward wage revision, and a deflationary threshold $K_D$, at which the fairness constraint ceases to bind and downward wage adjustment becomes admissible. Together, these thresholds define an inaction region within which nominal wages remain fixed despite fluctuations in economic conditions. As the capital stock evolves under stochastic investment dynamics, the economy repeatedly approaches and occasionally crosses these boundaries, generating asymmetric wage adjustments that shape the long-run behavior of the system.

This chapter brings together the analytical strands developed throughout the book to examine how stochastic investment dynamics, interacting with state-dependent wage-adjustment thresholds, shape the economy's long-run equilibrium. In this way, the long-run equilibrium emerges from the interaction between stochastic investment dynamics and the state-dependent wage-adjustment thresholds identified in the preceding chapter. The central object of interest is the invariant distribution of the capital stock and the associated neutral point that governs long-run macroeconomic behavior. The chapter asks how this distribution—and the equilibrium levels of employment, output, the interest rate, and the real wage it sustains—responds to changes in the model's exogenous parameters.

A natural starting point is the validity of the neutrality-of-money hypothesis in this stochastic environment. Specifically, we ask whether purely nominal parameters, such as the money supply and the nominal wage, influence real long-run

 
I. Takahashi, *Microfoundations of Keynesian Economics*, Evolutionary Economics and Social Complexity Science 37, https://doi.org/10.1007/978-981-95-9390-3_8

outcomes, or whether the stochastic steady state is instead pinned down entirely by real and structural features of the economy. Establishing the scope and limits of monetary neutrality provides a benchmark for the subsequent comparative-statics analysis of real parameters. The simulation results reported in this chapter are based on the stochastic transition structure developed in Chap. 5. They are presented in Sect. 8.4, while the baseline parameter values and implementation details are documented in Appendix 6.G.

The remainder of the chapter is organized as follows. Section 8.2 examines the neutrality of money in the stochastic steady state and explains why simple truncation of the capital distribution is not consistent with endogenous wage adjustment. Section 8.3 identifies the real anchors that determine the neutral point and shape the invariant distribution. Section 8.4 presents numerical comparative-statics results and explores how alternative institutional regimes and adjustment parameters affect the distribution of long-run outcomes.

## 8.2 Neutrality of Money in the Stochastic Steady State

We begin by establishing a benchmark result concerning the neutrality of money in the long run. The object of interest is the stochastic steady state characterized by the invariant distribution of the capital stock and its associated neutral point $K^*$. In this framework, neutrality is understood not in terms of short-run dynamics, but as a property of the long-run anchor around which stochastic fluctuations occur.

The central claim of this section is that the key real objects of the stochastic steady state—the neutral capital stock $K^*$, the corresponding interest rate $r^*$, the real wage $\omega^*$, employment $N^*$, and aggregate output $Q^*$—are independent of nominal parameters such as the money supply cap $M^S$ (hereafter referred to simply as the money supply and denoted by $M$) and the nominal wage $W$. Instead, they are pinned down entirely by real and structural features of the economy, including technology, preferences, and adjustment frictions. Establishing this result provides a reference point against which the subsequent analysis of invariant distributions and nominal adjustment mechanisms can be evaluated.

The SQSS analyzed in Chap. 4 describes a short-run equilibrium conditional on a given capital stock $K$. Firms revise investment decisions in response to expected profitability, so the economy evolves across a continuum of SQSS equilibria, each anchored at a different level of $K$ and associated with corresponding levels of employment, output, the interest rate, and the real wage. Perturbed by macroeconomic sentiment or mutations, firms update their confidence regarding prospective earnings, thereby expanding or contracting the capital stock over time. The neutral point can be characterized equivalently in terms of the capital stock $K^*$ or the associated neutral interest rate $r^*$.

Combining from Proposition 6.17 and Eq. (5.4.5) (Chap. 5), we obtain the following *neutral-point condition for the capital stock*:

$$1+\delta\zeta(K^*)=\frac{\alpha(c+\delta)}{r(K^*)+\delta}\cdot\frac{\eta-1}{\eta}=1+\delta\Big(\varphi-\frac{1+c_b}{2}\Big),\qquad \varphi=\frac{1}{1+\frac{1}{\alpha(\eta-1)}}. \tag{8.2.1}$$

Since $\zeta(K)$ is strictly decreasing in $K$, this condition pins down a unique stochastic equilibrium capital stock $K^*$. In the interior regime, the neutral-point condition admits an equivalent representation that is convenient for computing the neutral interest rate.

**Proposition 8.1 (Neutral-Point Condition in the Interior Regime)** *Assume that at the neutral point $K^*$ the clipping constraints in the transition probabilities do not bind (interior regime), so that $p_+(K^*)$, $p_-(K^*) \in (0, 1/2)$ and $\zeta(K^*) \in (\varphi-(1+c_b), \varphi)$. Let*

$$u(K) := \frac{K^d(K)}{K} = \frac{\alpha(c+\delta)}{r(K)+\delta}\cdot\frac{\eta-1}{\eta},\qquad \zeta(K) := \frac{u(K)-1}{\delta},$$

$$\varphi(\eta) := \frac{\alpha(\eta-1)}{1+\alpha(\eta-1)}.$$

*Then the neutral point defined by* drift$(K^*) = 0$ *satisfies*

$$\frac{\alpha(c+\delta)}{r^*+\delta}\cdot\frac{\eta-1}{\eta} = 1+\delta\Big(\varphi(\eta)-\frac{1+c_b}{2}\Big),\qquad r^* := r(K^*). \tag{8.2.2}$$

*Equivalently,*

$$r^*+\delta = \frac{\alpha(c+\delta)\,(\eta-1)/\eta}{1+\delta(\varphi(\eta)-(1+c_b)/2)}.$$

***Proof (Sketch)*** On $[0, 1/2]$, the function $f(x) = x(1-x/2)$ is strictly increasing. Hence, drift$(K) = 0$ is equivalent to $p_+(K) = p_-(K)$. In the interior regime, the definitions of $p_+$ and $p_-$ reduce to

$$p_+(\zeta) = \frac{1}{2} - b(\varphi-\zeta),\qquad p_-(\zeta) = \frac{1}{2} - b(\zeta+1+c_b-\varphi),$$

so $p_+(\zeta) = p_-(\zeta)$ implies

$$\zeta(K^*) = \varphi - \frac{1+c_b}{2}.$$

Using $\zeta = (u-1)/\delta$ yields $u(K^*) = 1+\delta(\varphi-(1+c_b)/2)$. Substituting $u = \alpha(c+\delta)\frac{\eta-1}{\eta}\frac{1}{r+\delta}$ and writing $r^* = r(K^*)$ give (8.2.2). □

Crucially, the neutral-point condition depends only on real and structural parameters—such as $\alpha$, $\eta$, $\delta$, $c_b$, and the propensity to consume $c$—and does not involve nominal variables such as the money supply $M$ or the nominal wage $W$.

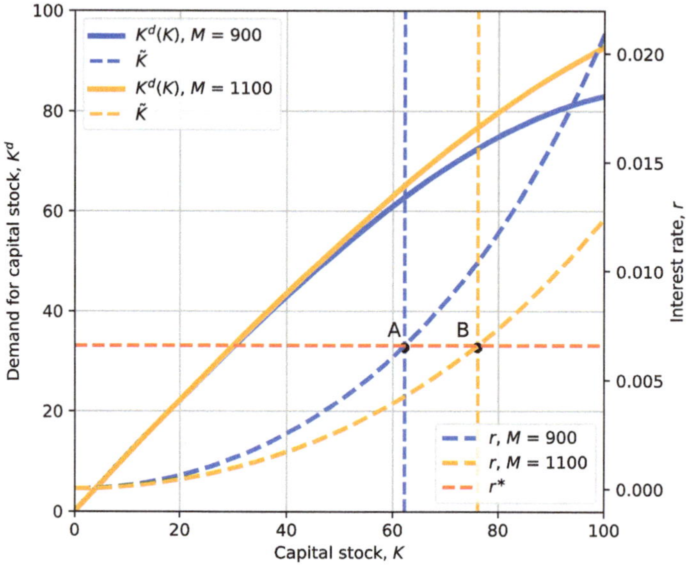


**Fig. 8.1** Neutral interest rate and capital-stock determination under different monetary conditions. The neutral interest rate $r^*$ is uniquely determined by real and structural parameters and does not depend on the monetary parameter $M$. Changes in $M$ shift the capital-demand schedule $K^d(\cdot; M)$, thereby altering the scale of operations $\tilde{K}$ consistent with a fixed $r^*$. The figure illustrates how monetary parameters affect the level of capital accumulation without changing the neutral rate of interest. The parameter $M$ in the figure legends denotes the exogenous settlement capacity of the payments system (i.e., the monetary supply $M^S$ in the model)

It follows that the associated equilibrium interest rate $r^*$ is likewise independent of $M$ and $W$.

Equivalently, the neutral-point condition can be written in interest rate form as Eq. (8.2.1) at $r = r^*$ yields the corresponding *neutral interest rate condition*:

$$\zeta(r^*) = \frac{\alpha(c+\delta)}{r^*+\delta} \cdot \frac{\eta-1}{\eta} = 1 + \delta\left(\frac{1}{1+\frac{1}{\alpha(\eta-1)}} - \frac{1+c_b}{2}\right). \tag{8.2.3}$$

To clarify why variations in the monetary parameter $M$ affect the steady-state capital stock $K^*$ without altering the neutral interest rate $r^*$, Fig. 8.1 provides a graphical representation of the neutral interest rate condition and the associated capital-demand schedules.

**Remark 8.1 (Neutral Interest Rate and Steady-State Capital)** Fix the neutral interest rate $r^*$. Then $h(r^*)$ is pinned down uniquely, and so is the steady-state capital stock $K^*$; see Eq. (4.D3) in Sect. 4.6.1. Under exponent matching, this relationship can be written as

$$K^* = \frac{\theta_{\mathrm{PM}}}{\theta_{\mathrm{LM}}\, h(r^*)}.$$

Hence, any parameter change that leaves $r^*$ unchanged affects $K^*$ only through the mapping from $h(r^*)$ to $K^*$. In particular, $K^*$ is inversely proportional to $\theta_{\mathrm{LM}}$.

Note that the neutral interest rate $r^*$ is pinned down by the neutral interest rate condition Eq. (8.2.3) alone and therefore depends only on real parameters (such as $\alpha, \eta, \delta, c, c_b$). In particular, it is independent of the LM scaling parameters (e.g., $\lambda$ and $W/M$) and does not rely on the exponent-matching assumption used elsewhere to obtain a closed-form mapping $r = r(K)$.[1]

Given a candidate stochastic steady state $(K^*, r^*)$, the PM relation uniquely determines the corresponding equilibrium real wage $\omega^*$. Consistency with the LM relation then requires

$$\omega^* = \left(\frac{W}{M}\, c\, \lambda^{1-\beta}\right) \cdot \left(\frac{r^*}{1+r^*}\right)^{-\beta} K^*.$$

This expression highlights the role of the nominal ratio $W/M$ in pinning down the real wage consistent with a given pair $(K^*, r^*)$. The question is therefore whether $(K^*, r^*)$ themselves depend on nominal magnitudes, or whether changes in $(W, M)$ that preserve $W/M$ amount to purely nominal rescalings.

**Proposition 8.2 (Neutrality of Money at the Stochastic Steady-State Anchor)** *Assume that the neutral point $K^*$ is defined by the neutral point condition Eq. (8.2.1) and that $\zeta(K)$ is strictly decreasing in $K$. Then the neutral point $K^*$ is uniquely determined by real and structural parameters and is independent of nominal variables such as $M$ and $W$.*

*Moreover, let $\omega^*$ denote the equilibrium real wage implied by the PM relation evaluated at $(K^*, r^*)$. If the LM relation can be written in the form*

$$\omega = \left(\frac{W}{M}\, c\, \lambda^{1-\beta}\right) \cdot \left(\frac{r}{1+r}\right)^{-\beta} K, \tag{8.2.4}$$

*then consistency of the PM–LM system at the neutral point implies that the ratio $W/M$ must be invariant when the real parameters $(\lambda, \beta, c)$ are held fixed (and hence $\alpha$ is fixed under exponent matching). Consequently, proportional rescalings of $(M, W)$ leave the stochastic steady-state anchor $(K^*, r^*, \omega^*)$ unchanged.*

***Proof*** Since $\zeta(K)$ is strictly decreasing, the neutral point condition Eq. (8.2.1) admits at most one solution. Because this condition depends only on real and structural parameters and does not involve $M$ or $W$, its unique solution $K^*$ is independent of nominal variables. It follows that $r^* := r(K^*)$ is likewise independent of $M$ and $W$.

[1] However, the distribution of $r$ can still change if the mapping $K \mapsto r(K)$ changes shape (e.g., if $\beta$ changes in the PM–LM mapping).

Given $(K^*, r^*)$, the PM relation pins down the corresponding equilibrium real wage $\omega^*$. Evaluating Eq. (8.2.4) at $(K^*, r^*, \omega^*)$ then shows that the ratio $W/M$ must take the constant value required for PM–LM consistency at the neutral point. This yields a long-run neutrality result at the stochastic steady-state anchor: Proportional rescalings of the nominal pair $(M, W)$ alter only the nominal scale, leaving the real anchor $(K^*, r^*, \omega^*)$ unchanged. Any endogenous quantities pinned down by $K^*$—such as $Q^* = Q(K^*)$ and $N^* = N(K^*)$—are likewise invariant to such nominal rescalings. □

This proposition formalizes long-run monetary neutrality at the stochastic steady-state anchor: Proportional rescalings of $(M, W)$ affect only the nominal scale (via $W/M$) and leave $(K^*, r^*, \omega^*)$ unchanged.[2]

This result, however, does not by itself pin down the *shape* of long-run fluctuations around the anchor. The invariant distribution may deform in response to nominal closures that suppress or permit wage and price adjustment, especially when the process is pushed against boundary constraints such as $K_F$ or $K_D$. We therefore turn next to a comparative study of invariant distributions—first under a fixed nominal wage closure, and then under nominal wage adjustment—to make transparent how monetary neutrality at the anchor coexists with substantial differences in the distributional profile of macroeconomic fluctuations.

**Remark 8.2 (Money in Wage Units)** In the present framework, the nominal scale relevant for real allocations is the money supply expressed in wage units, $M/W$, rather than the money supply or the price level taken in isolation. This formulation follows Keynes's insistence that monetary magnitudes should be measured in terms of wage units, reflecting the central role of nominal wages as the anchor for employment and production decisions (see Keynes, 1936, Ch. 4 for the original statement of the wage-unit concept).

**Remark 8.3 (Wage Rigidity and the Liquidity of Money)** Keynes repeatedly emphasized that the liquidity of money is not an intrinsic technological property but rests on the relative stability of money wages. If nominal wages were to adjust rapidly and continuously, the purchasing power of money in terms of labor would become unstable, undermining its role as a reliable store of value and unit of account. As Keynes argued, it is precisely the stickiness of money wages that anchors the value of money and sustains the liquidity premium associated with holding it (Keynes 1936, Chs. 17 and 19). From this perspective, wage rigidity should be understood not as a mere friction, but as a structural precondition for a monetary economy. In this sense, the neutrality of money established in this chapter is conditional on the existence of a stable wage anchor, a point already emphasized by Keynes in his analysis of money's liquidity.

[2] The neutrality result does not rely on the exponent-matching assumption. That assumption is imposed only to obtain closed-form expressions for employment, output, and wages in later sections and plays no role in pinning down the stochastic steady-state anchor $(K^*, r^*)$.

**Remark 8.4 (Interpretive Remarks on the Neutral Interest Rate)**

(*i*) *Wicksellian interpretation.* The neutral interest rate defined here is closely related to Wicksell's natural rate of interest, introduced in Wicksell (1936, Ch. 2), as the real rate consistent with the absence of cumulative disequilibrium. Unlike Wicksell's original formulation, however, neutrality is defined in terms of zero drift in a stochastic economy with nominal rigidities, rather than price-level stability in a deterministic setting.

(*ii*) *Relation to empirical natural-rate estimates.* In modern macroeconomics, the notation $r^*$ is often used to denote the natural rate of interest estimated from data, as in Laubach and Williams (2003). Unlike that empirical concept, the neutral rate derived here is structurally determined and governs the long-run drift of the capital stock in a stochastic economy.

(iii) *Borrower's risk and the servicing burden.* Under the 100% money regime with a fixed money wage, the contractionary role of the interest rate extends beyond its effect on new investment. Discounting the contractual service stream at rate $r$ implies that the long-run burden of capital commitments, measured in wage units, scales with the size of operations, $K/\omega$. However, the relevant constraint for borrower's risk is the period-by-period servicing flow, $(r + \delta)K/\omega$, which rises with the interest rate and is proportional to employment. An increase in $r$ therefore tightens financial conditions not only by reducing investment incentives but also by increasing the real servicing burden of existing commitments, thereby reinforcing the broader Keynesian contractionary effect of higher interest rates (see Appendix 8.C).

Having established monetary neutrality at the level of the stochastic steady-state anchor, we now turn to a comparative analysis of invariant distributions under alternative nominal closures, highlighting how fixed nominal wages and nominal wage adjustment differentially shape the distribution of long-run fluctuations.

## *8.2.1 Wage Adjustment and the Breakdown of Truncation*

This section reveals why truncation is not self-consistent. We begin by defining upper truncation and then show it triggers nominal wage adjustment, which feeds back through prices and real balances.

### 8.2.1.1 Erosion at the Upper Boundary: $K_F$

**Definition 8.1 (Upper Truncation at $K_F$)** Let $S = [1, K_F]$ and let $P$ be the transition kernel induced by the clipped update $K' = \min\{\Gamma(K\,\varepsilon), K_F\}$. An invariant distribution $\pi$ is said to be *truncated at $K_F$* if $\pi(K_F) > 0$. Equivalently, truncation holds if

$$\int_S \Pr(\Gamma(K,\varepsilon) \geq K_F \mid K)\,\pi(dK) > 0.$$

Formally, truncation corresponds to the case in which the clipped transition kernel assigns positive stationary probability to the upper boundary; a simple sufficient condition in terms of the maximal upward jump is given by the following lemma:

**Lemma 8.3 (Sufficient Condition for Upper Truncation at $K_F$)** *Let the state space be $S = [1, K_F]$, and suppose the capital-stock update rule is given by*

$$K' = \min\{K\gamma_+^{J_+},\ K_F\} \quad \text{on the up-branch,}$$

*where $J_+ \in \{0, 1, \ldots, J_{\max}\}$ is drawn with probabilities $\delta_+(K, j)$ and $\gamma_+ > 1$. Assume that $\delta_+(K, j)$ is continuous in $K$ and uniformly bounded away from zero on the interior domain $D = [K_{q0}, K_{p0}]$, and that $p_+(K) > 0$ for some $K \in D$.*

*If*

$$K_{p0}\,\gamma_+^{J_{\max}} \;\geq\; K_F,$$

*then the clipped transition kernel assigns positive probability to the upper boundary $K_F$ from a set of states with positive stationary measure. In particular, any invariant distribution $\pi$ of the process satisfies*

$$\int_S \Pr\big(K' = K_F \mid K\big)\,\pi(dK) \;>\; 0,$$

*so that the invariant distribution is truncated at $K_F$.*

***Proof (Deferred)*** See Appendix 8.A, Proof of Lemma 8.3.[3] □

**Lemma 8.4 (Adjustment-Induced Erosion of Truncation at $K_F$)** *Suppose the invariant distribution of the capital stock is truncated at the upper bound $K_F$, so that the economy repeatedly attains full employment. Then, under endogenous nominal wage adjustment and a fixed money supply $M$, such a truncated distribution cannot be invariant. Instead, the induced rise in the nominal wage leads to a leftward shift of the invariant distribution, eliminating the truncation at $K_F$.*

***Proof (Economic Mechanism)*** If the invariant distribution places positive mass near $K_F$, the economy repeatedly operates at full employment. Under endogenous nominal wage adjustment, these episodes generate persistent upward pressure on the nominal wage $W$. With the money supply $M$ held fixed, the induced increase in $W$ must ultimately be absorbed by a proportional increase in the price level $P$, so that the long-run real wage does not drift upward.

[3] In what follows, all proofs of the lemmas and propositions stated in the main text are provided in Appendix 8.A.

The resulting increase in $P$ reduces real money balances $M/P$, tightening monetary conditions and shifting effective demand downward at each capital stock. Investment incentives weaken accordingly, implying a downward shift in the drift of the capital-stock process. As a consequence, the invariant distribution shifts leftward, reducing the frequency with which the economy reaches the full-employment boundary $K_F$.

This adjustment continues until the invariant distribution assigns zero mass to the boundary region. Hence, a distribution truncated at $K_F$ is not self-consistent and cannot be invariant under fixed $M$ and endogenous nominal wage adjustment.

□

#### 8.2.1.2 Erosion at the Lower Boundary $K_D$

**Definition 8.2 (Lower Truncation at $K_D$)** Let $S = [1, K_F]$ and let $P$ be the transition kernel induced by the clipped update $K' = \max\{K_D, \min\{\Gamma(K, \varepsilon), K_F\}\}$. An invariant distribution $\pi$ is said to be *truncated at* $K_D$ if $\pi(\{K_D\}) > 0$. Equivalently, truncation holds if

$$\int_S \Pr(\Gamma(K, \varepsilon) \leq K_D \mid K)\, \pi(dK) > 0.$$

The next lemma provides a sufficient condition for lower truncation at $K_D$, that is, for the long-run capital-stock distribution to accumulate positive probability mass at the downward clip.

**Lemma 8.5 (Sufficient Condition for Lower Truncation at $K_D$)** *Let the state space be $S = [K_D, \infty)$, and suppose the capital-stock update rule on the down-branch is given by*

$$K' = \max\{K\gamma_-^{J_-},\ K_D\},$$

*where $J_- \in \{0, 1, \ldots, J_{\max}\}$ is drawn with probabilities $\delta_-(K, j)$ and $0 < \gamma_- < 1$. Assume that $\delta_-(K, j)$ is continuous in $K$ and uniformly bounded away from zero on the interior domain $D = [K_{q0}, K_{p0}]$, and that $p_-(K) > 0$ for some $K \in D$.*

*If*

$$K_{q0}\, \gamma_-^{J_{\max}} \ \leq\ K_D,$$

*then the clipped transition kernel assigns positive probability to the lower boundary $K_D$ from a set of states with positive stationary measure. In particular, any invariant distribution $\pi$ of the process satisfies*

$$\int_S \Pr\big(K' = K_D \mid K\big)\, \pi(dK)\ >\ 0,$$

*so that the invariant distribution is truncated at $K_D$.*

***Proof (Deferred)*** See Appendix 8.A, Proof of Lemma 8.5. □

**Lemma 8.6 (Adjustment-Induced Erosion of Truncation at $K_D$)** *Suppose the invariant distribution of the capital stock is truncated at the lower bound $K_D$, so that the economy repeatedly attains the lower clip. Then, under endogenous nominal wage adjustment and a fixed money supply $M$, such a truncated distribution cannot be invariant. Instead, the induced decline in the nominal wage leads to a rightward shift of the invariant distribution, eliminating the truncation at $K_D$.*

***Proof (Economic Mechanism)*** If the invariant distribution places positive mass near $K_D$, the economy repeatedly operates in a low-activity regime with slack labor markets. Under endogenous nominal wage adjustment, these episodes generate persistent downward pressure on the nominal wage $W$.

With the money supply $M$ held fixed, sustained movements in $W$ translate into proportional movements in the price level $P$ in the long run, so that the long-run real wage does not drift downward.

The resulting decline in $P$ increases real money balances $M/P$, loosening monetary conditions and shifting effective demand upward at each capital stock. Investment incentives strengthen accordingly, implying an upward shift in the drift of the capital -stock process. As a consequence, the invariant distribution shifts rightward, reducing the frequency with which the economy reaches the lower boundary $K_D$.

This adjustment continues until the invariant distribution assigns zero mass to the boundary region. Hence, a distribution truncated at $K_D$ is not self-consistent and cannot be invariant under fixed $M$ and endogenous nominal wage adjustment.

□

**Remark 8.5 (Qualification: Keynesian Asymmetry at the Lower Boundary)** Lemma 8.6 describes the symmetric adjustment mechanism under frictionless downward nominal wage adjustment. In practice, downward nominal wage rigidity (or explicit resistance to wage cuts) can prevent $W$ from falling sufficiently during slack episodes, thereby limiting the associated decline in the price level $P$ and the increase in real money balances $M/P$. Moreover, if monetary accommodation is constrained by an effective lower bound on the interest rate (for example, due to operational banking costs that imply an admissible floor for $r$), the economy may be unable to offset weak demand through further reductions in $r$.

These constraints break the symmetry: Lower-bound episodes need not self-correct as in Lemma 8.6, so mass near $K_D$ may persist and downward truncation can become long-run feasible. This formalizes Keynes's insight that a deflationary adjustment has no stable resting place below full employment unless some nominal magnitude is sticky or the interest rate becomes incapable of falling further (Keynes, 1936, Ch. 19, p. 304).

**Boundary-Induced Adjustment of the Stochastic Steady State**

The following corollary synthesizes the two truncation lemmas above.

**Corollary 8.7 (Boundary-Induced Adjustment of the Stochastic Steady State)** *Consider an economy with a fixed money supply $M$ and nominal wage adjustment driven by persistent labor-market imbalance. Suppose the capital-stock process admits a neutral point $K^*$ that is invariant to nominal parameters. Then:*

1. *If the invariant distribution is truncated at the upper boundary $K_F$, so that the economy repeatedly attains full employment, upward nominal wage pressure raises the price level $P$. The resulting fall in real money balances $M/P$ increases the interest rate and discourages investment, shifting the invariant distribution leftward and reducing the stationary probability of hitting $K_F$.*
2. *If the invariant distribution is truncated at the lower boundary $K_D$, so that the economy repeatedly experiences slack and unemployment, downward nominal wage pressure increases real money balances $M/P$. The resulting fall in the interest rate stimulates investment, shifting the invariant distribution rightward and reducing the stationary probability of hitting $K_D$.*

*In both cases, adjustment operates through a deformation of the invariant distribution rather than a displacement of the neutral point $K^*$, which remains determined by real structural parameters.*

***Proof*** The two claims follow directly from Lemmas 8.4 and 8.6. □

This stabilizing feedback operates as long as the economy remains on the stable (positive-rate) PM–LM branch and downward wage adjustment remains feasible.

Having characterized how boundary pressure interacts with wage adjustment, we now turn to numerical comparative statics of invariant distributions.

## 8.3 Real Anchors and the Shape of the Invariant Distribution

All closed-form SQSS objects are inherited from Chap. 4; here we focus on how these objects shape the invariant distribution. The neutrality result established above identifies a stochastic steady-state anchor for the economy, but it does not exhaust the long-run implications of the model. Beyond fixing the nominal invariance of the anchor, the structure of the economy determines how fluctuations are distributed around it. In particular, real parameters do not merely shift equilibrium relationships; they shape both the location of the neutral point and the geometry of the stochastic dynamics organized around it.

This section therefore shifts attention from nominal neutrality to the role of real anchors in shaping the invariant distribution. We examine how variations in real parameters—such as the capital share $\alpha$, the degree of market competitiveness $\eta$, and behavioral factors governing effective demand and liquidity preference—affect the long-run distribution of capital, employment, interest rates, and real wages. Changes in these parameters move the neutral point itself and, at the same time, alter the dispersion, skewness, and boundary behavior of the invariant distribution through their interaction with nominal closures and adjustment rules.

## 8.3.1 Real Anchors and the Neutral Point

Recall the (upward, downward) transition probabilities are incentivized by the excess replacement ratio $\zeta(K)$:

$$\begin{aligned} p_+(K) &= \text{clip}[\,0.5 - b\,(\varphi - \zeta(K))\,;\; 0,\; 0.5]\,, \\ p_-(K) &= \text{clip}[\,0.5 - b\,(\zeta(K) + 1 + c_b - \varphi\,);\; 0,\; 0.5]\,. \end{aligned} \tag{8.3.1}$$

The next paragraph summarizes the built-in stabilizing mechanism under the fixed-$M$ closure, operating through the interaction between the interest rate and the capital stock.

Along the stable PM–LM branch under a fixed money supply $M$, the capital stock $K$ and the interest rate $r(K)$ interact in a stabilizing way. When $K$ falls, transaction demand weakens and the equilibrium interest rate declines, lowering the user cost of capital. This raises the desired operating scale $K^d(K)$ relative to the existing stock $K$, increasing utilization and shifting the replacement ratio $\zeta(K)$ upward. As a result, the up-branch probability $p_+(K)$ rises, while the down-branch probability $p_-(K)$ falls, strengthening the incentive for recovery. Conversely, when $K$ becomes high, transaction demand tightens money-market conditions, pushing $r(K)$ upward and reducing $K^d(K)/K$; utilization and $\zeta(K)$ decline, upward revisions become less likely, and contractionary revisions become more likely. In this way, the fixed-$M$ closure generates a built-in mean-reverting force in the capital dynamics through the endogenous adjustment of the interest rate.

In short, under a fixed-$M$ closure, fluctuations in $K$ affect $r(K)$, which changes $\kappa(r)$ and therefore changes the desired operating scale $K^d$—this is the core stabilizing feedback loop in our model.

This stabilizing feedback shapes the drift of the capital-stock process around the neutral point. The location of the neutral point itself, however, is determined by real and structural parameters; see Proposition 8.2.

### 8.3.1.1 Comparative Statics of the Neutral Rate: $(\alpha, \eta, c_b, \lambda)$

Although the neutral point condition pins down $K^*$ (and hence $r^*$) uniquely from real parameters, the comparative statics with respect to deep parameters such as $\alpha$ and $\eta$ are not globally monotone and depend on the interaction between the profitability channel and the confidence threshold. For this reason, the numerical section below focuses on nominal closure experiments, which deliver sharper and more robust distributional implications.

The absence of globally monotone comparative statics with respect to $\alpha$ and $\eta$ partly reflects the exponent-matching assumption, which links technological curvature and monetary sensitivity; this assumption does not create the ambiguity but renders the offsetting channels explicit rather than concealed.

**Remark 8.6 (Competitive Pressure and SQSS Fragility)** The elasticity of substitution $\eta$ governs the degree of competitive pressure in each SQSS. As $\eta$ rises, markups shrink and the steady-state profit buffer above financing costs becomes thin. In this near-competitive region, even a small pessimistic disturbance—modeled here as a small mass of negative mutants that depress effective demand and expected profitability—can flip the sign of net investment incentives. Hence, large $\eta$ makes SQSS equilibria locally fragile: The economy becomes more sensitive to sentiment shocks, and the invariant distribution is more easily shifted toward contractionary states. While $\eta$ also enters the neutral point condition and therefore affects the stochastic steady-state anchor, its most transparent role is to govern the robustness of that anchor by controlling how much demand must fall before profitability drops below the financing cost.

To assess how the real structural parameters $\alpha$ and $\eta$ shape the invariant distribution of observable variables, it is useful to begin by examining how variations in these parameters shift the PM locus itself. $\theta_{\mathrm{PM}}$ governs the short-period PM schedule (how $\omega$ depends on $r$). These parameters do not merely shift equilibrium relationships; they determine the location of the neutral point around which the economy's stochastic dynamics are organized. The following lemma characterizes the comparative statics of the PM scale parameter $\theta_{\mathrm{PM}}$.[4]

**Lemma 8.8 (Comparative Statics of $\boldsymbol{\theta}_{\mathrm{PM}}$)** *Let $\alpha \in (0, 1)$ and $\eta > 1$. Define*

$$\theta_{\mathrm{PM}}(\alpha, \eta) \; := \; (1-\alpha)\Big(\frac{\eta-1}{\eta}\Big)^{\frac{1}{1-\alpha}} \alpha^{\frac{\alpha}{1-\alpha}}. \tag{8.3.2}$$

*Then:*

*(i) $\theta_{\mathrm{PM}}$ is strictly increasing in $\eta$:*

$$\frac{\partial \theta_{\mathrm{PM}}}{\partial \eta}(\alpha, \eta) \; > \; 0.$$

*(ii) For each fixed $\eta > 1$, the map $\alpha \mapsto \theta_{\mathrm{PM}}(\alpha, \eta)$ is strictly decreasing on $(0, 1)$:*

$$\frac{\partial \theta_{\mathrm{PM}}}{\partial \alpha}(\alpha, \eta) \; < \; 0.$$

***Proof** (Deferred)* The full proof is given in Appendix 8.A.3. □

**Remark 8.7 (Admissible Parameter Region for Comparative Statics)** All comparative statics statements with respect to $\eta$ are understood to hold on the admissible parameter region where the neutral point exists, i.e., where the endpoint sign conditions Eq. (6.6.5)–Eq. (6.6.6) in Sect. 6.6.1 remain satisfied. Outside this region

[4] $\theta_{\mathrm{PM}}$ is a scale parameter of the PM locus, not a behavioral parameter.

the defining equation for $K^*$ may fail to admit an interior solution, and the stochastic dynamics may instead be governed by boundary regimes.

Part (i) of Lemma 8.8 implies that a rise in $\eta$ increases the PM scale parameter $\theta_{\mathrm{PM}}$, shifting the PM locus upward in $(r, \omega)$-space for any given $r$. This is a PM-side effect and should be distinguished from the neutral interest rate condition, which determines how the neutral rate $r^*$ itself moves with $\eta$.

While changes in $\alpha$ and $\eta$ alter the PM locus through $\theta_{\mathrm{PM}}$, the implications for the neutral interest rate $r^*$ are more subtle, since $r^*$ is pinned down by the neutral interest rate condition rather than by the PM–LM intersection at a given capital stock.

The next lemma therefore examines how variations in $\alpha$ affect $r^*$ directly.

**Lemma 8.9 (Effect of $\alpha$ on the Neutral Interest Rate)** *Fix $\eta > 1$, $c > 0$, $\delta > 0$, and $c_b \geq 0$. Let*

$$d := \frac{1 + c_b}{2}, \qquad \varphi(\alpha) := \frac{\alpha(\eta - 1)}{1 + \alpha(\eta - 1)}.$$

*Suppose that $r^* = r^*(\alpha)$ is defined implicitly by the neutral interest rate condition*

$$\frac{\alpha(c + \delta)}{r^* + \delta} \cdot \frac{\eta - 1}{\eta} = 1 + \delta\big(\varphi(\alpha) - d\big), \tag{8.3.3}$$

*and assume*

$$B(\alpha) := 1 + \delta\big(\varphi(\alpha) - d\big) > 0 \tag{8.3.4}$$

*so that $r^* + \delta > 0$. If, in addition,*

$$\delta d < 1,$$

*then the neutral interest rate is strictly increasing in $\alpha$:*

$$\frac{\partial r^*}{\partial \alpha} > 0 \qquad \textit{for all } \alpha \in (0, 1) \textit{ satisfying (8.3.4).}$$

***Proof (Deferred)*** See Appendix 8.A.4. □

When $\alpha$ is small, the mapping from the neutral capital-labor ratio $\kappa^*$ into the real-wage schedule is highly leveraged.[5] Indeed, the neutrality condition pins down $\kappa^*$, and the real wage–interest rate relation Eq. (4.B12) in Appendix 4.B implies

[5] It follows from Eq. (4.D5) in Appendix 4.D together with the corrected neutral interest rate condition that $\kappa^* = \big(\frac{1+\delta(\varphi-(1+c_b)/2)}{c+\delta}\big)^{1/(1-\alpha)}$.

$$\omega = \frac{1-\alpha}{\alpha}\,\kappa^* \,(r+\delta),$$

so that $\partial\omega/\partial\kappa^* = \frac{1-\alpha}{\alpha}(r+\delta)$ becomes large as $\alpha \downarrow 0$. As a result, even a modest change in $\kappa^*$ induces a pronounced rotation of the capital intensity locus in the $(r, \omega)$-plane.

Lemma 8.9 shows that, under the mild condition $\delta d < 1$ (which is satisfied by the baseline calibration $(\delta, c_b) = (0.06, 0.05)$), the neutral interest rate $r^*$ is strictly increasing in $\alpha$ throughout the admissible range. Intuitively, an increase in $\alpha$ raises the capital share in production and therefore increases the level of utilization required to sustain neutrality. Because utilization enters the neutrality condition through the term $1 + \delta(\varphi - (1 + c_b)/2)$, this effect is only weakly attenuated by depreciation. Restoring neutrality therefore requires a higher cost of capital, which is achieved through an increase in the real interest rate $r^*$.

Although the effect of $\alpha$ on the neutral rate is ambiguous in general, it is sign definite under the baseline calibration. It is therefore useful to contrast this case with the role of $\eta$, which governs the degree of market power and enters the neutrality condition more directly.

**Lemma 8.10 (Effect of $\eta$ on the Neutral Interest Rate)** *Fix $\alpha \in (0, 1)$, $c > 0$, $\delta > 0$, and $c_b \geq 0$. Let*

$$d := \frac{1+c_b}{2}, \qquad \varphi(\eta) := \frac{\alpha(\eta-1)}{1+\alpha(\eta-1)}.$$

*Suppose the neutral interest rate condition*

$$\frac{\alpha(c+\delta)}{r^*+\delta}\cdot\frac{\eta-1}{\eta} = 1+\delta\left(\varphi(\eta)-d\right)$$

*admits a solution $r^* > 0$ for some $\eta > 1$. Then $r^*$ is differentiable in $\eta$, and*

$$\operatorname{sign}\left(\frac{\partial r^*}{\partial\eta}\right) = \operatorname{sign}\left[\frac{1}{\eta(\eta-1)} - \frac{\delta\,\varphi'(\eta)}{1+\delta(\varphi(\eta)-d)}\right], \qquad \varphi'(\eta) = \frac{\alpha}{\left(1+\alpha(\eta-1)\right)^2}.$$

*Proof (deferred)* See Appendix 8.A.5. □

**Corollary 8.11 (Monotonicity in $\eta$ under the baseline calibration)** *Under the baseline calibration $(\alpha, \delta, c_b) = (1/3, 0.06, 0.05)$, the bracketed term in Lemma 8.10 is strictly positive for $\eta$ in the range considered numerically (e.g., $\eta \in [6, 12]$). Hence,*

$$\frac{\partial r^*}{\partial\eta} > 0 \qquad \textit{throughout this range.}$$

*In particular, an increase in $\eta$—corresponding to tighter product-market competition and lower markups—unambiguously raises the neutral interest rate $r^*$ under the baseline parameter values.*

***Proof (Deferred)*** See Appendix 8.A.6. □

Economically, this result reflects the fact that greater competition compresses markups and raises the throughput required to sustain profitability; as a result, a given utilization disturbance must be absorbed by a larger adjustment in the real interest rate.

Although the effect of $\alpha$ on the neutral rate is ambiguous in principle, Lemma 8.9 and the calibration $\delta d < 1$ make it sign definite in the baseline economy: $r^*$ rises monotonically with $\alpha$. This relative simplicity of the $\alpha$-channel contrasts with the role of $\eta$, which governs the degree of market power and enters the neutrality condition more directly. Lemma 8.10 and Corollary 8.11 show that, over the empirically relevant range of $\eta$, increased competition (higher $\eta$) also raises $r^*$, but through a distinct mechanism operating via markups and effective utilization.

Next, we turn to transactions demand parameter $\lambda$. The parameter $\lambda$ is the transactions demand coefficient (a reduced-form measure of liquidity preference), governing how strongly real activity translates into desired money holdings. Although the parameter $\lambda$ enters the LM relation, it does not represent a nominal magnitude but rather the strength of liquidity preference, and therefore acts as a real behavioral parameter governing the sensitivity of the interest rate to fluctuations in economic activity.

**Remark 8.8 (Liquidity Preference $\lambda$ and the Stochastic Steady State)** *(i) Level effect.* The liquidity preference parameter $\lambda$ does not enter the neutral point condition and therefore does not affect the neutral interest rate $r^*$. Under exponent matching, however, $\lambda$ raises $\theta_{\mathrm{LM}}$ via $\theta_{\mathrm{LM}} \propto c\lambda^{1-\beta}$, so the steady-state identity $K^*h(r^*) = \theta_{\mathrm{PM}}/\theta_{\mathrm{LM}}$ implies $\partial K^*/\partial\lambda < 0$ whenever $\beta < 1$. Economically, a rise in precautionary money demand tightens the monetary side of the economy; holding the real anchor $r^*$ fixed, the long-run adjustment occurs through a lower capital stock rather than through a shift in the neutral interest rate. *(ii) Stabilization strength.* Under a fixed nominal money supply, the liquidity preference parameter $\lambda$ affects not only the level of the steady-state capital stock but also the strength of the interest-rate response to real fluctuations. In the PM–LM equilibrium under exponent matching,

$$h(r(K)) = \frac{\theta_{\mathrm{PM}}}{\theta_{\mathrm{LM}} K}, \qquad \theta_{\mathrm{LM}} \propto \lambda^{1-\beta},$$

so that an increase in $\lambda$ raises the equilibrium interest rate $r(K)$ for any given $K$. Differentiating the equilibrium condition implies

$$\frac{dr}{d\log K} = -\frac{1}{\frac{d}{dr}\log h(r)},$$

where $\frac{d}{dr}\log h(r) < 0$ on the stable (low-rate) branch. As $r(K)$ increases, the magnitude $\left|\frac{d}{dr}\log h(r)\right|$ typically declines, implying that a given fluctuation in $K$ requires a larger compensating change in the interest rate to restore monetary equilibrium.

Economically, a higher $\lambda$ amplifies the sensitivity of money demand to real activity. With nominal money fixed, variations in capital, output, and transactions generate larger shifts in desired liquidity holdings, which must be offset by stronger movements in the interest rate. In this sense, a high degree of liquidity preference strengthens the Keynes effect: Real fluctuations are more forcefully stabilized by interest-rate adjustments, but at the cost of heightened interest-rate volatility and a lower long-run capital stock.

We next turn to the role of the parameter $c_b$, which governs the benchmark normal profit requirement in the mutation-selection mechanism. Unlike $\alpha$, the effect of $c_b$ operates directly through the neutral interest rate condition and therefore admits a clear and unambiguous characterization in terms of the neutral rate $r^*$.

**Lemma 8.12 (Effect of the Downside Bias Parameter $c_b$ on $r^*$)** *Let $r^*$ denote the neutral interest rate determined by the neutral interest rate condition*

$$\frac{\alpha(c+\delta)}{r^*+\delta}\cdot\frac{\eta-1}{\eta} = 1+\delta\left(\varphi-\frac{1+c_b}{2}\right), \qquad \varphi = \frac{\alpha(\eta-1)}{1+\alpha(\eta-1)}.$$

*Holding all other parameters fixed, an increase in normal profit tolerance $c_b$ raises the neutral interest rate:*

$$\frac{\partial r^*}{\partial c_b} > 0.$$

***Proof*** An increase in $c_b$ lowers the right-hand side of the neutral interest rate condition (8.2.3) by construction, reflecting a higher required normal profit benchmark. Since the left-hand side is strictly decreasing in $r^*$, restoring equality requires an increase in $r^*$. The result follows. □

Since the LM-side relation is unaffected by $c_b$, the increase in $r^*$ implies a corresponding decline in the stochastic equilibrium capital stock $K^*$.

**Remark 8.9 (On the Effect of Depreciation $\delta$)** Unlike $c_b$, the depreciation rate $\delta$ enters the neutral point condition through multiple channels. First, holding $r$ fixed, it appears in the profitability wedge $(c+\delta)/(r+\delta)$, whose partial derivative with respect to $\delta$ is negative whenever $c > r$. Second, $\delta$ affects the equilibrium mapping $r = r(K)$ through the PM–LM system. Third, and specific to the corrected neutrality condition, $\delta$ enters the required utilization term on the right-hand side, $1+\delta(\varphi-(1+c_b)/2)$, which scales the utilization (or excess demand) threshold consistent with zero drift. In the interior regime $\varphi > (1+c_b)/2$, this requirement rises with $\delta$, pushing in the opposite direction of the direct profitability-wedge effect.

As a result, the comparative statics of $(K^*, r^*)$ with respect to $\delta$ are not globally monotone. When $c > r^*$ and the induced adjustment of $r(K)$ does not offset the direct wedge effect, an increase in $\delta$ tends to lower the neutral interest rate and, through the neutrality condition, raises the stochastic steady-state level of capital $K^*$. Absent such restrictions, however, the sign of $\partial K^*/\partial\delta$ remains parameter dependent.

**Remark 8.10 (Interpreting $c$ as Effective Demand and Confidence)** In the present framework, the parameter $c$ should not be interpreted narrowly as a technological marginal propensity to consume. Rather, it summarizes households' willingness to translate expected income into current expenditure and therefore governs effective demand and capacity utilization. Through this channel, variations in $c$ directly shape corporate profitability and expected returns on capital, which in turn discipline investment decisions over the business cycle. In this sense, the model formalizes the empirical insight emphasized by Ellis (2005), who argues that consumer spending is the primary driver of industrial production and corporate profits, with capital spending adjusting subsequently as profit expectations evolve.

**Remark 8.11 (The Propensity to Consume and the Neutral Point)** In the present specification, $c$ enters the neutral point condition through the utilization wedge and $\theta_{\mathrm{LM}}$. The neutrality condition implies that

$$\frac{\alpha(\eta-1)}{\eta}\cdot\frac{c+\delta}{r^*+\delta} = 1+\delta\Big(\varphi-\frac{1+c_b}{2}\Big), \qquad \Rightarrow \qquad \frac{\partial r^*}{\partial c} > 0. \tag{8.3.5}$$

Under exponent matching, the PM–LM equilibrium can be written as $K\,h(r) = \theta_{\mathrm{PM}}/\theta_{\mathrm{LM}}$ with $\theta_{\mathrm{LM}} \propto c$, hence

$$K^* = \frac{\theta_{\mathrm{PM}}}{\theta_{\mathrm{LM}}}\cdot\frac{1}{h(r^*)}.$$

Accordingly, $c$ affects $K^*$ through two opposing channels: Increasing $c$ raises $\theta_{\mathrm{LM}}$, which tends to lower $K^*$, but it also raises $r^*$ (by Eq. (8.3.5)), which tends to lower $h(r^*)$ on the low-rate branch and thereby raise $K^*$. The net sign of $\partial K^*/\partial c$ is therefore parameter dependent. For the baseline calibration used in the numerical section (where $r^*$ is small and $h(\cdot)$ is steep), the $r^*$-channel dominates so that an increase in $c$ shifts $K^*$ upward.

**Lemma 8.13 (A Sufficient Condition for $c$ Turning Point in $K^*(c)$)** *Maintain exponent matching and suppose that, at the neutral point, the neutral interest rate satisfies an affine relationship*

$$r^*(c) \;=\; a_c\,(c+\delta)-\delta, \qquad a_c > 0, \tag{8.3.6}$$

*where $a_c$ measures the pass-through from the user cost of capital into the neutral real interest rate, and that the stable PM–LM equilibrium mapping implies*

$$K^*(c) = \frac{\theta_{\text{PM}}}{\theta_{\text{LM}}(c)} \cdot \frac{1}{h(r^*(c))}, \qquad \theta_{\text{LM}}(c) \propto c, \qquad h(r) = \left[\frac{(r+\delta)(1+r)}{r}\right]^{\beta}, \tag{8.3.7}$$

*with* $\beta > 0$*. Then*

$$\operatorname{sgn}\left(\frac{dK^*}{dc}\right) = \operatorname{sgn}\left(-\frac{1}{c} - \frac{d}{dc}\log h(r^*(c))\right) = \operatorname{sgn}\left(-\frac{1}{c} - \frac{h'(r^*)}{h(r^*)}a_c\right). \tag{8.3.8}$$

*Moreover, there exist constants* $0 < c_{\text{low}} < c_{\text{high}}$ *(within the admissible region) such that*

$$\frac{dK^*}{dc} > 0 \quad \textit{for} \quad c \in (0, c_{\text{low}}), \qquad \frac{dK^*}{dc} < 0 \quad \textit{for} \quad c > c_{\text{high}}.$$

***Proof (Deferred)*** See Appendix 8.A.7. □

**Remark 8.12 (Self-stabilization, Confidence, and Macroeconomic Performance)** Lemma 8.13 highlights a non-monotone role of the propensity to consume in shaping the stochastic steady-state level of capital. When $c$ is low, aggregate demand is weak and the neutral interest rate $r^*$ is correspondingly small. In this region, an increase in $c$ primarily raises utilization and expected profitability, shifting the stochastic steady-state capital stock upward. Because $h(r)$ is steep at low interest rates, this profitability channel dominates the direct scaling effect through $\theta_{\text{LM}}$.

As $c$ rises further, however, the economy approaches real capacity constraints. The neutral interest rate increases, the function $h(r)$ flattens, and additional demand pressure is increasingly absorbed by higher financing costs rather than higher real activity. Beyond this region, the expansionary effect of confidence weakens and the direct $\theta_{\text{LM}}$ channel dominates, so that further increases in $c$ crowd out investment and reduce $K^*$.

This mechanism clarifies two broader points. First, with a fixed money supply, the economy possesses an inherent self-stabilizing force: Excessively optimistic demand conditions are checked endogenously by rising interest rates, while pessimistic conditions leave room for profitability-driven recovery. Second, macroeconomic performance depends not only on structural parameters but also on collective sentiment. A sufficiently high and sustained level of confidence supports investment and capital accumulation, whereas either chronic pessimism or overheated optimism undermines long-run performance through distinct channels.

In the present framework, $c$ is treated not as a reduced-form regression coefficient, but as a structural summary of effective-demand formation—capturing households' willingness to translate expected income into current expenditure, and hence disciplining profitability and investment incentives in the long run.

**Remark 8.13 (Why Focus on *c*?)** Modern macroeconomic analysis has often regarded consumption-function parameters as non-deep objects whose values may shift with policy regimes. The present analysis does not require $c$ to be policy invariant in that strong sense. Rather, the objective is to treat $c$ as a disciplined demand-side shifter that governs long-run profitability, utilization, and investment incentives. In this role, $c$ operates as a parsimonious summary of confidence and spending norms and becomes central precisely because it affects the stochastic equilibrium through channels that are distinct from those driven by technology or market structure.

The advantage of focusing on $c$ is that, unlike $(\alpha, \eta)$, it shifts the profitability scale $A$ without simultaneously relocating the confidence threshold $\varphi$, thereby isolating a purely demand-driven channel of long-run feasibility and stochastic stability.

The next remark presents a larger $\beta$ strengthens the Keynes effect in the sense that smaller interest-rate adjustments are sufficient to stabilize real fluctuations.

**Remark 8.14 (The Role of $\beta$ in the Steady State and the Keynes Effect)** The parameter $\beta$ does not enter the neutral point condition and therefore does not affect the neutral interest rate $r^*$ directly. Under exponent matching, however, the PM–LM equilibrium implies

$$K\,h(r) = \frac{\theta_{\mathrm{PM}}}{\theta_{\mathrm{LM}}}, \qquad h(r) = \left[\frac{(r+\delta)(1+r)}{r}\right]^{\beta}, \qquad \theta_{\mathrm{LM}} \propto c\,\lambda^{\,1-\beta}.$$

Evaluated at the neutral rate $r^*$, this yields $K^* = \theta_{\mathrm{PM}}\big/\big(\theta_{\mathrm{LM}}(\beta)\,h(r^*;\beta)\big)$. If $0 < \lambda < 1$ and $(r^*+\delta)(1+r^*)/r^* > 1$, then both $\theta_{\mathrm{LM}}(\beta)$ and $h(r^*;\beta)$ are increasing in $\beta$, implying $\partial K^*/\partial\beta < 0$.

The effect of $\beta$ on the strength of the Keynes effect can be read from the sensitivity of the equilibrium interest rate to fluctuations in $K$ under a fixed-$M$ closure. Writing the equilibrium as $h(r(K)) = \theta_{\mathrm{PM}}/(\theta_{\mathrm{LM}}K)$ and differentiating in logs give

$$\frac{dr}{d\log K} = -\frac{1}{\frac{d}{dr}\log h(r)} = -\frac{1}{\beta\,\frac{d}{dr}\log\!\left(\frac{(r+\delta)(1+r)}{r}\right)}.$$

On the stable branch $\frac{d}{dr}\log h(r) < 0$, so $dr/d\log K > 0$, and its magnitude is proportional to $1/\beta$. Thus a higher $\beta$ reduces the interest-rate movement required to offset a given real disturbance, dampening interest-rate volatility while lowering the long-run capital stock through the $\theta_{\mathrm{LM}}$ and $h(r^*)$ channels. Finally, a larger $\beta$ makes the PM–LM linkage "stiffer" in $(r, \omega)$-space, so a smaller interest-rate movement is needed to offset a given real disturbance.

## 8.4 Numerical Comparative Statics: Invariant Distribution

In this section, we illustrate how these real anchors shape the location and dispersion of invariant distributions under alternative nominal closures.

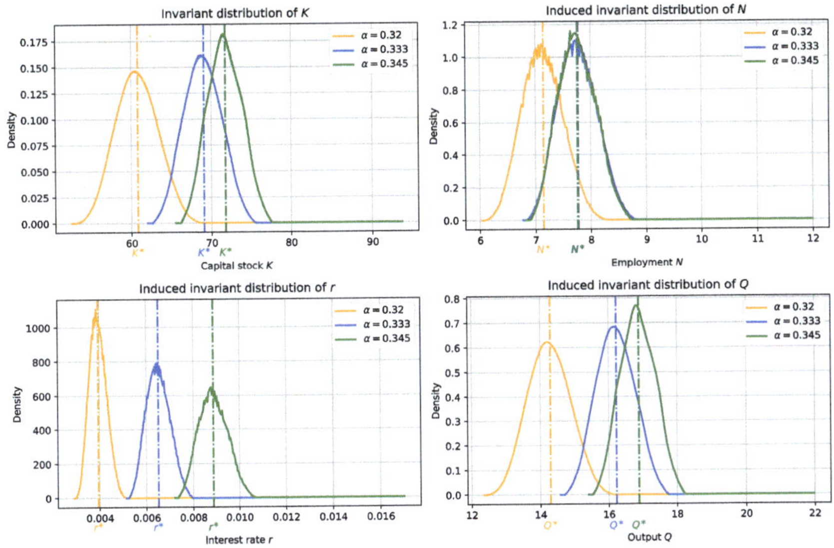


**Fig. 8.2** Effect of the capital income share $\alpha$ on the invariant distributions.
For readability, the fourth panel reports the induced invariant distribution of output $Q$ rather than the real wage $\omega$. The implications for real wages are discussed in the text

We focus on the roles of $\alpha$ and $\eta$ (technology and market structure), $c$ and $c_b$ (effective demand and profitability benchmarks), $b$ (adjustment sensitivity), $\lambda$ (transactions demand and liquidity preference parameter), and $\sigma_0$ (the scaling of nominal disturbances).

**Capital Share $\alpha$** We begin by examining how changes in the capital income share $\alpha$ affect the shape of the invariant distribution. Although the analytical results establish that $\alpha$ influences the neutral interest rate $r^*$ in a systematic way, the stochastic equilibrium turns out to be highly sensitive to $\alpha$, so that only a relatively narrow intermediate range yields economically meaningful invariant distributions.

Figure 8.2 shows that, as the capital income share $\alpha$ increases, the neutral interest rate $r^*$ rises, as predicted by the analytical results, shifting the invariant distribution of $r$ to the right. At the same time, the higher capital share strengthens investment incentives and raises the stochastic equilibrium capital stock $K^*$, which in turn supports higher output. Interpreted structurally, a larger $\alpha$ corresponds to a more roundabout production system with higher capital intensity and greater aggregate productivity.

At the same time, an increase in $\alpha$ lowers the real wage by redistributing income toward capital, even as the capital-labor ratio rises. These opposing forces largely offset each other in their effect on employment, so that the invariant employment distributions associated with medium and high values of $\alpha$ are nearly indistinguishable.

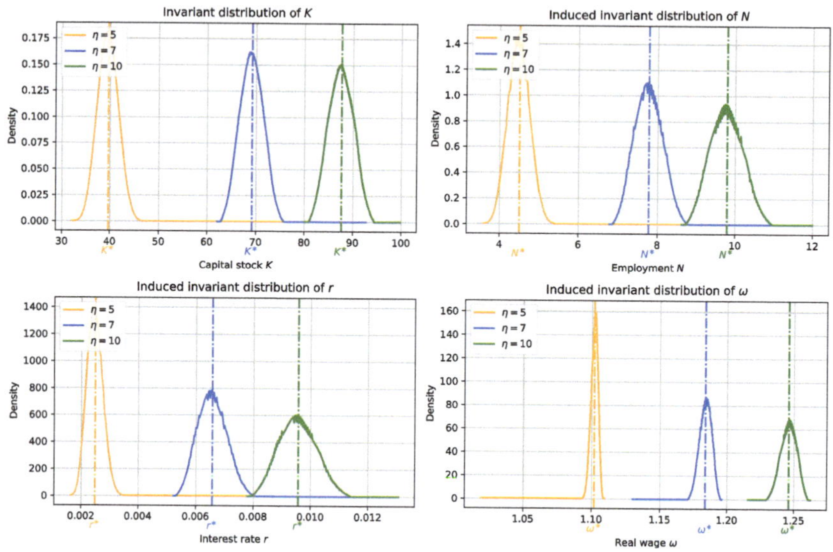


**Fig. 8.3** Effect of market competitiveness on the invariant distributions.
The figure displays the invariant distribution of the capital stock $K$ and the induced distributions of employment $N$, the interest rate $r$, and the real wage $\omega$ for three values of the elasticity of substitution $\eta$

Finally, the simulations indicate that higher $\alpha$ is associated not only with higher mean output but also with a more concentrated invariant distribution of $Q$. This suggests that, over the relevant range, increasing capital intensity moves the economy deeper into the interior region and strengthens mean-reverting forces in the capital-stock dynamics, thereby reducing the frequency of boundary excursions that generate additional volatility. This volatility compression is calibration and range dependent: If higher $\alpha$ were to raise the intensity of investment shocks or to increase financial fragility endogenously, the dispersion of $Q$ could instead widen.

**Competitive Price Elasticity $\eta$** An increase in $\eta$ corresponds to a more competitive environment with lower markups. In the SQSS, a higher $\eta$ shifts the PM locus outward, relaxing the joint constraint on the real wage and the interest rate. Figure 8.3 shows that, in the admissible interior regime, this pro-competitive shift raises the neutral interest rate $r^*$ and increases the stochastic equilibrium capital stock $K^*$, thereby supporting higher employment. Moreover, greater competition improves allocative efficiency and raises the equilibrium real wage $\omega^*$. For a given nominal wage, the identity $P = W/\omega$ then implies a lower price level, while higher capital utilization and employment translate into higher output $Q$. As a result, the invariant distributions of $(K, N, r, \omega)$ shift coherently to the right, reflecting improved long-run macroeconomic performance under increased competitiveness.

At the same time, higher $\eta$ can increase the sensitivity of the equilibrium mapping $K \mapsto (r, \omega)$, so fluctuations in activity translate into larger movements in prices and interest rates. In this sense, intensified competition may be associated with greater fragility even when average performance improves.

**Remark 8.15 (Competition, Safety Margins, and Depressive Instability)** In the present framework, stronger competition—captured by a higher effective elasticity $\eta$—compresses markups and raises static efficiency in normal times. At the same time, it erodes the profitability buffer that sustains investment. As $\eta$ increases, the confidence threshold

$$\varphi(\eta) = \frac{\alpha(\eta - 1)}{1 + \alpha(\eta - 1)}$$

moves toward unity, shrinking the margin above the normal profit benchmark (see Lemma 5.3). Once $\varphi(\eta)$ approaches the critical level $1 - c_b$, even small pessimistic disturbances suffice to eliminate net investment incentives.

This mechanism is particularly relevant in depressive episodes. When demand is weak and excess capacity is pervasive, firms and workers are forced into intensified price competition in order to secure sales and employment. In reduced form, this corresponds to a rise in effective competition, even absent changes in market structure. The resulting compression of safety margins makes the economy more susceptible to tipping into low-investment regimes, thereby amplifying unemployment, firm exit, and deflationary pressure. In this sense, heightened competition stabilizes outcomes in normal times but undermines macroeconomic stability in downturns.

This interpretation echoes Keynes's emphasis on confidence and margins of safety in investment decisions (Keynes, 1936, 1937, Chs. 11, 12), Kalecki's principle of increasing risk (Kalecki, 1937), and Minsky's notion of endogenous fragility arising from compressed safety margins (Minsky, 1986, Ch. 7, 8).

**Investment Confidence Parameter $c$**[6] To illustrate how the propensity to consume shapes the stochastic steady state, Fig. 8.4 compares the invariant distribution of the capital stock $K$ and the induced invariant distributions of employment $N$, the interest rate $r$, and the real wage $\omega$ for three nearby values of $c$. As $c$ increases, the invariant distribution of $K$ shifts rightward, and the induced distribution of employment shifts upward accordingly, reflecting higher average activity and more frequent operation near capacity. At the same time, the induced distribution of the interest rate $r$ moves to the right and becomes noticeably wider, consistent with the expansion of the admissible interest-rate window predicted by Proposition 8.14. Finally, the induced distribution of the real wage $\omega$ shifts leftward, indicating that stronger demand is accommodated primarily through higher utilization and higher

[6] The parameter $c$ should be interpreted as a structural investment confidence parameter—capturing the responsiveness of investment to prospective profitability—rather than as a time-varying state variable. Endogenizing the evolution of confidence is an important topic but lies beyond the scope of this book.

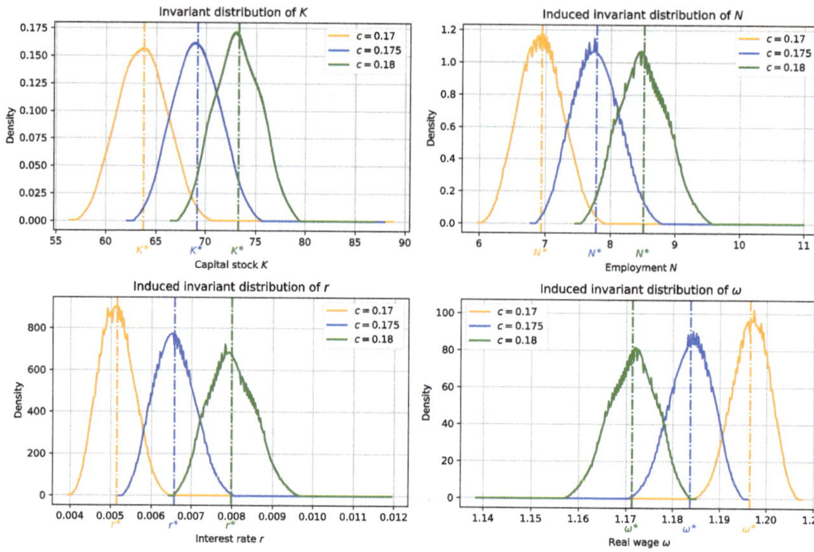


**Fig. 8.4** Effect of the propensity to consume $c$ on the invariant distributions.
Invariant distributions of the capital stock $K$ (left panel) and the interest rate $r$ (right panel) for $c = 0.17$, $c = 0.175$, and $c = 0.18$, holding all other parameters fixed

interest rates rather than through sustained increases in real wages. Although the right tail of the capital distribution for $c = 0.18$ approaches the full-employment boundary $K_F$, the distribution remains largely interior, and no collapse is observed in this range.

**Liquidity preference λ** Because $\lambda$ does not enter the drift condition that pins down $r^*$, its main effect operates through the scaling of the PM–LM mapping. In the numerical experiments, this channel shifts the distribution of $K$ while leaving the distribution of $r$ essentially unchanged. In the limit as the elasticity of substitution between consumption and money tends to zero, money demand takes a cash-in-advance form, $M^d = \lambda PC$, so that $\lambda$ measures the nominal liquidity required per unit of consumption.

The parameter $\lambda$ does not enter the drift condition $\zeta$ that determines the admissible interest-rate window $[r_-, r_+]$, and, as shown in Fig. 8.5, the invariant distribution of $r$ is therefore unchanged. Applying Remark 8.1, changes in $\lambda$ affect the stochastic dynamics around the neutral point without altering the value of $r^*$, shifting the invariant distribution of $K$ through tighter financing conditions captured by the scaling factor $\sigma_0 = \theta_{\mathrm{LM}}/\theta_{\mathrm{PM}}$. The resulting contraction in capital reduces transaction demand, offsetting the initial pressure on interest rates and leaving the long-run distribution of $r$ unchanged.

We report only the distributions of $K$ and $r$ here, since employment and the real wage move monotonically with the capital stock under the present closure and therefore do not add independent information in this comparative exercise.

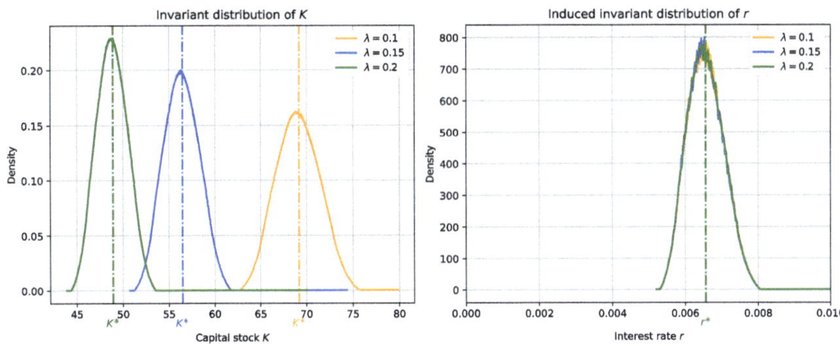


**Fig. 8.5** Effect of transaction demand coefficient λ on the invariant distributions.
The figure displays the invariant distribution of the capital stock $K$ and the interest rate $r$ for three values of the strength of transactions demand λ

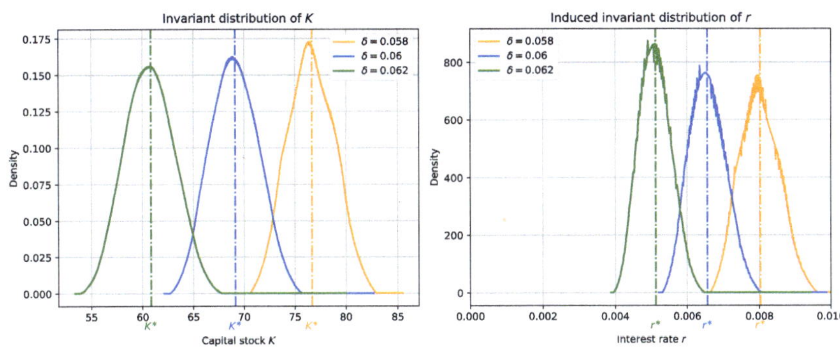


**Fig. 8.6** Effect of depreciation rates $\delta$ on the invariant distributions.
The figure displays the invariant distribution of the capital stock $K$ and the induced distribution of the interest rate $r$ for three values of the depreciation rate $\delta$. A higher depreciation rate raises the effective user cost of capital, reducing the profitability of investment and shifting the invariant distribution of $K$ leftward. Consistent with this contraction in capital accumulation, the corresponding distribution of the interest rate also shifts downward

**Depreciation Rate $\delta$** We next examine the role of capital depreciation, summarized by the parameter $\delta$, which governs the effective user cost of capital and therefore directly influences long-run investment incentives. Figure 8.6 shows that an increase in $\delta$ raises the replacement burden on firms and weakens profitability, suggesting a contraction in the long-run scale of production. Although not shown in the figure, lower capital intensity implies reduced employment and a lower real wage in the stochastic steady state. Overall, faster depreciation compresses the long-run scale of economic activity and weakens profit prospects across the distribution.

**Remark 8.16 (The Elasticity of Substitution Between Consumption and Money)** $\beta$ enters the model only through composite parameters such as $\theta_{\mathrm{LM}}$. Under the exponent-matching condition, $\beta$ cannot vary independently and is therefore treated as a structural parameter defining the monetary regime rather than a margin for comparative statics.

Accordingly, varying $\beta$ does not constitute a meaningful comparative statics exercise within the present framework, but instead corresponds to a change in the underlying monetary regime.

### 8.4.1 Nominal Closures and Distributional Deformation

To illustrate how nominal closures affect the long-run stochastic equilibrium, Figs. 8.7 and 8.8 compare the invariant distributions of the capital stock and its induced real variables under fixed nominal wages and under nominal wage adjustment, respectively.[7] The invariant distributions are computed directly from the transition kernel rather than from simulated sample paths; once the kernels coincide, the distributions are identical up to numerical precision. Note that holding $W$ fixed while increasing $M$ changes $W/M$ and therefore changes the real monetary environment; this is not a neutrality experiment.

Figure 8.7 displays the invariant distribution of the capital stock $K$ and the induced invariant distributions of employment $N$, the interest rate $r$, and the real wage $\omega$ for two money supplies, $M = 1000$ and $M = 1050$, holding the nominal wage fixed. An increase in $M$ raises real money balances mechanically, lowers the interest rate, and shifts the invariant distribution rightward. When the shift is sufficiently strong, the distribution places more mass near the upper boundary $K_F$. As the employment panel shows, the invariant distribution assigns substantial mass to $N(K_F)$, implying that the economy repeatedly enters and remains in states of full employment for extended durations, indicating recurrent full-employment pressure.

Figure 8.8 shows the same invariant and induced distributions as in Fig. 8.7, now allowing nominal wages to adjust proportionally with the money supply so as to keep the real monetary environment unchanged. Under this neutrality-consistent closure, the invariant distributions for $M = 1000$ and $M = 1050$ coincide and are therefore visually indistinguishable.

When nominal wages are allowed to adjust, proportional changes in the nominal pair $(M, W)$ correspond to purely nominal rescalings. In that case, the ratio $W/M$ is invariant, the real monetary environment is unchanged, and the invariant distribution of the capital stock—together with its induced distributions for employment, the interest rate, and the real wage—coincides across nominal scales. The visual overlap in Fig. 8.8 therefore provides a distributional counterpart to Proposition 8.2: Once nominal adjustment is admitted, purely nominal rescalings affect only the units in which prices and wages are measured, not the stochastic long-run equilibrium in real terms.

[7] The numerical illustrations employ a reduced-form Markov kernel that collapses the industry-level jump-size distribution to a law-of-large-numbers (mean-jump) update conditional on the up/down branch. This approximation preserves the state dependence of branch probabilities and the implied drift, while avoiding a high-dimensional jump-size state in the transition matrix. The theoretical results in the main text are stated for the full discrete-jump specification.

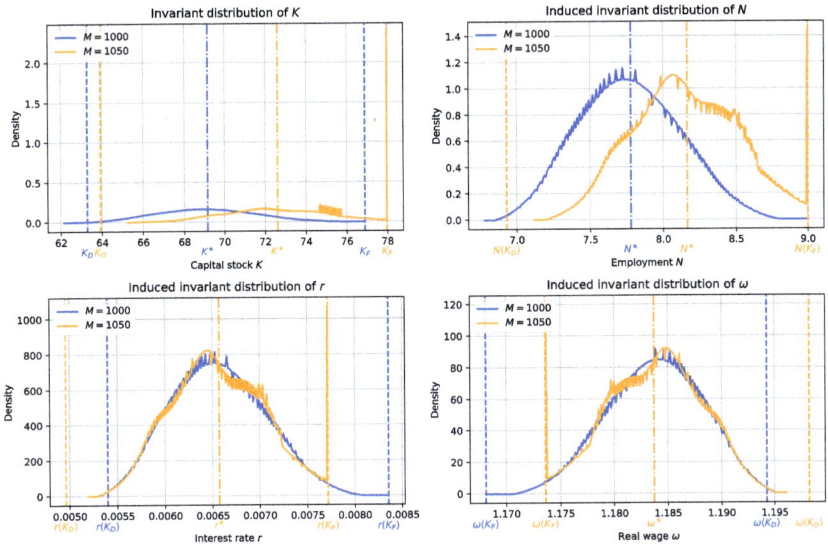


**Fig. 8.7** Invariant and induced distributions under a fixed nominal wage.
This configuration is not self-consistent in the long run, since persistent boundary pressure would trigger nominal wage and price adjustment

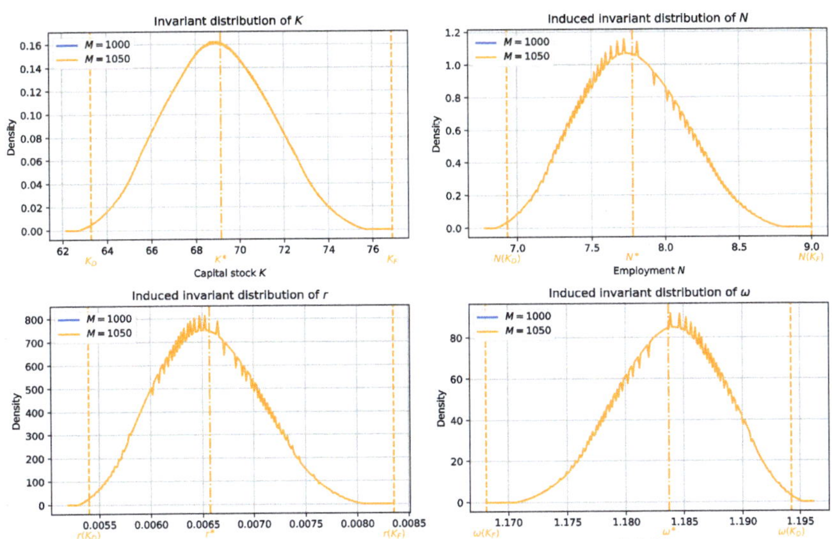


**Fig. 8.8** Invariant and induced distributions under nominal wage adjustment.
The overlap confirms that proportional nominal rescalings leave the stochastic steady state unchanged: They do not affect the neutral point or the long-run distributions of real variables, only the nominal scale of prices and wages

The shape of the invariant distribution in Fig. 8.8 resonates strongly with a classic Keynesian insight. As Skidelsky (2010, p. 179) emphasizes, Keynes argued that under laissez-faire, full-employment levels of investment are typically attained only during episodes of exceptional excitement, strong enough to overcome the uncertainty surrounding future returns. In the present framework, this observation acquires a precise stochastic interpretation: Although the upper boundary $K_F$ is reachable and repeatedly touched, it is visited only through rare realizations of sufficiently strong aggregate sentiment. The economy therefore fluctuates endogenously within its stability corridor, with full-utilization configurations appearing as tail events rather than as typical long-run outcomes.[8]

By contrast, changes in $M$ or $W$ that alter the ratio $W/M$ are not neutrality experiments. Such shifts modify the effective LM scaling and can deform the invariant distribution of $K$ even though the neutral interest rate $r^*$ remains pinned down by the real neutrality condition.

A contraction of the money supply raises distinct issues: When the economy is pushed toward the lower boundary $K_D$, downward adjustment of nominal wages may amplify deflationary pressure rather than neutralize it. This case is therefore deferred to the subsequent analysis of weak economies and deflationary fragility.

### 8.4.2 Boundary Proximity and Real Adjustment Parameters

The interaction of the excess-demand mapping $\zeta(r)$ with the interior probability bounds implies that admissible interest rates are confined to a finite interval $[r_-, r_+]$. The width of this admissible interest-rate window depends on real structural parameters governing profitability, competition, and adjustment asymmetries and is invariant to nominal rescalings. Moreover, this interest-rate window induces a corresponding interior window for the capital stock, whose absolute width scales with the nominal parameter $\sigma_0^{-1}$, while its relative width remains unchanged. A formal statement and proof of these results are given in Proposition 8.14 in Appendix 8.B.

Before deferring the formal derivation to the Appendix, we pause to state the main economic insight behind the admissible-window construction and to interpret the composite scale parameter $A$ that governs the mapping from utilization shocks into interest-rate adjustment.

[8] The analysis suggests that strict full employment emerges primarily at the upper boundary of the invariant distribution, where nominal pressures begin to accumulate. By contrast, the economy tends to stabilize within the interior of the stability corridor. The apparent tension between full employment and price stability therefore reflects not a short-run trade-off, but a structural property of the stochastic equilibrium: Policy that anchors nominal stability implicitly selects an interior operating point rather than the full-employment frontier.

**Key Technical Insight: Admissible Windows and the Meaning of $A$**
A central message of Proposition 8.14 is that the long-run stochastic dynamics admit an *interior viability region* governed by a pair of admissible windows: an interest-rate window $[r_-, r_+]$ in ($r$-space) and an induced capital window $[K_{q0}, K_{p0}]$ in ($K$-space). The width of the interest-rate window is pinned down by *real* and *behavioral* parameters: The sentiment amplitude $V$ determines how widely the common macroshifter $\nu$ can move, $c_b$ governs the degree of status quo bias in replacement decisions, and $\varphi$ locates the confidence threshold at which investment incentives flip sign. Within this structure, the scale parameter

$$A := \alpha(c+\delta)\frac{\eta-1}{\eta}$$

plays a distinct role: $A$ scales the sensitivity of utilization $\zeta$ to changes in the user cost of capital $r+\delta$, via

$$\zeta(r) = \frac{1}{\delta}\Big(\frac{A}{r+\delta} - 1\Big).$$

In particular, $A$ summarizes how technology ($\alpha$), market structure ($\eta$), and effective-demand intensity ($c$) jointly scale the sensitivity of utilization to financing costs. In this sense, Proposition 8.14 cleanly separates behavioral feasibility (the admissible $\zeta$-window) from the monetary adjustment required to accommodate it (the mapping into $r$-space), scaled by $A$.

**Interpretation of $(A, V, c_b, \varphi)$ and the Role of $c$**
First, $V$ governs the amplitude of exogenous sentiment innovations and therefore the raw range of macroeconomic pushes the economy may be exposed to. Second, $c_b$ is a status quo wedge that captures reluctance to revise operating scale, and thus pins down the width of the admissible utilization window in $\zeta$-space. Third, $\varphi$ determines the location of the confidence threshold at which net investment incentives flip between expansion and contraction. Finally, $A = \alpha(c+\delta)\frac{\eta-1}{\eta}$ scales how strongly the user cost of capital $r+\delta$ compresses utilization and hence determines how a fixed admissible range of $\zeta$ maps into the required variability of the interest rate. This decomposition is particularly useful for studying the propensity to consume $c$: For given technological and market-structure parameters ($\alpha$, $\eta$), variations in $c$ shift the sensitivity parameter $A$ while leaving the confidence threshold $\varphi$ unchanged. This isolates a transparent demand-side channel that is otherwise difficult to separate, since changes in $\alpha$ and $\eta$ typically move both $A$ and $\varphi$ simultaneously.

**Result (Admissible Interest-Rate Window)** Under the interior regime of the mutation–selection dynamics, the admissible range of real interest rates is given by an interval $[r_-, r_+]$ whose width satisfies

$$\Delta r := r_+ - r_- = \frac{2A}{\delta}\cdot\frac{D}{\varphi^2 - D^2}, \quad A = \alpha(c+\delta)\frac{\eta-1}{\eta}, \quad D = \frac{V}{2c_b} - \frac{1+c_b}{2},$$

where $\varphi = \frac{\alpha(\eta-1)}{1+\alpha(\eta-1)}$ and $\bar{\varphi} = \varphi + 1/\delta - (1+c_b)/2$. Assume $D > 0$, equivalently $V/c_b > 1 + c_b$, so the interior window is nondegenerate.

To understand the result intuitively, it is useful to distinguish between (i) the width of the admissible excess-demand window in $\zeta$-space and (ii) the mapping from that window into the admissible interest-rate range.

Define

$$\zeta_- := \zeta(K_{q0}) = \varphi + \frac{V}{2c_b} - (1+c_b), \qquad \zeta_+ := \zeta(K_{p0}) = \varphi - \frac{V}{2c_b}.$$

The admissible $\zeta$-window is therefore

$$\Delta\zeta := \zeta_- - \zeta_+ = \frac{V}{c_b} - (1+c_b) = 2D.$$

As noted earlier, the mapping from $\zeta$ to the interest rate is governed by

$$\zeta(r) = \frac{1}{\delta}\Big(\frac{A}{r+\delta} - 1\Big), \qquad A := \alpha(c+\delta)\frac{\eta-1}{\eta}.$$

This window determines the maximal range of real fluctuations that can occur without activating boundary regimes. Its width is governed entirely by real and structural parameters and is invariant to nominal rescalings. Importantly, the admissible capital window admits a scale-free representation. In particular, its relative width is independent of nominal parameters:

$$\frac{K_{p0} - K_{q0}}{K^*} = \frac{x(r_+)^\beta - x(r_-)^\beta}{x(r^*)^\beta}, \qquad x(r) := (\sigma_0 K)^{\frac{1}{\beta}}.$$

Thus, proportional rescalings of $(M, W)$ do not affect the relative size of the long-run viability region. See Appendix 8.B.1 for the full derivation.

Figure 8.9 provides a schematic overview of how behavioral constraints on utilization translate into interest-rate volatility and, in turn, into fluctuations of the capital stock.

**Remark 8.17 (Competition, Profitability, and the Required Variability of the Interest Rate)** An increase in the elasticity of substitution $\eta$ compresses markups and reduces the profit margin earned at any given level of activity. As a result, firms must operate at a higher throughput—that is, at a higher level of utilization—to sustain a given rate of profitability. In such an environment, a fixed disturbance to utilization translates into a larger required adjustment of the user cost of capital $r + \delta$ in order to restore investment incentives. A higher $\eta$ raises $A$ and therefore amplifies the interest-rate movement needed to accommodate a given admissible fluctuation in $\zeta$. The admissible interest-rate window thus expands, even though the underlying behavioral thresholds determining the $\zeta$-window remain unchanged.

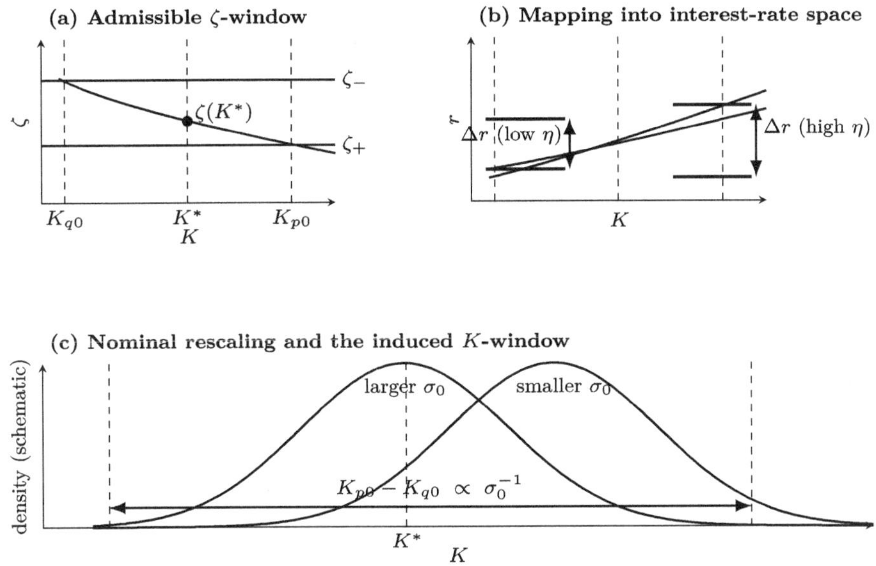


**Fig. 8.9** From behavioral thresholds to interest-rate volatility (schematic).
Panel (**a**) shows how behavioral bounds pin down the admissible $\zeta$-window. Panel (**b**) illustrates how the same $\zeta$-window maps into different admissible interest-rate ranges $[r_-, r_+]$, with higher competitiveness ($\eta$) requiring a wider adjustment range. Panel (**c**) shows how nominal rescaling shifts the induced $K$-window without changing relative dispersion

This mechanism clarifies an observation that has appeared repeatedly in macroeconomic thought. In Wicksell's analysis of cumulative processes, greater competitive pressure lowers the normal profit rate and makes the economy more sensitive to deviations between the market rate and the natural rate of interest (Wicksell, 1936, Chs. 2–4). Keynes emphasized the same point from the demand side: When profit margins are thin, small changes in expected sales require substantial interest-rate movements to stabilize investment (Keynes, 1936, Chs. 11, 17).

In Minsky's language, market power provides a cushion of safety by stabilizing revenues when demand falls; intense competition erodes that cushion and makes debt-financed investment more fragile (Minsky, 1986, p. 181).

The present framework provides a precise stochastic formulation of this idea. Competition does not merely affect equilibrium levels; it alters the economy's tolerance for interest-rate variability. Highly competitive economies require a wider range of interest-rate movement to absorb utilization shocks, rendering monetary conditions more volatile even when long-run real anchors remain unchanged.[9]

[9] This mechanism should be distinguished from the standard New Keynesian channel in which greater price flexibility (or a flatter Phillips curve) alters the transmission of monetary policy. By contrast, the channel emphasized here operates through profitability and capital utilization rather than price adjustment. When markups are compressed, firms require higher throughput to sustain

As long as the state remains within this interior region, branch probabilities stay strictly inside $(0, 1/2)$ and the stochastic dynamics operate under a stable adjustment regime. Boundary effects arise only as the state approaches the edges, where probability clipping and regime-specific real adjustments become relevant. In this sense, $[K_{q0}, K_{p0}]$ provides a precise notion of proximity to the boundary against which subsequent mechanisms—such as the money supply, the nominal wage, and (when treated as an LM scale shifter) liquidity preference—should be interpreted.

Two implications of Proposition 8.14 are particularly important. First, the width of the interest-rate window $\Delta r := r_+ - r_-$ is determined entirely by real parameters:

$$\Delta r = \frac{2A}{\delta} \cdot \frac{D}{\bar{\varphi}^2 - D^2},$$

where $A$, $D$, and $\bar{\varphi}$ are functions of technological and market-structure parameters such as $\alpha$, $\eta$, $\delta$, and the behavioral asymmetry parameter $c_b$. Purely monetary variables do not enter this determination.

Second, for a given interest-rate window $[r_-, r_+]$, the induced range of admissible capital stocks satisfies

$$K_{p0} - K_{q0} = \frac{x(r_+)^\beta - x(r_-)^\beta}{\sigma_0},$$

so that the absolute width of the capital window scales inversely with $\sigma_0 = \theta_{\mathrm{LM}}/\theta_{\mathrm{PM}}$. Monetary parameters entering $\theta_{\mathrm{LM}}$—such as the money supply, the nominal wage, and liquidity preference—thus affect the *scale* of capital fluctuations, but not the underlying interest-rate window itself. Real demand parameters that influence both $\sigma_0$ and $A$ (notably $c$ and $\beta$) operate through multiple channels, whereas purely nominal parameters act only through this rescaling effect.

The width of the admissible interest-rate window plays a central role in the stability of the stochastic steady state. Since $\Delta r$ governs how far real fluctuations can proceed before boundary regimes are activated, it is useful to record its comparative statics with respect to the underlying structural parameters.

**Remark 8.18 (Comparative Statics of the Admissible Interest-Rate Window)** Maintain the interior regime of Proposition 8.14, so that $\zeta_- - \zeta_+ = 2D > 0$ and $\bar{\varphi}^2 - D^2 > 0$. Recall

normal profits, so a given utilization disturbance must be offset by a larger movement in the user cost of capital. Competition therefore amplifies the required variability of the interest rate even when prices remain flexible and long-run real anchors are unchanged. The two mechanisms thus have opposite implications for interest-rate volatility despite both being associated with increased competition.

$$\Delta r = \frac{2A}{\delta} \cdot \frac{D}{\bar{\varphi}^2 - D^2}, \qquad A := \alpha(c+\delta)\frac{\eta - 1}{\eta}, \qquad D := \frac{1}{2b} - \frac{1+c_b}{2},$$

$$\varphi := \frac{\alpha(\eta-1)}{1+\alpha(\eta-1)}, \qquad \bar{\varphi} := \varphi + \frac{1}{\delta} - \frac{1+c_b}{2}.$$

Then the following statements hold.

**(i) Profitability scale:** Holding $(\delta, D, \bar{\varphi})$ fixed, the window width is strictly increasing in the profitability scale:

$$\frac{\partial \Delta r}{\partial A} = \frac{2}{\delta} \cdot \frac{D}{\bar{\varphi}^2 - D^2} > 0.$$

In particular, if $\varphi$ is treated as fixed (so only the mechanical $A$-channel is active) when varying $\alpha$ and $\eta$, then

$$\frac{\partial \Delta r}{\partial \alpha} > 0, \qquad \frac{\partial \Delta r}{\partial c} > 0, \qquad \frac{\partial \Delta r}{\partial \eta} > 0,$$

since $A$ is increasing in each of $\alpha$, $c$, and $\eta$ through $A = \alpha(c+\delta)(\eta-1)/\eta$.

**(ii) Endogenous $\varphi$-channel:** Allowing $\varphi = \varphi(\alpha, \eta)$ to vary as defined above, the elasticity of $\Delta r$ with respect to $\alpha$ admits the decomposition

$$\frac{\partial}{\partial \alpha} \log \Delta r = \frac{1}{\alpha} - \frac{2\bar{\varphi}}{\bar{\varphi}^2 - D^2} \frac{\partial \varphi}{\partial \alpha}, \qquad \frac{\partial \varphi}{\partial \alpha} = \frac{\eta - 1}{(1+\alpha(\eta-1))^2} > 0.$$

Hence, $\partial/\partial\alpha\, \Delta r > 0$ holds whenever

$$\frac{1}{\alpha} > \frac{2\bar{\varphi}}{\bar{\varphi}^2 - D^2} \cdot \frac{\eta - 1}{(1+\alpha(\eta-1))^2}.$$

Analogously, the elasticity with respect to $\eta$ is

$$\frac{\partial}{\partial \eta} \log \Delta r = \frac{1}{\eta(\eta-1)} - \frac{2\bar{\varphi}}{\bar{\varphi}^2 - D^2} \frac{\partial \varphi}{\partial \eta}, \qquad \frac{\partial \varphi}{\partial \eta} = \frac{\alpha}{(1+\alpha(\eta-1))^2} > 0,$$

so $\partial/\partial\eta\, \Delta r > 0$ holds whenever

$$\frac{1}{\eta(\eta-1)} > \frac{2\bar{\varphi}}{\bar{\varphi}^2 - D^2} \cdot \frac{\alpha}{(1+\alpha(\eta-1))^2}.$$

**(iii) Probability-bound parameters:** Holding $(A, \delta, \bar{\varphi})$ fixed, the window width is strictly increasing in $D$:

$$\frac{\partial \Delta r}{\partial D} = \frac{2A}{\delta} \cdot \frac{\bar{\varphi}^2 + D^2}{(\bar{\varphi}^2 - D^2)^2} > 0.$$

Consequently, a larger sentiment shock amplitude $V$ increases $D$, hence widens the admissible windows, while a larger wedge cost $c_b$ reduces $D$ and therefore tends to shrink the admissible window:

$$\frac{\partial D}{\partial V} = \frac{1}{2c_b} > 0, \qquad \frac{\partial D}{\partial c_b} = -\frac{V}{2c_b^2} - \frac{1}{2} < 0.$$

At the same time, $c_b$ also lowers $\bar{\varphi}$ via $\partial/\partial c_b \bar{\varphi} = -\frac{1}{2}$, which tends to *widen* the window by reducing the denominator $\bar{\varphi}^2 - D^2$. Thus the net sign of $\partial/\partial c_b \Delta r$ is generally ambiguous: $c_b$ simultaneously tightens the interior probability span (through $D$) and relaxes the interest-rate mapping (through $\bar{\varphi}$).

(iv) **Depreciation:** The depreciation rate $\delta$ enters $\Delta r$ through $A = \alpha(c + \delta)(\eta - 1)/\eta$, through the prefactor $1/\delta$, and through $\bar{\varphi} = \varphi + 1/\delta - (1 + c_b)/2$. Accordingly, the sign of $\partial/\partial\delta \Delta r$ is parameter dependent, and $\Delta r$ need not vary monotonically with depreciation.

**Remark 8.19 (Competition and Interest-Rate Volatility)** A central implication of the admissible interest-rate window is that greater product-market competition necessitates *larger* interest-rate movements in order to stabilize real activity. As competition intensifies (higher $\eta$), markups are compressed and profit margins thin. Firms must therefore operate at higher throughput to sustain normal profitability, making capital utilization more sensitive to financing conditions.

In such an environment, a given disturbance to utilization requires a larger adjustment of the user cost of capital $r + \delta$ to restore investment incentives. This mechanism operates through the scale parameter $A = \alpha(c+\delta)\frac{\eta-1}{\eta}$, which amplifies the mapping from utilization shocks into interest-rate movements. While behavioral thresholds determine the admissible range of utilization fluctuations, competition governs how large an interest-rate response is required to absorb them.

This is closely related to Minsky's "cushions of safety" mechanism: When cash-flow buffers are compressed, financial conditions become a stronger amplifier of macroeconomic fluctuations (Minsky, 1986, Ch. 7). The present framework formalizes this insight in stochastic terms. Interest-rate volatility is not a policy failure in competitive economies, but a structural implication of low markups and tight profitability constraints—even when long-run real anchors remain unchanged.

One immediate implication of Proposition 8.14 is its relevance to the effectiveness of monetary expansion, summarized in the following remark.

**Remark 8.20 (Effective Monetary Expansion)** When the upper labor-supply constraint $\bar{N}$ is sufficiently large and the lower truncation $K_D$ is not binding, a monetary expansion under downward nominal wage rigidity reduces $\sigma_0$ (through the LM scaling via $W/M$) and therefore shifts both the neutral point $K^*$ and the admissible capital window $[K_{q0}, K_{p0}]$ proportionally to the right. The relative dispersion of the invariant distribution remains unchanged, since the admissible interest-rate window is pinned down by real parameters. In this regime, monetary expansion raises the long-run scale of economic activity without altering the underlying stabilizing structure of the economy.

This mechanism is particularly relevant when the labor endowment $\bar{N}$ increases, shifting the full-employment frontier outward and thereby creating room for monetary expansion without immediately activating the upper boundary of the invariant distribution.

From the perspective of boundary proximity, the primary role of monetary expansion is therefore not to fine-tune interior dynamics, but to move the economy away from regions in which real adjustment rules—such as downward wage revision or capacity scrapping—are activated. The subsequent analysis of wage bargaining and deflationary responses should be read precisely in this state-contingent sense.

### 8.4.3 *Wage Adjustment Under a Fixed Money Supply*

Figure 8.10 indicates that, under a fixed money supply $M$, the economy is remarkably resilient to deflationary pressure induced by wage cuts.

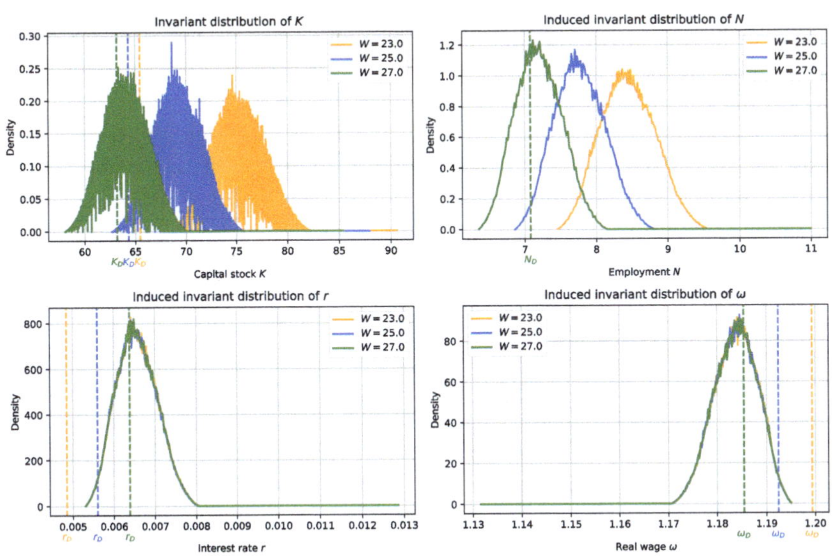


**Fig. 8.10** Wage cuts under a fixed money supply and boundary proximity.
Invariant distributions of $K$, $N$, $r$, and $\omega$, under a fixed money supply $M$ and varying nominal wage $W$. As $W$ drops near the lower boundary $K_D$, $K$ shifts away from $K_D$, $N$ moves right, and $\omega$ and $r$ stay near neutral. Downward wage adjustment triggers real adjustments absorbed by the price level, not a deflationary spiral

When the economy weakens and the capital stock falls below the bargaining threshold $K_D$ (around $K \simeq 61$ when $W = 27$), downward nominal wage adjustment is triggered. Suppose the wage is cut discretely from $W = 27$ to $W = 25$. Holding the money supply fixed, this wage cut lowers the price level proportionally, so that the real wage $\omega = W/P$ and the real interest rate remain tightly anchored around their neutral values, as shown in the lower panels of the figure. At the same time, for a given capital stock $K$, the lower nominal wage reduces transaction demand, induces a decline in the interest rate $r(K)$ along the PM–LM locus (not directly visible in the invariant distribution), and thereby stimulates investment. As a result, the invariant distribution of $K$ shifts to the right and away from the lower boundary, and employment rises correspondingly. This stabilizing feedback is most clearly seen in the $\omega$-panel: The dashed vertical lines indicate the threshold real wage $\omega_D$ that triggers a wage cut. As $W$ is reduced from 27 to 25, and further to 23, the implied threshold $\omega_D$ shifts upward relative to the tightly concentrated distribution of $\omega$, eventually eliminating the incentive for further wage cuts. In this way, under a fixed money supply (100% reserve) regime, nominal wage flexibility operates as a boundary-triggered adjustment that mitigates deflationary pressure and helps the economy move back toward the interior region, rather than initiating a cumulative deflationary spiral.

**Remark 8.21 (Self-correction Under Fixed Money and Its Fragility Under Endogenous Money)** The stabilizing role of downward nominal wage adjustment under a fixed money supply can be interpreted in Keynes–Wicksell terms as an endogenous realignment between the prevailing (market) interest rate and the natural (neutral) rate.

Under the fixed-$M$ closure, a fall in the capital stock raises the utilization ratio and the excess replacement incentive $\zeta(K)$, thereby strengthening the up-branch probability and weakening the down-branch probability. In this sense, the mutation-selection dynamics already embed an interior self-correction mechanism: Depressed states tend to generate stronger incentives for recovery even before any nominal adjustment is triggered.

If the recession becomes sufficiently severe that the process reaches the wage-bargaining boundary $K_D$, downward nominal wage adjustment provides an additional stabilizing channel. A wage cut lowers the price level proportionally under fixed $M$, increases real money balances, and reduces the market interest rate along the PM–LM relation for a given $K$. This stimulates investment and shifts the invariant distribution away from the distress region, helping restore the economy toward its neutral long-run anchor.

This interpretation also clarifies a fragility of the mechanism: Under a fractional reserve regime in which the money supply contracts endogenously in slumps, the increase in real balances and the associated interest-rate relief may be weakened or reversed, so the recovery forces described above need not operate with the same strength.

**Nominal Wage Adjustment as Self-correction Under Fixed Money**
In this sense, a wage cut operates primarily as a nominal rescaling: The price level falls roughly in proportion to $W$, leaving the key real anchors largely intact. Moreover, once employment rises, the economy becomes less likely to trigger further wage reductions, because the wage-adjustment rule is hit less frequently at the lower boundary. This mechanism creates a stabilizing feedback under fixed $M$: Nominal deflation absorbs the wage adjustment, while real activity is supported through higher utilization, higher employment, and a shift of the long-run capital distribution away from the distress region near $K_D$. Accordingly, the model does not generate a self-reinforcing deflationary spiral solely from a nominal wage cut when $M$ is fixed, because the real anchors $\omega$ and $r$ remain stable, while the lower boundary shifts upward relative to the invariant distribution, reducing the likelihood of further wage adjustments.

This mechanism resembles what Axel Leijonhufvud (1968, p. 340) calls the "Keynes-effect" proper: under a fixed nominal money supply, a fall in money wages increases real balances and tends to reduce the interest rate. In the present framework, the mechanism operates through the LM funding constraint and additionally shifts the invariant distribution away from the distress boundary $K_D$.

**Behavioral Fragility and the Wage-Cut Boundary**
The preceding robustness result should be read as conditional on fixed behavioral parameters. More precisely, the stabilizing mechanism relies on three conditions: (i) sufficient nominal adjustment margins, (ii) stability of behavioral parameters, and (iii) the presence of a sufficient buffer from the lower boundary. If one allows for psychological feedback effects, the stabilizing mechanism can weaken. In particular, a deterioration of economic conditions may reduce the propensity to consume ($c$) (precautionary saving) and raise liquidity preference $\lambda$, both of which depress effective demand and tighten financing conditions, increasing the frequency with which the economy visits the low-$K$ region.

The deflationary threshold $K_D$, defined Sect. 7.4.2, marks the point at which downward nominal wage adjustment becomes admissible. The robustness results above should therefore be read as conditional on fixed behavioral parameters. Second, if one allows for psychological feedback effects, the stabilizing role of wage adjustment may weaken. Thus wage cuts become admissible only when the implied "Keynesian wage pressure" $W_0/W$ is sufficiently large. This logic yields a deflation threshold $\text{Def\_ratio} = 1 - \frac{(1-\theta)v(1-g)}{\theta g(1-b_-)}$, and the associated boundary $K_D$ is defined implicitly by $W_0(K_D)/W = \text{Def\_ratio}$. Increasing firms' bargaining power $\theta$ (or raising the asymmetry $b_-$) relaxes the rigidity inequality and shifts $K_D$ upward, so that the economy reaches the wage-cut regime at less depressed levels of $K$. The key point is not that wage cuts mechanically generate a deflationary spiral under fixed $M$, but rather that shifts in $c$ and $\lambda$ can move the state distribution toward $K_D$ and make downward wage adjustment more relevant for real activity. Monetary expansion in this sense does not eliminate real adjustment boundaries, but mitigates the likelihood that the economy persistently operates near them.

These considerations highlight that real adjustment boundaries are shaped jointly by behavioral responses and by the economic environment in which they operate.

**Competition, Profit Buffers, and Institutional Breakdown**

A third qualification concerns not behavioral feedback per se, but the economic environment in which behavioral and institutional commitments can be sustained.

Viewed from this perspective, the sustainability of nominal wage commitments depends on the availability of a sufficient profit buffer. When margins are ample, employers can credibly commit to a stable nominal wage in order to preserve workers' morale and effort. As economic conditions deteriorate, price competition intensifies and the effective elasticity of substitution $\eta$ rises, compressing markups and eroding this safety margin. In such an environment, survival considerations dominate morale preservation, making downward wage revisions more likely even in the presence of strong effort asymmetries.

Because this erosion of wage discipline operates at the institutional level, it may precede the transmission of monetary stimulus through the price channel, allowing effective demand to collapse before the stabilizing $P$-effect takes hold.

While wage adjustment under a fixed money supply leaves the stochastic steady state largely intact and exhibits substantial resilience to deflationary pressure, this robustness need not survive once the money supply itself becomes state dependent.

**Discussion: Neutrality, Self-Correction, and Keynesian Asymmetry**

This proposition clarifies the sense in which monetary neutrality holds in the present framework. Nominal variables such as the money supply and the nominal wage do not determine the location of the neutral point $K^*$, nor the long-run real levels of employment, output, or the real wage. These real magnitudes are pinned down by structural features of technology, preferences, and adjustment behavior. Monetary neutrality therefore holds at the level of the stochastic balance point.

At the same time, nominal adjustment plays a crucial role in shaping the *distributional* properties of the stochastic steady state. When the invariant distribution becomes truncated at either boundary, nominal wage and price adjustments alter real money balances and interest rates, thereby deforming the invariant distribution itself. This mechanism ensures that boundary truncation is not self-sustaining: Persistent full employment triggers inflationary pressure that erodes real balances and dampens investment, while persistent slack triggers deflationary pressure that raises real balances and encourages recovery.

This distinction between the invariance of the neutral point and the malleability of the invariant distribution provides a precise formulation of Keynes's well-known asymmetry between inflation and deflation. With a fixed money supply, inflation necessarily reduces real balances and must eventually come to a halt. Deflation, by contrast, may persist so long as downward nominal wage adjustment remains feasible and the interest rate stays above its lower bound. Once these channels are exhausted—through a zero lower bound on interest rates or downward rigidity of nominal wages—the self-correcting mechanism may fail, allowing deflationary stagnation to endure.

The stochastic framework developed in this book thus reconciles long-run monetary neutrality with a rich set of Keynesian adjustment dynamics. Neutrality holds in the sense of long-run real anchoring, while nominal institutions and rigidities govern the shape, dispersion, and boundary behavior of macroeconomic fluctuations around that anchor.

These considerations raise a broader question: Even when local self-correcting mechanisms operate and nominal anchors remain in place, do they suffice to ensure that the economy's invariant distribution is centered on full employment, or can substantial mass persist away from it?

### *8.4.4 Revisiting Tobin's Question: Full Employment and Invariant Distributions*

Tobin (1993) raised a fundamental question concerning whether "natural market adjustment mechanisms" reliably restore full-employment equilibrium after a negative demand shock:

> The big issue between Keynes and his "old classical" opponents was the efficacy of the economy's natural market adjustment mechanisms in restoring full-employment equilibrium, once a negative real demand shock had pushed the economy off that equilibrium (Tobin, 1993, p. 47).

In the present framework, however, the relevant object is not a pointwise "full-employment equilibrium," but the long-run invariant distribution and its boundary dynamics. In the stochastic steady state, the economy generically spends most of its time away from the full-employment boundary; full employment may occur, but it is not the typical outcome.

Interpreted through a welfare lens, the analysis suggests that the central normative issue is therefore not whether unemployment can be eliminated in principle, but whether the stationary level and dispersion of unemployment are socially tolerable. This observation naturally motivates discussion of insurance or redistribution mechanisms, although such institutional design lies beyond the scope of the present book.

Tobin's question can thus be reformulated in stochastic terms. Rather than asking whether adjustment mechanisms restore a pointwise full-employment equilibrium, the appropriate analogue concerns the behavior of the invariant distribution and its interaction with boundary regimes. Lemma 8.4, together with the simulated invariant distributions, indicates that full employment is not a persistent attractor: When the process approaches the full-employment boundary $K_F$, endogenous nominal wage adjustment shifts the long-run distribution leftward, eliminating truncation at $K_F$ and restoring interior stochastic stability. The economically relevant question is therefore the magnitude and variability of unemployment in the stationary regime, rather than the existence of full employment as a long-run outcome.

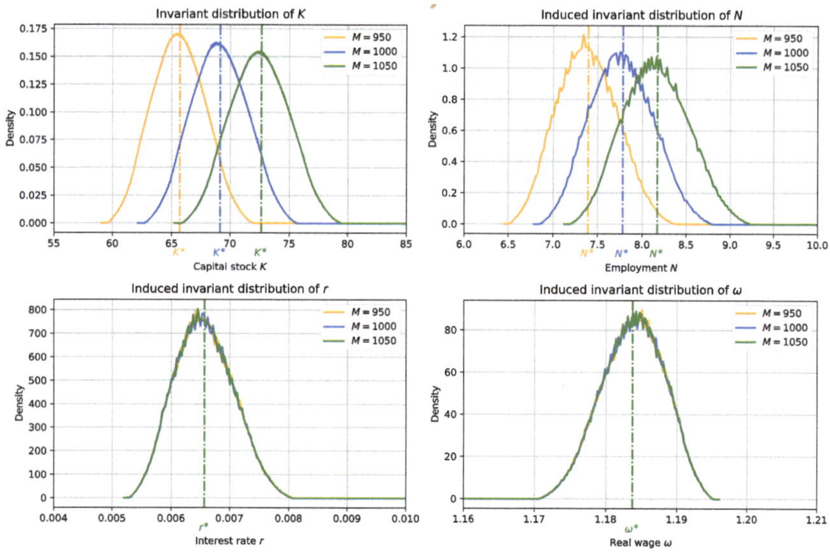


**Fig. 8.11** Invariant distributions under alternative money supplies.
Variations in the money supply shift the invariant distributions of capital and employment leftward or rightward, so that under a fractional reserve regime monetary contraction near the lower edge of activity and expansion near the upper edge act as state-dependent amplification mechanisms rather than stabilizers

### *8.4.5 State-Dependent Monetary Amplification Under a Fractional Reserve Regime*

We now turn to the case of a variable money supply under a fractional reserve banking regime and examine how state-dependent monetary expansion and contraction affect the economy's stochastic steady state.

Figure 8.11 illustrates how variations in the money supply $M$ under a fractional reserve regime shift the invariant distribution of capital and the induced distributions of employment, the interest rate, and the real wage, holding the nominal wage fixed. It demonstrates that a lower money supply shifts the invariant distributions of capital and employment leftward, while a higher money supply shifts them to the right. Under a fractional reserve regime, such state-dependent monetary contraction near the lower edge of activity and expansion near the upper edge act as amplification mechanisms, deepening slumps and reinforcing booms rather than stabilizing the economy around its interior region.

Under a fractional reserve regime, variations in the money supply introduce a state-dependent amplification mechanism at both ends of the activity spectrum. When the economy operates near the lower edge of capital and employment, a contraction of bank credit shifts the invariant distributions of $K$ and $N$ further leftward, deepening and prolonging the slump. Conversely, when activity approaches the upper edge, credit expansion shifts the stochastic steady state further to the right, reinforcing the boom and increasing the risk of overheating rather than stabilizing the economy around its interior region. These effects arise even in the absence of endogenous banking behavior, reflecting the comparative properties of the invariant distributions: Variable money supply acts as a destabilizing force by withdrawing liquidity in downturns and supplying it in upturns, thereby amplifying fluctuations instead of dampening them.

This asymmetric amplification mechanism echoes Irving Fisher's debt-deflation hypothesis (Fisher, 1933), according to which contractions of credit during downturns and expansions during booms mechanically intensify fluctuations rather than stabilizing economic activity.

### 8.4.6 The Regime Comparison: 100% Reserve vs. Pure Credit

This chapter has shown that long-run macroeconomic stability in a monetary economy is best understood in terms of real anchors and boundary proximity rather than pointwise deterministic equilibrium. Monetary neutrality holds at the level of the stochastic anchor: The neutral point $K^*$ and the associated real variables $(r^*, N^*, Q^*, \omega^*)$ are pinned down by technology, preferences, and adjustment behavior, not by the nominal scale of money and wages. At the same time, nominal institutions and rigidities shape the *distribution* of outcomes around that anchor, determining how frequently the economy visits regions of slack or full utilization and how strongly boundary regimes feed back into real activity. When the state remains within the interior capital window $[K_{q0}, K_{p0}]$, interest-rate adjustment and mutation-selection dynamics generate a self-stabilizing stochastic equilibrium; when the state is pushed toward the boundaries $K_D$ or $K_F$, wage bargaining, liquidity preference, and psychological responses of consumption become decisive for macroeconomic performance. The framework developed here therefore reconciles long-run monetary neutrality with a rich Keynesian asymmetry between inflation and deflation and provides a coherent benchmark for evaluating policy regimes that aim to keep the economy away from its real boundaries while preserving the stabilizing role of the interest rate.

The comparison between a 100% reserve regime and a pure-credit regime thus turns on whether nominal institutions reinforce or counteract the economy's intrinsic tendency to drift toward its real boundaries.

**Remark 8.22 (Policy Interpretation Under a 100% Reserve Regime)** Under a 100% reserve regime with a fixed nominal money supply, the adjustment mechanisms highlighted in this chapter admit a natural policy interpretation. Because discretionary monetary expansion is absent by design, stabilization must rely on institutional and contractual mechanisms that become operative as the economy approaches its real boundaries.

Downward nominal wage adjustment can play precisely this boundary role when it remains feasible. When the capital stock falls toward the lower threshold $K_D$, a wage cut reduces the price level proportionally, raises real money balances, lowers transaction demand, and induces a decline in the market interest rate for a given state. This interest-rate response stimulates investment and shifts the invariant distribution away from the boundary, thereby reducing the likelihood of further wage cuts.

From a policy perspective, the key objective in a 100% reserve system is therefore not fine-tuning aggregate demand, but maintaining sufficient distance from boundary regimes in which real adjustment rules are activated. Institutional arrangements that preserve predictable financial conditions and stable wage-setting norms thus serve as primary stabilizing devices in the absence of active monetary intervention.

**Remark 8.23 (Policy Interpretation Under a Pure-Credit Regime)** Under a pure-credit regime with an endogenous money supply, the same adjustment mechanisms operate in a fundamentally different institutional environment. Because bank lending and deposit creation respond endogenously to economic conditions, the money supply becomes state dependent, expanding in upturns and contracting in downturns.

When credit is procyclical, such state-dependent monetary adjustment can work as an amplification mechanism by shifting the invariant distributions of capital and employment toward the boundaries. Near the lower threshold $K_D$, credit contraction tightens financing conditions and suppresses investment, reinforcing the slump; near the upper threshold $K_F$, credit expansion lowers financing costs and validates optimistic expectations, amplifying the boom.

In contrast to the 100% reserve regime, stabilization under a pure-credit system therefore requires institutional constraints that counteract endogenous credit dynamics, since interest-rate adjustment alone need not provide a reliable stabilizing feedback toward the interior region of the stochastic equilibrium.

Taken together, the regime comparison suggests that macroeconomic stability in a monetary economy is less a matter of activating discretionary policy instruments than of suppressing state-dependent amplification mechanisms. A 100% reserve regime achieves this by fixing the nominal anchor and allowing stabilization to operate primarily through interest-rate adjustment and boundary-triggered real mechanisms, whereas a pure-credit regime requires additional institutional constraints to prevent endogenous money creation and destruction from pushing the economy persistently toward its real boundaries. The analysis therefore shifts the focus of policy evaluation from short-run demand management to the long-run design of monetary institutions that preserve interior stochastic stability.

**Remark 8.24 (Scope: Endogenous Credit and Horizontalist Closures)** The analysis in this chapter treats the money supply as an institutional parameter (fixed under a 100% reserve regime or varied exogenously in comparative exercises) and therefore abstracts from endogenous credit creation and destruction in a pure-credit or horizontalist framework. In such environments, downturns may be accompanied by rising risk premia, debt repayment, and contraction of bank balance sheets, so that the effective financing cost need not decline even as real activity weakens. Incorporating this mechanism requires an explicit banking block with endogenous money and risk premia and is left for future work.

With these scope qualifications in place, the concluding chapter draws together the book's main results to clarify how neutral anchors, boundary mechanisms, and institutional regimes jointly shape long-run macroeconomic stability.

# Appendix

## 8.A Proofs of Lemmas and a Corollary

This appendix collects the proofs deferred from the main text, grouped in the order in which the results appear in Chap. 8.

### *8.A.1 Proof of Lemma 8.3 (Sufficient Condition for Upper Truncation at $K_F$)*

***Proof (Sketch)*** For $K \in S$, define the (one-step) upper-clipping probability

$$\Pr\big(K' = K_F \mid K\big) = p_+(K) \sum_{j=0}^{J_{\max}} \delta_+(K, j)\, \mathbf{1}\{K\gamma_+^j \geq K_F\}.$$

Under $K_{p0}\gamma_+^{J_{\max}} \geq K_F$, we have $K_{p0}\gamma_+^{J_{\max}} \geq K_F$ and hence $\mathbf{1}\{K_{p0}\gamma_+^{J_{\max}} \geq K_F\} = 1$. Therefore,

$$\Pr\big(K' = K_F \mid K_{p0}\big) \;\geq\; p_+(K_{p0})\, \delta_+(K_{p0}, J_{\max}).$$

By continuity of $p_+(\cdot)$ and $\delta_+(\cdot, J_{\max})$ and the uniform interior lower bound on $\delta_+$ over $D$, there exist an $\varepsilon > 0$ and a constant $\underline{q} > 0$ such that $\Pr\big(K' = K_F \mid K\big) \geq \underline{q}$ for all $K \in (K_{p0} - \varepsilon, K_{p0}] \subset D$. Since any invariant distribution $\pi$ assigns positive mass to every neighborhood that is visited with positive long-run frequency (in particular, to $(K_{p0} - \varepsilon, K_{p0}]$ under irreducibility on $D$), it follows that

$$\int_S \Pr\big(K' = K_F \mid K\big)\, \pi(dK) \;\geq\; \underline{q}\, \pi\big((K_{p0} - \varepsilon, K_{p0}]\big) \;>\; 0,$$

which establishes truncation at the upper boundary in the stated sense. □

### 8.A.2 Proof of Lemma 8.5 (Sufficient Condition for Lower Truncation at $K_D$)

***Proof (Sketch)*** For $K \in S$, define the (one-step) lower-clipping probability

$$\Pr\big(K' = K_D \mid K\big) = p_-(K) \sum_{j=0}^{J_{\max}} \delta_-(K, j)\, \mathbf{1}\{K\gamma_-^j \le K_D\}.$$

Under $K_{q0}\gamma_-^{J_{\max}} \le K_D$, we have $\mathbf{1}\{K_{q0}\gamma_-^{J_{\max}} \le K_D\} = 1$, and hence

$$\Pr\big(K' = K_D \mid K_{q0}\big) \;\ge\; p_-(K_{q0})\, \delta_-(K_{q0}, J_{\max}).$$

By continuity of $p_-(\cdot)$ and $\delta_-(\cdot, J_{\max})$ and the uniform interior lower bound on $\delta_-$ over $D$, there exist an $\varepsilon > 0$ and a constant $\underline{q} > 0$ such that $\Pr\big(K' = K_D \mid K\big) \ge \underline{q}$ for all $K \in [K_{q0}, K_{q0} + \varepsilon) \subset D$. Since any invariant distribution $\pi$ assigns positive mass to every neighborhood that is visited with positive long-run frequency (in particular, to $[K_{q0}, K_{q0} + \varepsilon)$ under irreducibility on $D$), it follows that

$$\int_S \Pr\big(K' = K_D \mid K\big)\, \pi(dK) \;\ge\; \underline{q}\, \pi\big([K_{q0}, K_{q0} + \varepsilon)\big) \;>\; 0,$$

which establishes truncation at the lower boundary in the stated sense. □

### 8.A.3 Proof of Lemma 8.8 (Comparative Statics of $\theta_{\mathrm{PM}}$)

***Proof (Sketch)*** Work with $\log \theta_{\mathrm{PM}}$. From (8.3.2),

$$\log \theta_{\mathrm{PM}} = \log(1-\alpha) + \frac{1}{1-\alpha} \log\left(\frac{\eta - 1}{\eta}\right) + \frac{\alpha}{1-\alpha} \log \alpha. \tag{8.A1}$$

**Step 1 (derivative w.r.t. $\eta$).** Log-differentiating (8.A1) with respect to $\eta$ (holding $\alpha$ fixed) gives

$$\begin{aligned}\frac{\partial}{\partial \eta} \log \theta_{\mathrm{PM}} &= \frac{1}{1-\alpha} \frac{\partial}{\partial \eta} \log\left(\frac{\eta-1}{\eta}\right) = \frac{1}{1-\alpha}\left(\frac{1}{\eta - 1} - \frac{1}{\eta}\right) \\ &= \frac{1}{(1-\alpha)\eta(\eta-1)} \;>\; 0.\end{aligned}$$

Hence, $\partial_\eta \theta_{\mathrm{PM}} = \theta_{\mathrm{PM}}\, \partial_\eta \log \theta_{\mathrm{PM}} > 0$, proving (i).

**Step 2 (derivative w.r.t. $\alpha$).** Differentiate (8.A1) with respect to $\alpha$ (holding $\eta$ fixed). Using $\frac{d}{d\alpha}\frac{1}{1-\alpha} = \frac{1}{(1-\alpha)^2}$ and $\frac{d}{d\alpha}\frac{\alpha}{1-\alpha} = \frac{1}{(1-\alpha)^2}$, we obtain

$$\frac{\partial}{\partial\alpha}\log\theta_{\mathrm{PM}} = -\frac{1}{1-\alpha} + \frac{1}{(1-\alpha)^2}\log\left(\frac{\eta-1}{\eta}\right) + \frac{1}{(1-\alpha)^2}\log\alpha + \frac{1}{1-\alpha}.$$

The first and last terms cancel, leaving

$$\frac{\partial}{\partial\alpha}\log\theta_{\mathrm{PM}} = \frac{1}{(1-\alpha)^2}\Big[\log\alpha + \log\left(\frac{\eta-1}{\eta}\right)\Big] = \frac{1}{(1-\alpha)^2}\log\left(\alpha\cdot\frac{\eta-1}{\eta}\right).$$

Since $\alpha \in (0, 1)$ and $(\eta-1)/\eta \in (0, 1)$, their product lies in $(0, 1)$, so the logarithm is negative. Therefore $\partial_\alpha \log\theta_{\mathrm{PM}} < 0$, hence $\partial_\alpha\theta_{\mathrm{PM}} < 0$, proving (ii). □

### 8.A.4 *Proof of Lemma 8.9 (Effect of α on the Neutral Interest Rate)*

***Proof*** Let $k := \eta - 1 > 0$ and

$$A \;:=\; (c+\delta)\frac{k}{\eta} \;>\; 0.$$

Rearranging (8.3.3) gives

$$r^* + \delta = \frac{\alpha A}{B(\alpha)}, \qquad B(\alpha) = 1 + \delta\Big(\frac{\alpha k}{1+\alpha k} - d\Big).$$

Since $A > 0$ and $B(\alpha) > 0$ by (8.3.4), differentiation yields

$$\frac{\partial r^*}{\partial\alpha} = A\,\frac{B(\alpha) - \alpha B'(\alpha)}{B(\alpha)^2}. \tag{8.A2}$$

Thus $\mathrm{sign}(\partial r^*/\partial\alpha)$ is the sign of

$$S(\alpha) \;:=\; B(\alpha) - \alpha B'(\alpha).$$

Next compute

$$\varphi'(\alpha) = \frac{k}{(1+\alpha k)^2}, \qquad B'(\alpha) = \delta\,\varphi'(\alpha) = \frac{\delta k}{(1+\alpha k)^2}.$$

Substituting into $S(\alpha)$ gives

$$S(\alpha) = 1 - \delta d + \delta\,\frac{\alpha k}{1+\alpha k} - \delta\,\frac{\alpha k}{(1+\alpha k)^2}.$$

Let $x := \alpha k > 0$. Then

$$\frac{\alpha k}{1+\alpha k} - \frac{\alpha k}{(1+\alpha k)^2} = \frac{x}{1+x} - \frac{x}{(1+x)^2} = \frac{x^2}{(1+x)^2},$$

so

$$S(\alpha) = 1 - \delta d + \delta \frac{x^2}{(1+x)^2}.$$

Since $x^2/(1+x)^2 \in [0, 1)$ for all $x > 0$, we obtain the bounds

$$1 - \delta d \leq S(\alpha) < 1 - \delta d + \delta.$$

Under the parameter restriction $\delta d < 1$, the lower bound is strictly positive, so $S(\alpha) > 0$ for all $\alpha \in (0, 1)$ satisfying (8.3.4). By (8.A2), this implies $\partial r^*/\partial\alpha > 0$ on the admissible range. □

### 8.A.5 Proof of Lemma 8.10 (Effect of $\eta$ on the Neutral Interest Rate)

***Proof*** From (8.2.2),

$$r^*+\delta = \frac{\alpha(c+\delta)\,N(\eta)}{B(\eta)}, \qquad N(\eta) := \frac{\eta-1}{\eta}, \qquad B(\eta) := 1+\delta\Big(\varphi(\eta) - \frac{1+c_b}{2}\Big).$$

Taking logs and differentiating,

$$\frac{d}{d\eta}\ln\big(r^*+\delta\big) = \frac{N'(\eta)}{N(\eta)} - \frac{B'(\eta)}{B(\eta)} = \frac{1}{\eta(\eta-1)} - \frac{\delta\,\varphi'(\eta)}{B(\eta)}.$$

Since $r^*+\delta > 0$, $\partial/\partial\eta r^*$ has the same sign as the bracketed expression, yielding the claim. □

### 8.A.6 Proof of Corollary 8.11 (Monotonicity in $\eta$ Under the Baseline Calibration)

***Proof*** For $\eta \geq 6$ and $\alpha = 1/3$,

$$\varphi'(\eta) = \frac{\alpha}{(1+\alpha(\eta-1))^2} \leq \frac{1/3}{(1+(1/3)\cdot 5)^2} = \frac{1/3}{(8/3)^2} = \frac{3}{64}.$$

Moreover, since $\varphi(\eta) \in (0, 1)$ and $(1 + c_b)/2 > 0$, we have $B(\eta) = 1 + \delta(\varphi - (1 + c_b)/2) \geq 1 - \delta(1 + c_b)/2$. With $\delta = 0.06$ and $c_b = 0.05$, this gives $B(\eta) \geq 1 - 0.06 \cdot 1.05/2 = 0.9685$. Hence,

$$\frac{\delta\,\varphi'(\eta)}{B(\eta)} \leq \frac{0.06}{0.9685} \cdot \frac{3}{64} < 0.00291.$$

On the other hand, for $\eta \in [6, 12]$,

$$\frac{1}{\eta(\eta - 1)} \geq \frac{1}{12 \cdot 11} = \frac{1}{132} \approx 0.00758.$$

Therefore, the bracket in Lemma 8.10 is strictly positive on $[6, 12]$, implying $\partial/\partial\eta r^* > 0$ on this interval. □

### 8.A.7 *Proof of Lemma 8.13 (A Sufficient Condition for c Turning Point in $K^*(c)$)*

***Proof (Sketch)*** From Eq. (8.3.7), $\log K^*(c) = \text{const} - \log c - \log h(r^*(c))$, hence Eq. (8.3.8). Next note that

$$\frac{d}{dr} \log h(r) = \beta\Big(\frac{1}{r + \delta} + \frac{1}{1 + r} - \frac{1}{r}\Big) = -\beta\Big(\frac{1}{r} - \frac{1}{r + \delta} - \frac{1}{1 + r}\Big) < 0,$$

and its magnitude satisfies the asymptotics

$$-\frac{d}{dr} \log h(r) \sim \frac{\beta}{r} \quad (r \downarrow 0), \qquad -\frac{d}{dr} \log h(r) \sim \frac{\beta}{r^2} \quad (r \to \infty).$$

Since $r^*(c)$ increases linearly in $c$ by Eq. (8.3.6), the term $-\frac{h'(r^*)}{h(r^*)}a = -\frac{d}{dr} \log h(r^*)\, a$ becomes arbitrarily large as $c \downarrow 0$ (along sequences with $c \downarrow 0$ for which $r^*(c) \to 0$), whereas it decays like $O(1/c^2)$ as $c \to \infty$. Therefore, for small $c$, the positive contribution $-\frac{d}{dc} \log h(r^*(c))$ dominates the term $1/c$, yielding $dK^*/dc > 0$, while for sufficiently large $c$ the $1/c$ term dominates, yielding $dK^*/dc < 0$. □

## 8.B Admissible Windows and Long-Run Feasibility

Behavioral parameters are interpreted in Sect. 5.2.2; here we use them only to characterize feasibility.

### 8.B.1 Formal Statement and Proof: Admissible Interest-Rate Window and Induced Capital Windows

**Proposition 8.14 (Admissible Interest-Rate Window and Induced Capital Window)** *Fix real parameters $(\delta, A, V, c_b, \varphi)$ with $\delta > 0$, $A > 0$, and define the excess-demand map*

$$\zeta(r) := \frac{1}{\delta}\Big(\frac{A}{r+\delta} - 1\Big), \qquad r > 0. \tag{8.B1}$$

*Let the admissible excess-demand bounds be*

$$\zeta_+ := \varphi - \frac{V}{2c_b}, \qquad \zeta_- := \varphi + \frac{V}{2c_b} - (1 + c_b),$$

*and assume $\zeta_- > \zeta_+ > 0$ (equivalently $V/c_b > 1 + c_b$ and $\varphi > V/2c_b$). Let*

$$D := \frac{V}{2c_b} - \frac{1+c_b}{2},$$

*and assume $D > 0$, equivalently $V/c_b > 1 + c_b$, so the interior window is nondegenerate.*

***(i) Admissible interest-rate window.*** *There exists a unique interval $[r_-, r_+]$ such that*

$$\zeta(r) \in [\zeta_+, \zeta_-] \iff r \in [r_-, r_+],$$

*with endpoints*

$$r_\pm := \frac{A}{1 + \delta\zeta_\pm} - \delta, \qquad r_- < r_+. \tag{8.B2}$$

*Moreover, on the stable PM–LM branch of interest, parameters are such that $r_- > 0$.*

***(ii) Width of the admissible interest-rate window.*** *Define*

$$\bar{\varphi} := \varphi + \frac{1}{\delta} - \frac{1+c_b}{2}.$$

*Then $\zeta_- - \zeta_+ = 2D$, and the width $\Delta r := r_+ - r_-$ admits the exact expression*

$$\Delta r = \frac{2A}{\delta} \cdot \frac{D}{\bar{\varphi}^2 - D^2}, \tag{8.B3}$$

*with the small-$D$ approximation*

$$\Delta r = \frac{2A}{\delta}\left(\frac{D}{\bar{\varphi}^2} + O(D^3)\right), \qquad (D \to 0).$$

***(iii) Induced capital window and monetary scaling.*** *Fix* $(\beta, \sigma_0)$ *with* $\beta > 0$ *and* $\sigma_0 > 0$, *and define*

$$x := \sigma_0^{1/\beta} K^{1/\beta}, \qquad K = \frac{x^\beta}{\sigma_0}. \tag{8.B4}$$

*This normalization absorbs the nominal scale* $\sigma_0$ *into the state variable. Let* $x(r)$ *denote the PM–LM consistent value of* $x$ *at interest rate* $r$. *Then the admissible capital window is*

$$[K_{q0}, K_{p0}] = \left[\frac{x(r_-)^\beta}{\sigma_0}, \frac{x(r_+)^\beta}{\sigma_0}\right], \tag{8.B5}$$

*so that*

$$K_{p0} - K_{q0} = \frac{x(r_+)^\beta - x(r_-)^\beta}{\sigma_0}, \tag{8.B6}$$

$$\frac{K_{p0} - K_{q0}}{K^*} = \frac{x(r_+)^\beta - x(r_-)^\beta}{x(r^*)^\beta}. \tag{8.B7}$$

*In particular, the* absolute *width of the capital window scales with* $\sigma_0^{-1}$, *while the* relative *width is independent of nominal parameters.*

***Proof*** **Step 1 (monotonicity and inversion).** From (8.B1),

$$\zeta'(r) = -\frac{A}{\delta(r+\delta)^2} < 0,$$

so $\zeta$ is strictly decreasing on $(0, \infty)$ and hence invertible. Solving $\zeta(r_\pm) = \zeta_\pm$ gives (8.B2). Since $\zeta_- > \zeta_+$ and $\zeta$ is decreasing, it follows that $r_- < r_+$, proving (i).

**Step 2 (an exact expression for $\Delta r$).** Write

$$\delta\zeta(r) + 1 = \frac{A}{r+\delta} \quad \Longrightarrow \quad \frac{1}{r+\delta} = \frac{\delta\zeta(r)+1}{A}.$$

Evaluating at $r_\pm$ yields

$$\frac{1}{r_- + \delta} - \frac{1}{r_+ + \delta} = \frac{\delta}{A}(\zeta_- - \zeta_+).$$

Because $\zeta_- > \zeta_+$, the left-hand side is positive, consistent with $r_- < r_+$.

On the other hand,

$$\frac{1}{r_- + \delta} - \frac{1}{r_+ + \delta} = \frac{r_+ - r_-}{(r_- + \delta)(r_+ + \delta)} = \Delta r \cdot \frac{1}{(r_- + \delta)(r_+ + \delta)}.$$

Combining gives

$$\Delta r = \frac{\delta}{A}(\zeta_- - \zeta_+)\,(r_- + \delta)(r_+ + \delta). \tag{8.B8}$$

Using again $r_\pm + \delta = A/(1 + \delta\zeta_\pm)$,

$$(r_- + \delta)(r_+ + \delta) = \frac{A^2}{(1 + \delta\zeta_-)(1 + \delta\zeta_+)}.$$

Substitute into (8.B8) to obtain the exact formula

$$\Delta r = A\delta\,\frac{\zeta_- - \zeta_+}{(1 + \delta\zeta_-)(1 + \delta\zeta_+)}.$$

Since

$$\zeta_- - \zeta_+ = \frac{V}{c_b} - (1 + c_b) = 2D, \qquad \zeta_\pm + \frac{1}{\delta} = \bar{\varphi} \mp D,$$

we have

$$(1 + \delta\zeta_-)(1 + \delta\zeta_+) = \delta^2\Big(\zeta_- + \frac{1}{\delta}\Big)\Big(\zeta_+ + \frac{1}{\delta}\Big) = \delta^2(\bar{\varphi}^2 - D^2).$$

Substituting these identities yields (8.B3). The Taylor expansion follows because $\frac{D}{\bar{\varphi}^2 - D^2} = \frac{D}{\bar{\varphi}^2} \cdot \frac{1}{1-(D/\bar{\varphi})^2} = \frac{D}{\bar{\varphi}^2}\big(1 + O(D^2)\big)$.

**Step 3 (mapping the $r$-window into a $K$-window).** By (8.B4), any admissible value of $x$ corresponds to a unique capital stock $K = x^\beta/\sigma_0$. Evaluating at $r_\pm$ and using $x(r_\pm)$ give (8.B5). The scaling relation (8.B6) is immediate. Finally, $K^* = x(r^*)^\beta/\sigma_0$ implies (8.B7), completing (iii). □

### 8.B.2 *Formal Statement and Proof: Feasibility Implications*

**Corollary 8.15 (Effect of $c_b$ with Fixed Sentiment Range $V$)** *Assume the setting of Proposition 8.14. Then*

$$\Delta\zeta \;:=\; \zeta_- - \zeta_+ \;=\; \frac{V}{c_b} - (1 + c_b).$$

*Thus sentiment shock amplitude $V$ enlarges both the admissible $\zeta$- and $r$- windows. $c_b$ shrinks the admissible $\zeta$-window, while the induced effect on $\Delta r$ is generally ambiguous because $c_b$ also shifts $\bar{\varphi}$.*

## 8.C Lender's and Borrower's Risk and Equilibrium Admissibility (Kalecki)

**Lender's and Borrower's Risk (Kalecki)**[10] A company's decision to invest depends on the size of its current debt burden. A company with current physical capital $K$ has a periodic payment of $(r+\delta)P^S K$. The total repayment amount for that capital stock is

$$(r+\delta)PK\{1+(1-\delta)+(1-\delta)^2+\dots\} = \frac{(r+\delta)PK}{\delta}.$$

Since $P^S \equiv 1, \omega = W/P$, and $P = P^S$, under the fixed wage assumption, we know

$$BS = \frac{(r+\delta)P^S K}{\delta W} = \frac{(r+\delta)K}{\delta\omega}.$$

The real burden measured in wage units $BS$ equals

$$\kappa = \frac{\alpha\omega}{(1-\alpha)(r+\delta)} \Rightarrow \frac{r+\delta}{\omega} = \frac{\alpha}{(1-\alpha)\kappa}, \qquad N^d = \frac{K}{\kappa}.$$

We can see that the magnitude of this debt burden is proportional to the employment level.

$$BS = \frac{(r+\delta)K}{\delta\omega} = \frac{K}{\delta}\cdot\frac{r+\delta}{\omega} = \frac{K}{\delta}\cdot\frac{\alpha}{(1-\alpha)\kappa} = \frac{\alpha}{(1-\alpha)\delta}\cdot\frac{K}{\kappa} = \frac{\alpha}{(1-\alpha)\delta}N^d.$$

Thus the real debt-service burden is proportional to employment; in this sense, tighter balance-sheet positions can amplify the Keynesian interest-rate channel under the fixed-$M$ regime.

[10] Kalecki's *principle of increasing risk* implies that investment financed beyond internal funds carries a rising external-finance premium; see Kalecki (1937). We abstract from such endogenous risk premia here. See also Minsky (1986, Ch.8).

## References

Ellis, J. H. (2005). *Ahead of the curve: A commonsense guide to forecasting business and market cycles*. Harvard Business School Press.

Fisher, I. (1933). The debt-deflation theory of great depressions. *Econometrica, 1*(4), 337–357. https://doi.org/10.2307/1907327

Kalecki, M. (1937). The principle of increasing risk. *Economica, 4*(16), 440–447. https://doi.org/10.2307/2626879

Keynes, J. M. (1936). *The general theory of employment, interest and money*. Macmillan

Keynes, J. M. (1937). The general theory of employment. *The Quarterly Journal of Economics, 51*(2), 209–223. https://doi.org/10.2307/1882087

Laubach, T., & Williams, J. C. (2003) Measuring the natural rate of interest. *Review of Economics and Statistics, 85*(4), 1063–1070. https://doi.org/10.1257/000282803321946246

Leijonhufvud, A. (1968). *On Keynesian economics and the economics of Keynes*. New York: Oxford University Press.

Minsky, H. P. (1986). *Stabilizing an unstable economy*. Yale University Press.

Skidelsky, R. (2010). *Keynes: the return of the master*. Public Affairs. https://doi.org/10.1057/eej.2010.42

Tobin, J. (1993). Price flexibility and output stability: An old Keynesian view. *The Journal of Economic Perspectives, 7*(1), 45–65. https://doi.org/10.1257/jep.7.1.45

Wicksell, K. (1936). Interest and prices: A study of the causes regulating the value of money. Macmillan for the Royal Economic Society, London, translated by R. F. Kahn; introduction by Bertil Ohlin. First published in German (1898) as *Geldzins und Güterpreise*.

# Chapter 9
# Conclusion: Boundary Proximity, Neutral Anchors, and Macroeconomic Stability

*The object of our analysis is, not to provide a machine, or method of blind manipulation, but to provide ourselves with an organized and orderly method of thinking.*

— John Maynard Keynes, *The General Theory*, Ch. 21

This book has advanced a methodological alternative to equilibrium-centered macroeconomics by treating the economy as a stochastic system governed by decentralized decision rules, institutional constraints, and boundary-triggered adjustments. Rather than deriving macroeconomic outcomes from global optimization or representative-agent dynamics, the analysis has focused on how real anchors, nominal closures, and financial structure jointly shape the invariant distribution of economic states—and, equivalently, the stability corridor within which those states evolve. The conclusion returns to this perspective—not to restate results—but to clarify what they imply for macroeconomic stability, institutional design, and policy.

## 9.1 Neutral Anchors, Invariant Distributions, and Stochastic Stability

A central result of the book is that long-run macroeconomic stability is organized around a *neutral anchor* rather than a unique equilibrium path. Neutrality, established in Chap. 8, holds at the level of *location*, fixing the center of the long-run distribution; institutions, by contrast, govern its *dispersion* and the *frequency* with which the economy approaches its *boundaries*. Under broad conditions, the interaction of investment behavior, interest-rate adjustment, and mutation-selection dynamics generates a stochastic steady state centered on a neutral point $(K^*, r^*)$. This anchor is pinned down by real parameters—technology, preferences, market structure, and adjustment behavior—and is invariant to purely nominal rescalings that leave the real monetary environment unchanged.

 
I. Takahashi, *Microfoundations of Keynesian Economics*, Evolutionary Economics and Social Complexity Science 37, https://doi.org/10.1007/978-981-95-9390-3_9

At the same time, neutrality in this sense does not imply that monetary institutions are irrelevant. While the anchor itself is neutral, the *distribution* of macroeconomic states around it is not. Nominal closures, liquidity conditions, and wage-setting institutions shape the dispersion of outcomes and, crucially, determine how frequently the economy approaches regions in which standard adjustment mechanism cease to apply. The distinction between neutral anchors and non-neutral distributions provides a natural reconciliation between long-run neutrality and the pervasive asymmetries emphasized in Keynesian analysis.

This perspective naturally leads to an organizing role for boundary proximity in characterizing macroeconomic stability.

## 9.2 Boundary Proximity as an Organizing Principle

The analysis has shown that macroeconomic dynamics are best understood in terms of *boundary proximity*. Interior regions of the state space, characterized by strictly interior adjustment probabilities, admit smooth stochastic dynamics governed by interest-rate adjustment and investment responses. Proposition 8.14 formalizes this notion by characterizing an admissible interest-rate window $[r_-, r_+]$ and its induced capital window $[K_{q0}, K_{p0}]$. As long as the economy remains within this window, the stabilizing forces of the monetary economy operate without invoking regime switches or real discontinuities.

Boundary regimes arise only when the economy approaches the edges of this window. Near these boundaries, probability clipping, capacity constraints, and bargaining thresholds become active, and real adjustment mechanisms—such as wage bargaining, labor-supply constraints, or capacity scrapping—play a decisive role. The economy becomes fragile not when it deviates from the neutral point, but when it spends too much time near boundaries at which clipping, bargaining constraints, or scrapping decisions become binding. From this perspective, macroeconomic instability is not the result of small deviations from an equilibrium path, but of repeated or persistent visits to boundary regions where the usual adjustment mechanism no longer applies.

## 9.3 Asymmetry, Wage Adjustment, and Deflation

The book has emphasized the fundamental asymmetry between expansion and contraction that emerges naturally once boundary proximity is taken seriously. In particular, downward nominal wage adjustment has been analyzed as a boundary-triggered response rather than as a continuously operative flexibility. Under *100% reserve regime* (with a fixed money-supply), wage cuts are shown to be largely absorbed through proportional price-level adjustment, leaving the real wage and the neutral interest rate effectively unchanged. By lowering transaction demand and inducing a decline in the market interest rate for a given real state $K$ *in the interior region*, wage cuts can stimulate investment and shift the invariant distribution away from the lower boundary.

This mechanism provides a concrete explanation for why nominal wage flexibility need not generate a cumulative deflationary spiral in a monetary economy with a stable interest-rate anchor. At the same time, the analysis makes clear that this robustness is conditional. Behavioral responses—such as increases in liquidity preference or declines in the propensity to consume—can push stationary probability mass in the stationary distribution back toward the boundary and reactivate deflationary regimes. Deflation is therefore neither mechanically stabilizing nor inherently destabilizing; its effects depend on the interaction between nominal adjustment rules and the economy's distance from its real boundaries.

## 9.4 Policy Under Institutional Regimes

A central implication of the analysis is that macroeconomic policy should be understood as a problem of institutional design rather than discretionary fine-tuning. Under a 100% reserve regime, discretionary monetary expansion is absent by construction, and stabilization must operate through the preservation of a well-defined interest-rate reaction mechanism and the management of boundary proximity. In this context, nominal wage flexibility, stable bargaining norms, and predictable financial conditions act as automatic stabilizers that help keep the economy within its interior region.

More generally, the analysis suggests that the primary objective of policy is not to control aggregate demand at each point in time, but to design institutions that minimize the likelihood of persistent boundary contact. Monetary regimes, wage-setting institutions, and financial structures should be evaluated by how effectively they preserve the interest-rate anchor and prevent the economy from becoming trapped in regimes where real adjustment rules dominate. This boundary-based criterion provides a unifying framework for comparing alternative monetary and financial regimes.

## 9.5 Confidence, the Propensity to Consume, and Long-Run Stability

Among the parameters governing long-run macroeconomic performance, the propensity to consume $c = c_w/(1 - c_y)$ plays a distinctive stabilizing role. In the model, $c$ can be interpreted narrowly as a reduced-form summary of households' expenditure behavior, but it also admits a broader interpretation as a proxy for

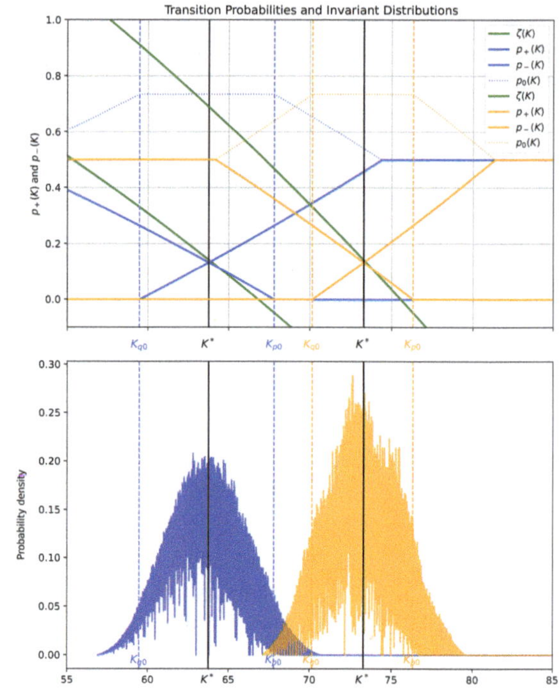


**Fig. 9.1** Effect of an increase in the propensity to consume on transition probabilities and the invariant distribution. The upper panel shows the transition probabilities $p_+(K)$ and $p_-(K)$ for $c = 0.17$ and $c = 0.18$. A higher $c$ shifts both schedules rightward. The lower panel displays the corresponding invariant probability densities, which move toward higher $K$ while exhibiting similar dispersion. Vertical lines indicate the thresholds $K_{q0}$, $K_{p0}$, and the neutral point $K^*$

macroeconomic confidence—the willingness to convert expected income into current spending rather than precautionary saving.

Figure 9.1 illustrates this mechanism. An increase in $c$ shifts the transition probabilities $p_+(K)$ and $p_-(K)$ rightward, thereby shifting the invariant distribution of capital toward higher levels of $K$ while leaving its dispersion broadly unchanged. To complement the invariant-distribution comparison above, we also report in Fig. 9.2 a representative sample trajectory of the capital-stock process under the same calibration. Consistent with the *stability corridor* concept introduced in Chap. 1, we interpret the admissible capital interval $[K_{q0}, K_{p0}]$ as the operational stability corridor within which the discrete dynamics remain locally viable, and we illustrate how its location and width vary with the propensity to consume $c$.

A central lesson of this book is that not all parameters governing long-run macroeconomic performance are equally amenable to policy influence. Structural parameters such as the capital income share $\alpha$ and the degree of market competitiveness $\eta$ are shaped by technology, industrial organization, and institutional design and therefore tend to evolve only slowly over time. By contrast, the propensity to consume $c$, which summarizes households' willingness to translate expected income into current expenditure, is inherently behavioral and state dependent. For this reason, $c$ emerges as a particularly important margin through which policy and institutions can influence long-run outcomes without altering the underlying technological structure of the economy.

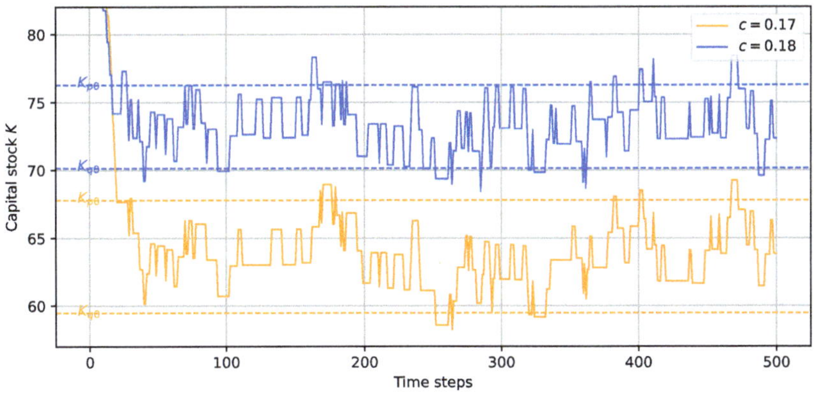


**Fig. 9.2** Sample trajectories of $K_t$ and the stability corridor $[K_{q0}, K_{p0}]$ for two values of $c$. Representative sample trajectories of the capital stock $K_t$ for two values of $c$ under the same shock realization. The stability corridor $[K_{q0}, K_{p0}]$ shifts upward and narrows as $c$ rises, so the realized path is both higher on average and less volatile

Unlike in standard saving-driven frameworks where a higher propensity to consume reduces capital accumulation via $S = I$, in the present model an increase in $c$ raises effective demand and utilization, which in turn supports higher output and capital formation; see Sect. 8.4.2 for the underlying mechanism.

The stochastic transition mechanism developed in Chap. 6 captures the state-dependent nature of decentralized spending behavior. Rather than representing exogenous randomness imposed upon an otherwise homogeneous economy, it summarizes the heterogeneous responses of industries and households operating under differing local conditions, financial positions, utilization rates, and expectations. Even under a common macroeconomic environment, sectors and households do not respond uniformly.

A simple illustration helps clarify the intuition. Consider an unusually severe summer in which temperatures rise substantially above normal levels. In an expanding economy, households may respond by purchasing more ice cream and beverages, traveling to cooler regions, or increasing leisure expenditures more broadly. The same weather shock therefore propagates through a wide range of industries, stimulating demand across multiple sectors. By contrast, during a sharp economic downturn, households facing income uncertainty and financial stress may respond very differently. Rather than increasing discretionary expenditures, they may simply avoid costly activities and seek low-cost ways of adapting to the heat. The physical shock is identical in both cases, yet its economic consequences differ markedly because the surrounding macroeconomic conditions alter the economy's responsiveness. The stochastic component of the model therefore represents not arbitrary noise, but a reduced-form expression of unresolved industrial and behavioral heterogeneity beneath the aggregate economy.

The probabilistic structure introduced in this book is intended to capture precisely this feature of decentralized capitalist economies. Aggregate fluctuations emerge not because all firms and households move mechanically together, but because common macroeconomic conditions interact with heterogeneous local circumstances and differing propensities to respond. The resulting stochastic dynamics are therefore interpreted not as arbitrary randomness, but as a reduced-form representation of heterogeneous and state-dependent economic responsiveness.

Within the present framework, the parameter $c$ plays a dual role. First, it enters the neutrality condition through the utilization wedge, thereby affecting the location of the stochastic steady-state anchor $(K^*, r^*)$. As illustrated in Fig. 9.1, an increase in $c$ from 0.17 to 0.18 induces a pronounced rightward shift in the transition probability schedules $p_+(K)$ and $p_-(K)$, reflecting the expansion of aggregate demand and the associated upward pressure on capital accumulation.

Second, changes in $c$ reshape the geometry of the admissible interest-rate window and the corresponding capital window. While the relative dispersion of the invariant distribution remains broadly similar, the entire distribution shifts toward higher levels of capital and moves closer to the upper boundary of the admissible region. In this way, variations in $c$ primarily affect the location of the invariant distribution and its proximity to boundaries, rather than their intrinsic stochastic spread.

An economy characterized by persistently low $c$ operates with weak effective demand, low utilization, and a heightened probability of approaching the lower boundary $K_D$, where real adjustment mechanisms such as downward wage revision become relevant. As noted by Gesell (1958), the asymmetry between costless money holding and costly goods storage creates a bias toward liquidity, which in our framework can be interpreted as a lower transactions intensity $c$. Conversely, an increase in $c$ shifts the invariant distribution away from distress regions, reducing the likelihood that the economy persistently operates near binding constraints.

Crucially, raising $c$ need not be understood as the result of direct intervention in consumption decisions. Rather, $c$ reflects the degree of economic security and confidence with which households form expenditure plans. Policies and institutions that reduce income uncertainty, mitigate unemployment risk, and stabilize expectations operate indirectly on $c$ by lowering precautionary motives and strengthening effective demand. In this sense, the model formalizes a Keynesian insight: Macroeconomic performance depends not only on prices, wages, and interest rates, but also on the social and institutional conditions under which households assess the safety of spending out of income.[1]

[1] Keynes emphasized that the aggregate propensity to consume depends not only on the level of income but also on its distribution. Since lower-income households typically exhibit higher propensities to consume, redistributive policies can raise aggregate consumption even when total income is unchanged. See Keynes (1936, Chs. 8 and 24), where income distribution, underconsumption, and unemployment are treated as closely related macroeconomic phenomena rather than as separate issues of equity and efficiency.

The analysis cautions against viewing increases in $c$ as unambiguously expansionary. When confidence becomes excessive, the resulting pressure on utilization is increasingly absorbed by higher interest rates rather than by sustained increases in real activity, and long-run capital accumulation may eventually decline. The stabilizing role of confidence is therefore inherently state contingent. Sustainable macroeconomic performance is achieved not by maximizing optimism, but by maintaining the economy within an interior region in which stochastic dynamics remain organized around real anchors and boundary adjustments remain dormant.

## 9.6 Implications and Outlook

The framework developed in this book reorients macroeconomic analysis along several fundamental dimensions. It replaces pointwise equilibrium with distributional stability, local comparative statics with global geometry, and shock-by-shock policy responses with institutional design at the regime level. In doing so, it provides a coherent way to reconcile Keynesian asymmetries, Wicksellian interest-rate adjustment, and long-run monetary neutrality within a unified stochastic framework.

In particular, the present analysis may shed new light on the relationship between competition and macroeconomic stability. The standard efficiency results associated with competitive markets primarily concern static allocative efficiency under idealized equilibrium conditions. The framework developed here highlights an additional dimension: the dynamic stability of decentralized monetary production economies operating under uncertainty and heterogeneous adjustment.

Greater competitive pressure may compress profit margins and increase the throughput required to sustain profitability, thereby increasing the economy's sensitivity to fluctuations in demand and utilization. In such an environment, firms may become more vulnerable to temporary downturns, and the economy as a whole may require stronger stabilizing mechanisms to avoid cumulative contractionary dynamics. From this perspective, institutional arrangements that appear less efficient from a static allocative standpoint may nevertheless contribute to macroeconomic robustness and long-run stability.

The present argument should not be interpreted as a rejection of competition itself, but rather as a reminder that allocative efficiency and dynamic stability are distinct properties of an economic system. A decentralized capitalist economy must ultimately be evaluated not only by its static efficiency, but also by its capacity to maintain coordinated and sustainable activity through time under uncertain conditions.

Several avenues for further research remain open. Empirically, the emphasis on invariant distributions and boundary frequencies suggests new objects of interest beyond average growth rates or impulse responses. Conceptually, the comparison of alternative monetary and financial regimes—such as pure-credit systems or hybrid reserve structures—can be analyzed within the same boundary-based framework.

Finally, the interaction between expectations, learning, and boundary proximity remains a promising direction for further exploration.

The central message of this book is therefore both modest and ambitious. Macroeconomic stability does not require perfect foresight, continuous price flexibility, or discretionary fine-tuning by an omnipotent policymaker. It requires institutions that preserve real anchors, discipline nominal adjustment, and keep the economy away from regions in which boundary constraints dominate behavior. When such conditions are in place, a monetary economy can remain robust—not despite uncertainty, heterogeneity, and decentralized decision-making, but because these features are accommodated within a stable institutional framework rather than suppressed by fragile equilibrium constructions.

## References

Keynes, J. M. (1936). *The general theory of employment, interest and money*. Macmillan.

Gesell, S. (1958). *The natural economic order*. London: Peter Owen. Originally published as Die natürliche wirtschaftsordnung (1916).

# Index




I. Takahashi, *Microfoundations of Keynesian Economics*, Evolutionary Economics and Social Complexity Science 37, https://doi.org/10.1007/978-981-95-9390-3

FSC
www.fsc.org
MIX
Papier | Fördert
gute Waldnutzung
FSC® C083411

Kolibri 360 GmbH
Ferdinand-Jühlke-Straße 7
99095 Erfurt, Deutschland
produktsicherheit@kolibri360.de